D0866913

Considerable attention is now being paid to the use of molecular evidence in studies of human diversity and origins. Much of the early work was based on evidence from mitochondrial DNA, but this has now been supplemented by important new information from nuclear DNA from both the Y chromosomes and the autosomes. The bulk of the material available is also from living populations, but this is now being extended by the study of DNA from archaic populations. The underlying models used in interpreting this evidence are developments of the neutral theory of molecular evolution, but also consider the possible role of selections. This volume brings together this new evidence and methodology from an international group of research workers. It will be an important reference for researchers in human biology, molecular biology and genetics alike.

SOCIETY FOR THE STUDY OF HUMAN BIOLOGY SYMPOSIUM SERIES: 38

Molecular biology and human diversity

PUBLISHED SYMPOSIA OF THE
SOCIETY FOR THE STUDY OF HUMAN BIOLOGY

10 Biological Aspects of Demography ED. W. BRASS
11 Human Evolution ED. M.H. DAY
12 Genetic Variation in Britain ED. D.F. ROBERTS AND E. SUNDERLAND
13 Human Variation and Natural Selection ED. D.F. ROBERTS
(*Penrose Memorial Volume reprint*)
14 Chromosome Variation in Human Evolution ED. A.J. BOYCE
15 Biology of Human Foetal Growth ED. D.F. ROBERTS
16 Human Ecology in the Tropics ED. J.P. GARLICK AND R.W.J. KEAY
17 Physiological Variation and its Genetic Base ED. J.S. WEINER
18 Human Behaviour and Adaptation ED. N.J. BLURTON JONES AND V. REYNOLDS
19 Demographic Patterns in Developed Societies ED. R.W. HIORNS
20 Disease and Urbanisation ED. E.J. CLEGG AND J.P. GARLICK
21 Aspects of Human Evolution ED. C.B. STRINGER
22 Energy and Effort ED. G.A. HARRISON
23 Migration and Mobility ED. A.J. BOYCE
24 Sexual Dimorphism ED. F. NEWCOMBE *et al.*
25 The Biology of Human Ageing ED. A.H. BITTLES AND K.J. COLLINS
26 Capacity for Work in the Tropics ED. K.J. COLLINS AND D.F. ROBERTS
27 Genetic Variation and its Maintenance ED. D.F. ROBERTS AND G.F. DE STEFANO
28 Human Mating Patterns ED. C.G.N. MASCIE-TAYLOR AND A.J. BOYCE
29 The Physiology of Human Growth ED. J.M. TANNER AND M.A. PREECE
30 Diet and Disease ED. G.A. HARRISON AND J. WATERLOW
31 Fertility and Resources ED. J. LANDERS AND V. REYNOLDS
32 Urban Ecology and Health in the Third World ED. L.M. SCHELL, M.T. SMITH AND A. BILSBOROUGH
33 Isolation, Migration and Health ED. D.F. ROBERTS, N. FUJIKI AND K. TORIZUKA
34 Physical Activity and Health ED. N.G. NORGAN
35 Seasonality and Human Ecology ED. S.J. ULIJASZEK AND S.S. STRICKLAND
36 Body Composition Techniques in Health and Disease ED. P.S.W. DAVIES AND T.J. COLE
37 Long-term Consequences of Early Environment ED. J.H.K. HENRY AND S.J. ULIJASZEK

Numbers 1–9 were published by Pergamon Press, Headington Hill Hall, Headington, Oxford OX3 0BY. Numbers 10–24 were published by Taylor & Francis Ltd, 10–14 Macklin Street, London WC2B 5NF. Further details and prices of back-list numbers are available from the Secretary of the Society for the Study of Human Biology.

Molecular biology and human diversity

EDITED BY

A. J. BOYCE

Department of Biological Anthropology, Oxford University

and

C. G. N. MASCIE-TAYLOR

Department of Biological Anthropology, Cambridge University

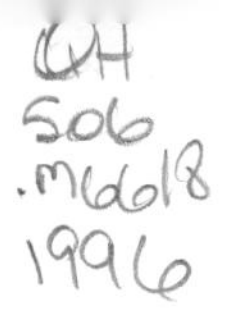

Published by the Press Syndicate of the University of Cambridge
The Pitt Building, Trumpington Street, Cambridge CB2 1RP
40 West 20th Street, New York, NY 10011-4211, USA
10 Stamford Road, Oakleigh, Melbourne 3166, Australia

First published 1996

Printed in Great Britain at the University Press, Cambridge

A catalogue record for this book is available from the British Library

Library of Congress cataloguing in publication data

Molecular biology and human diversity / edited by A. J. Boyce and C. G. N. Mascie-Taylor.
p. cm. – (Society for the Study of Human Biology symposium series; 38)
Includes index.
ISBN 0 521 56086 1
1. Molecular biology. Molecular evolution. 3. Human evolution.
I. Boyce, A. J. (Anthony J.) II. Mascie-Taylor, C. G. N.
III. Series.
QH506.M6618 1996
574.87′328–dc20 96-22586 CIP

ISBN 0 521 56086 1 hardback

VN

Contents

List of contributors

G. Barbujani
Dipartimento di Scienze Statistiche, via Belle Arti 41, I-40126 Bologna, Italy

J. Bertranpetit
Laboratori d'Antropologia, Facultat de Biologia, Universitat de Barcelona, Diagonal 645, 08028 Barcelona, Catolonia, Spain; and Institut de Salut Pública de Catalunya, Spain

G. Bertorelle
Dipartimento di Biologia, via Trieste 75, I-35121 Padova, Italy

J. Bond
MRC Molecular Haematology Unit, Institute of Molecular Medicine, Oxford University, Oxford OX3 9DU, UK

N. Bouzekri
Department of Genetics, University of Leicester, University Road, Leicester LE1 7RH, UK

K. B. Bulayeva
Institute of General Genetics, Russian Academy of Sciences, Moscow 117809, Russia

F. Calafell
Laboratori d'Antropologia, Facultat de Biologia, Universitat de Barcelona, Diagonal 645, 08028 Barcelona, Catolonia, Spain

J. B. Clegg
MRC Molecular Haematology Unit, Institute of Molecular Medicine, Oxford University, Oxford OX3 9DU, UK

D. Comas
Laboratori d'Antropologia, Facultat de Biologia, Universitat de Barcelona, Diagonal 645, 08028 Barcelona, Catolonia, Spain

H. Côrte-Real
Department of Cellular Medicine, Institute of Molecular Medicine, Oxford University, Oxford OX3 9DU, UK

G. A. Dover
Department of Genetics, University of Leicester, University Road, Leicester LE1 7RH, UK

L. Excoffier
Genetics and Biometry Laboratory, Department of Anthropology and Ecology, University of Geneva CP 511, 1211 Geneva 24, Switzerland

J. Flint
MRC Molecular Haematology Unit, Institute of Molecular Medicine, Oxford University, Oxford OX3 9DU, UK

N. Fretwell
Department of Genetics, University of Leicester, University Road, Leicester LE1 7RH, UK

S. M. Fullerton
Department of Anthropology, University of Durham, UK

E. Hagelberg
Department of Biological Anthropology, University of Cambridge, UK

R. M. Harding
MRC Molecular Haematology Unit, Institute of Molecular Medicine, Oxford University, Oxford OX3 9DU, UK

H. C. Harpending
Department of Anthropology, Pennsylvania State University, University Park, PA 16802, USA

R. Hewitt
Department of Human Genetics, The South African Institute for Medical Research and University of the Witwatersrand, Johannesburg, South Africa

A. V. S. Hill
Wellcome Trust Centre for Human Genetics & Institute of Molecular Medicine, Nuffield Department of Medicine, University of Oxford, Oxford OX3 9DU, UK

A. J. Jeffreys
Department of Genetics, University of Leicester, University Road, Leicester LE1 7RH, UK

T. Jenkins
Department of Human Genetics, The South African Institute for Medical Research and University of the Witwatersrand, Johannesburg, South Africa

M. A. Jobling
Department of Genetics, University of Leicester, University Road, Leicester LE1 7RH, UK

O. V. Kalnina
Institute of Gene Biology, Russian Academy of Sciences, Moscow 117334, Russia

V. V. Kalnin
Institute of Gene Biology, Russian Academy of Sciences, Moscow 117334, Russia

I. M. Khidiatova
Department of Biochemistry and Cytochemistry, Bashkir Science Center of the Ural Branch of the Russian Academy of Sciences, Ufa 450054, Russia

E. K. Khusnutdinova
Department of Biochemistry and Cytochemistry, Bashkir Science Center of the Ural Branch of the Russian Academy of Sciences, Ufa 450054, Russia

J. R. Kidd
Department of Genetics, School of Medicine, Yale University, 333 Cedar Street, PO Box 3333, New Haven, CT 06510-8005, USA

K. K. Kidd
Department of Genetics, School of Medicine, Yale University, 333 Cedar Street, PO Box 3333, New Haven, CT 06510-8005, USA

A. Krause
Department of Human Genetics, The South African Institute for Medical Research and University of the Witwatersrand, Johannesburg, South Africa

N. S. Kupriyanova
Institute of Gene Biology, Russian Academy of Sciences, Moscow 117334, Russia

A. Langaney
Laboratoire d'Anthropologie Biologique, CNRS, Musée de l'Homme, Paris, France

S. A. Limborska
Institute of Molecular Genetics, Russian Academy of Sciences, Moscow 123182, Russia

J. J. Martinson
MRC Molecular Haematology Unit, Institute of Molecular Medicine, Oxford University, Oxford OX3 9DU, UK

E. Mateu
Laboratori d'Antropologia, Facultat de Biologia, Universitat de Barcelona, Diagonal 645, 08028 Barcelona, Catolonia, Spain

N. E. Morton
Human Genetics Department, University of Southampton, Princess Anne Hospital, Southampton SO16 5YA, UK

A. Pérez-Lezaun
Laboratori d'Antropologia, Facultat de Biologia, Universitat de Barcelona, Diagonal 645, 08028 Barcelona, Catolonia, Spain

E. S. Poloni
Genetics and Biometry Laboratory, Department of Anthropology and Ecology, University of Geneva, CP 511, 1211 Geneva 24, Switzerland

M. I. Prosnyak
Institute of Molecular Genetics, Russian Academy of Sciences, Moscow 123182, Russia

J. H. Relethford
Department of Anthropology, State University of New York, Oneonta, NY 13820, USA

M. Richards
Department of Cellular Medicine, Institute of Molecular Medicine, Oxford University, Oxford OX3 9DU, UK

D. F. Roberts
17 Montague Court, Gosforth, Newcastle upon Tyne NE3 4JL, UK

A. P. Ryskov
Institute of Gene Biology, Russian Academy of Sciences, Moscow 117334, Russia

N. Saitou
Laboratory of Evolutionary Genetics, National Institute of Genetics, 1111 Yata, Mishima-shi, Shizuoka-ken, 411, Japan

S. Santachiara-Benerecetti
Dipartimento di Genetica e Microbiologia, Università di Pavia, Italy

O. Semino
Dipartimento di Genetica e Microbiologia, Università di Pavia, Italy

S. T. Sherry
Department of Anthropology, Pennsylvania State University, University Park, PA 16802, USA

H. Soodyall
Department of Anthropology, The Pennsylvania State University, University Park, PA 16802, USA

M. Stoneking
Department of Anthropology, The Pennsylvania State University, University Park, PA 16802, USA

B. Sykes
Department of Cellular Medicine, Institute of Molecular Medicine, Oxford University, Oxford OX3 9DU, UK

J. W. Teague
Human Genetics Department, University of Southampton, Princess Anne Hospital, Southampton SO16 5YA, UK

R. H. Ward
Institute of Biological Anthropology, Oxford University, Oxford OX2 6QS, UK

Preface

This volume is based on papers presented at the 38th annual symposium of the Society for the Study of Human Biology held on 3 and 4 April 1995 at Corpus Christi College and the Department of Biological Anthropology, Cambridge. The sessions were chaired by Dr John Clegg, Professor Lord Renfrew, Dr Derek Roberts and Dr Chris Stringer.

The Editors wish to record their thanks to Dr Tracey Sanderson, Commissioning Editor, Biological Sciences, Ms Susan Bishop, Production Controller, and Ms Mary Sanders, Copy Editor, of Cambridge University Press for their help during the production of this volume.

The Committee of the Society acknowledges with gratitude the generous financial support for the meeting provided by the Boise Fund of Oxford University, the International Union of Biological Sciences, and the Wellcome Trust.

1 *Mitochondrial DNA in ancient and modern humans*

E. HAGELBERG

Introduction

The study of the patterns of genetic diversity in modern human populations can help reconstruct human evolutionary history, including ancient migrations, population expansions, and bottlenecks. However, the interpretation of early events based on modern genetic patterns may easily be obscured by more recent developments, for instance, multiple migrations, invasions, or demographic collapse resulting from infectious diseases or genocide. Additional information in the form of linguistic analyses, historical reports or the archaeological record can be used to build some kind of approximate picture of our past. Moreover, the development of powerful techniques for DNA analysis, notably the polymerase chain reaction, enables us to gain direct genetic information about past peoples in the form of DNA amplified from bones or mummified soft tissues.

In recent years, mitochondrial DNA (mtDNA) has achieved a prominent place in the study of human evolutionary history because it is simple, easy to study and evolves quickly. The human mtDNA genome is a closed circular molecule of approximately 16 569 base pairs in length (Anderson *et al.*, 1981) located in the cellular cytoplasm. It contains only 37 genes, no introns, and little other non-coding DNA besides the 1 kb (kilobase) control region. The mtDNA genome is inherited in a maternal fashion, without recombination, and evolution occurs by the accumulation of mutations through time. The rapid rate of evolution (mtDNA evolves on average about 10 times faster than nuclear DNA) means that deleterious mtDNA mutations are important in human disease and ageing (Wallace, 1995). It also means that there are numerous harmless mutations, either silent or in non-coding regions, which can provide convenient genetic markers. Lastly, and most importantly for ancient DNA research, the existence of thousands of copies of mtDNA in each cell means that some mtDNA sequences are likely to persist for a long time in compromised biological samples such as those found in archaeological or forensic contexts.

Mitochondrial DNA and human evolution

Possibly the most notable example of the use of genetic markers, and specifically mtDNA, for the reconstruction of human evolutionary history was the 1987 study by the late Allan Wilson and colleagues at the University of California in Berkeley (Cann, Stoneking & Wilson, 1987). Using high resolution restriction mapping of mtDNA from living humans of different geographical origins, these workers generated data to test the two contrasting models for the evolution of anatomically modern humans. The first of these models, known as the regional continuity model, places the ultimate origin of archaic *Homo erectus* in Africa, but states that archaic humans evolved into modern humans in different parts of the world, with enough gene flow between populations of different continents to ensure that modern *Homo sapiens* evolved into a single biological species. This hypothesis contrasts with the 'Out of Africa' model that proposes that anatomically modern humans originated in Africa only, and eventually replaced the archaic humans throughout the Old World in a recent expansion from Africa. Before the mtDNA study was published, palaeontologists generally supported a middle view with elements of both theories, but the mtDNA data, frequently misunderstood or overinterpreted, provoked a deep polarization of opinions (Wolpoff, 1989).

Wilson and colleagues observed little variation in the mtDNA types of humans from different parts of the world, suggesting a recent common origin for present-day people. Moreover, they detected substantially more variation in African mtDNA types than anywhere else, consistent with the view that African lineages were the oldest and had more time to accumulate differences. Phylogenetic analysis by maximum parsimony suggested that all modern mtDNAs could be traced to a single individual, known as the African Eve, who lived in Africa about 200 000 years ago. The concept of an African Eve, suggesting a single female ancestor for modern humans, undoubtedly contributed to some of the controversy that greeted the results (Lewin, 1987; Wainscoat, 1986). Additional studies based on sequencing of the hypervariable control region of mtDNA, confirmed a date of about 200 000 years for the common ancestor and showed that the deepest branches of the most parsimonious phylogenetic tree were in Africa, giving further support to the Out of Africa model (Vigilant *et al.*, 1991). This model also agreed with the conclusion reached somewhat earlier by the analysis of β-globin gene variation in different human populations (Wainscoat *et al.*, 1986).

Although later studies have questioned the statistical validity of the mitochondrial DNA maximum parsimony tree (Templeton, 1992), there does seem to be more genetic variation in Africa than elsewhere. A recent restriction enzyme analysis of mtDNA variation revealed the most ancient of all the continent-specific lineages in African populations (Chen *et al.*, 1995). On balance, although there are still many problems in the interpretation of phylogenetic data, the genetic evidence seems to point to Africa as the place of origin of modern humans. The emphasis of the debate on recent human evolution has shifted away from the place of origin of modern humans to the date of the common ancestor. Ways to calibrate the mtDNA clock range from taking the sequence difference between humans and chimpanzee mtDNAs and dividing by the presumed time of divergence of the two species (too inaccurate to be of real value) to the application of coalescence techniques to recently diverged human groups such as native Americans (Ward *et al.*, 1993). Depending on which mutation rate is used to calculate the date of the common human ancestor, the mtDNA data could satisfy either the multiregional origin model, assuming that our most recent common ancestor lived about a million years ago, or the Out of Africa model, if the common ancestor lived about 100 000 to 200 000 years ago (Stoneking, 1993). The debate rages on.

Mitochondrial DNA polymorphisms as anthropological markers

Although these, often contentious, studies have focused on the general patterns of recent human evolution, more detailed analyses of populations of different geographic locations are needed to address specific anthropological questions such as the route and time of migrations of humans in regions such as Africa, Europe, the Pacific and the Americas.

The Pacific area was one of the first regions of the world to be studied in detail by molecular anthropologists as it was settled relatively recently and provides an excellent scenario for testing models of human colonization. Although Australia and Papua New Guinea (PNG) were probably occupied as early as 50 000 or 60 000 years ago, the remote archipelagos of eastern Polynesia were settled for the first time by humans only in the last 1000 years (Bellwood, 1989). The Wilson group used the date of settlement of PNG as a convenient measure to calibrate the mtDNA tree by analysing mtDNA diversity in present-day populations of PNG and assuming that this diversity had accumulated since the first settlement of the island. Unfortunately, this calibration disregarded the effects of multiple migrations into the regions studied, a problem inherent in most attempts to determine the rate of evolution of any DNA locus, including mtDNA. The very wide margin of error in the estimates of the rate of mutation of mtDNA in humans causes significant problems in the interpretation of relatively recent events, such as the colonization of the Pacific and the New World (Stoneking, 1993).

Nevertheless, analyses of mtDNA diversity in humans of different geographical regions promise to provide new insights into the patterns of past migrations. One of the first anthropologically useful mtDNA markers to be identified was a deletion of 9 base pairs (9 bp) in a small non-coding region between the genes for cytochrome oxidase II and lysil transfer RNA (Wrischnik *et al.*, 1987). The 9 bp deletion was observed at relatively high frequencies (5–40%) in individuals of Asian origin and was later found to be present in Polynesians at frequencies reaching fixation (100%). The occurrence of the 9 bp deletion throughout Asia and the Pacific, and its fixation in Polynesians, was presented as evidence of the ultimately Southeast Asian ancestry of the Polynesians. This interpretation of the data is supported by the fact that the deletion is absent or virtually absent in Australia and in the highlands of PNG, areas which are presumably inhabited by the descendants of the Australoid and Papuan people who first settled the Pacific during the Pleistocene (Hertzberg *et al.*, 1989; Stoneking & Wilson, 1989).

Although the 9 bp deletion was originally thought to occur only in populations of Asian origin, including native Americans, it has subsequently been observed in African populations. It is now known that the deletion occurred independently in different parts of the world, and that this part of the mtDNA genome is particularly sensitive to mutation. We have detected the deletion of one of the 9 bp motifs as well as a 9 bp triplication in several European individuals. Nevertheless, the 9 bp deletion has proved to be very useful in studies of Pacific human migrations, particularly in conjunction with sequence analyses of the highly variable mtDNA control region.

Research by us and others has revealed several mtDNA polymorphisms that seem to be present exclusively in Polynesia and island Melanesia (Hagelberg & Clegg, 1993; Lum *et al.*, 1994; Melton *et al.*, 1995). This 'Polynesian mtDNA haplotype' is characterized by the 9 bp deletion and by base substitutions at positions 16 189, 16 217, 16 247 and 16 261 of the control region of the mtDNA genome. The Polynesian mtDNA type seems to derive from an ancestral type present in Asia, characterized by the 16 189 substitution and the 9 bp deletion (Type i in Table 1.1). An additional mutation at position 16 217 occurred later on this background. This mtDNA type (Type ii) is present in Taiwan and Borneo and also in the Americas (Schurr *et al.*, 1990; Horai *et al.*, 1991; Ballinger *et al.*, 1992; our unpublished observations), reflecting the Asian origin of the populations of the New World. The additional two substitutions, 16 247 and 16 261, appear eastwards in the Pacific, with the Polynesian type (Type iv)

Table 1.1. *Mitochondrial type*

Type	Deletion			
i	16 189			
ii	16 189	16 217		
iii	16 189	16 217	16 261	
iv	16 189	16 217	16 247	16 261

accounting for about 80% of individuals sampled, and Type iii for about 15%. The remaining (non-deleted) mtDNA types found in individuals of eastern Polynesia can usually be traced back to types found in either Melanesia or Southeast Asia. Interestingly, the Polynesian mtDNA motif is also present at high frequency in the Malagasy, showing that migrations from Southeast Asia did not only lead to the colonization of Oceania in the east, but also of Madagascar in the west (Soodyall, Jenkins & Stoneking, 1995 and see Chapter 12).

MtDNA control region polymorphisms associated with the 9-bp deletion

The Polynesian mtDNA type (Type iv) derives from an ancestral type that can be traced back to Asia. We have observed the Polynesian mtDNA type in bone samples from prehistoric sites in eastern Polynesia. The Polynesian mtDNA type was also detected in skeletal remains of prehistoric inhabitants of Easter Island, confirming the Polynesian ancestry of the original inhabitants of this most remote of inhabited Pacific islands (Hagelberg *et al*., 1994). However, we failed to detect the polymorphisms in several bones from sites in Melanesia and the central Pacific that were occupied by supposedly proto-Polynesian Lapita settlers (Hagelberg & Clegg, 1993). This would tend to argue against the simple scenario of a recent expansion of people from island Southeast Asia into Polynesia and would support the contention that the earlier settlers of island Melanesia may have expanded gradually into the central Pacific before the arrival of the Polynesians (Terrell, 1986). Further research on ancient bone samples is needed before this question can be answered. This and other problems relating to the migration patterns of past peoples can only really be addressed using direct genetic evidence based on ancient DNA.

Despite fairly unanimous agreement on the Southeast Asian ancestry of the Pacific islanders, evidence from archaeology and botany seems to suggest a certain amount of contact between Polynesia and the New World. There is extremely little variability in the non-coding (hypervariable) mtDNA region of modern and ancient Polynesians, reflecting the recent population expansion of the proto-Polynesian colonizers. In addition, we also observe very little diversity in native Americans belonging to the lineage associated with the 9 bp deletion (haplogroup B). The low sequence divergence in haplogroup B Amerinds, and the absence of the 9 bp deletion in aboriginal Siberians, Beringians and northern North Americans (Shields *et al.*, 1992, 1993; Torroni *et al.*, 1993b) have led to the suggestion that haplogroup B people colonized the Americas through a route other than the Bering Strait, and possibly much more recently than the original migrations from Asia of the other native American haplogroups (Torroni *et al.*, 1993*a*). Our evidence from coding and non-coding mtDNA sequences of living individuals (Thomas & Hagelberg, unpublished observations) confirms that the Polynesians and haplogroup B Amerinds can be traced back to a relatively recent group of people who lived in mainland Asia or island Southeast Asia. We are now embarked on a new ancient DNA project to investigate the genetic origins of the prehistoric inhabitants of the Pacific coast of South America and it will be interesting to determine if there is evidence to support the idea of recent trans-Pacific contacts.

Ancient DNA studies

The field of ancient DNA was born in 1984 with the report of the cloning of DNA from the skin of an extinct quagga (Higuchi *et al.*, 1984). Initially, developments were slow due to the technical problems associated with the analysis of ancient DNA sequences, but the subject has enjoyed rapid growth in the last few years (three international conferences on ancient DNA studies have taken place, and the first issue of a journal dedicated to ancient DNA research is being published in Spring 1996), as well as an unprecedented amount of attention from the news media.

From early on, the polymerase chain reaction (Saiki *et al.*, 1985) became the technique of choice in ancient DNA studies, as it permits the amplification of specific DNA fragments from degraded organic samples, even in the presence of vast amounts of microbial DNA and other contaminants, and avoids some of the sequence artefacts associated with DNA cloning (Pääbo & Wilson, 1988; Pääbo *et al.*, 1989). Ancient DNA studies have been burdened from the start by technical problems, including the difficulty of extracting usable DNA from some types of tissues, contamination by modern DNA, and PCR inhibition by unknown compounds in the tissue extracts. Despite this, DNA has been extracted from plant and insect tissues of considerable antiquity (DeSalle & Grimaldi, 1994).

Because of the interest generated by ancient DNA research, scientists are pressed to produce exciting results on very old and exotic materials, and much of the research has been of a headline-catching type, rather than of real scientific interest. Some research groups have shied away from work on human remains because of the contamination problems. Samples can become contaminated by people handling the bones or by small amounts of human DNA in common dust, as well as by the sequences generated in previous PCR reactions. Whereas in animal studies one can rely on phylogenetic inference for checking the validity of the results, in the case of human studies it may be difficult to distinguish between a genuine DNA sequence and one arising from contamination (Hagelberg & Clegg, 1991; Hagelberg, 1994).

An interesting illustration of the dangers associated with over-optimistic analyses of ancient remains is the claimed recovery of a mtDNA sequence from 80 million year-old dinosaur bones (Woodward, Weyand & Bunnell, 1994). The authors took heroic precautions to avoid contamination by modern DNA and succeeded in amplifying a very short fragment of the mtDNA cytochrome b gene. Although they did not carry out a phylogenetic analysis of the sequence, the amplified fragment was different to the orthologous vertebrate sequences lodged in current DNA databases, so the authors assumed that a genuine ancient sequence had been amplified from the dinosaur bones. However, recent phylogenetic comparison between the 'dinosaur' sequence and number of vertebrate sequences showed that the putative dinosaur sequence aligned with humans rather than birds, the closest living relatives of the dinosaurs, suggesting that the amplified sequences were from minuscule amounts of contaminating human DNA (Hedges & Schweizer, 1995).

However, despite the technical problems, human skeletal and soft tissue remains have been analysed with success and have yielded important genetic information. In addition to the studies on ancient Pacific islanders described above, an excellent survey of skeletal remains from a pre-Columbian site in Illinois showed no appreciable difference in the levels of genetic variation in ancient and present-day populations (Stone & Stoneking, 1993). The recent analysis of mtDNA extracted from a sample of muscle from the Tyrolean Iceman helped to confirm that the 5000 year-old man was an ancient European, rather than an Amerind or Egyptian mummy placed in the ice as an elaborate hoax (Handt *et al.*, 1994).

The forensic identification of human remains became an early target of ancient DNA techniques, as well as providing an excellent way to test the validity of the analyses by comparing DNA extracted from the remains with DNA from close relatives of a presumed victim. Bone DNA typing by analysis of polymorphic nuclear microsatellite DNA has been used in a number of forensic cases (Hagelberg, Gray & Jeffreys, 1991; Jeffreys *et al.*, 1992; Gill *et al.*, 1994). In addition, despite the difficulty in calculating likelihood ratios based on mtDNA data, mtDNA typing is an extremely useful tool for forensic identification as it can be used for difficult samples such as single hairs, or skeletal remains that are too old or decayed to permit the amplification of single-copy nuclear sequences (Higuchi *et al.*, 1988; Stoneking *et al.*, 1991; Ginther *et al.*, 1992; Holland *et al.*, 1993).

To conclude, mtDNA provides an excellent tool for the study of the genetic relationships of ancient and modern human populations, as well as for forensic identification. Although the strictly maternal inheritance of mtDNA means that it is highly susceptible to genetic drift and any results must be interpreted with caution, the relative simplicity of mtDNA and its usefulness in ancient DNA research undoubtedly will help maintain its important role in the study of human evolutionary history.

Note

A similar paper was presented at the 16th International Congress of the International Society for Forensic Haemogenetics, Santiago de Compostela, Spain, 12–16 September 1995.

References

Anderson, S., Bankier, A. T., Barrell, B. G., de Bruijn, M. H. L., Coulson, A. R., Drouin, J., Eperon, I. C., Nierlich, D. P., Roe, B. A., Sanger, F., Schreier, P. H., Smith, A. J. H., Staden, P. & Young, I. G. (1981). Sequence organisation of the human mitochondrial genome. *Nature*, **290**, 457–65.

Ballinger, S. W., Schurr, T. G., Torroni, A., Gan, Y. -Y., Hodge, J. A., Hassan, K., Chen, K. -H. & Wallace, D. C. (1992). Southeast Asian mitochondrial DNA analysis reveals genetic continuity of ancient mongoloid migrations. *Genetics*, **130**, 139–52.

Bellwood, P. S. (1989). The colonization of the Pacific: some current hypotheses. In *The Colonization of the Pacific: A Genetic Trail*, ed. A. V. S.Hill and S. W. Serjeantson, pp. 1–59. Oxford: Clarendon Press.

Cann, R. L., Stoneking, M. & Wilson, A. C. (1987). Mitochondrial DNA and human evolution. *Nature*, **325**, 31–6.

Chen, Y. -S., Torroni, A., Excoffier, L., Santachiara-Benerecetti, A. S. & Wallace, D. C. (1995). Analysis of mtDNA variation in African populations reveals the most ancient of all human continent-specific haplogroups. *American Journal of Human Genetics*, **57**, 133–49.

DeSalle, R. & Grimaldi, D. (1994). Very old DNA. *Current Opinion in Genetics and Development*, **4**, 810–15.

Gill, P., Ivanov, P. L., Kimpton, K., Piercy, L., Benson, L., Tully, G., Evett, I., Hagelberg, E. & Sullivan, K. (1994). Identification of the remains of the Romanov family by DNA analysis. *Nature Genetics*, **6**, 130–5.

Ginther, C., Issel-Tarver, L. & King, M. C. (1992). Identifying individuals by sequencing mitochondrial DNA from teeth. *Nature Genetics*, **2**, 135–8.

Hagelberg, E. (1994). Mitochondrial DNA from ancient bones. In *Ancient DNA*, ed. B. Herrmann and S. Hummel, pp.195–204. New York: Springer-Verlag.
Hagelberg, E. & Clegg, J. B. (1991). Isolation and characterisation of DNA from archaeological bone. *Proceedings of the Royal Society of London Series B*, **244**, 45–50.
Hagelberg, E. & Clegg, J. B. (1993). Genetic polymorphisms in prehistoric Pacific islanders determined by analysis of ancient bone DNA. *Proceedings of the Royal Society of London Series B*, **252**, 163–70.
Hagelberg, E., Gray, I. C. & Jeffreys, A. J. (1991). Identification of the skeletal remains of a murder victim by DNA analysis. *Nature*, **352**, 427–9.
Hagelberg, E., Quevedo, S. & Turbon, D. & Clegg, J. B. (1994). DNA from prehistoric Easter Islanders. *Nature*, **369**, 25–6.
Handt, O., Richards, M., Trommsdorff, M., Kilger, C., Simanainen, J., Georgiev, O., Bauer, K., Stone, A., Hedges, R., Schaffner, W., Utermann, G., Sykes, B. & Pääbo, S. (1994). Molecular genetic analyses of the Tyrolean Iceman. *Science*, **264**, 1775–8.
Hedges, S. B. & Schweizer, M. H. (1995). Detecting dinosaur DNA. *Science*, **268**, 1191.
Hertzberg, M., Mickleson, K. P. N., Serjeantson, S. W., Prior, J. F. & Trent, R. J. (1989). An Asian-specific 9-bp deletion of mitochondrial DNA is frequently found in Polynesians. *American Journal of Human Genetics*, **44**, 504–10.
Higuchi, R., Bowman, B., Freiberger, M., Ryder, O. A & Wilson, A. C. (1984). DNA sequences from the quagga, an extinct member of the horse family. *Nature*, **312**, 282–4.
Higuchi, R., von Beroldingen, C. H., Sensabaugh, G. F. & Erlich, H. A. (1988). DNA typing from single hairs. *Nature*, **332**, 543–6.
Holland, M. M., Fisher, D. L., Mitchell, L. G., Rodriguez, W. C., Canik, J. J., Merril, C. R. & Weedn, V. W. (1993). Mitochondrial DNA sequence analysis of human skeletal remains from the Vietnam War. *Journal of Forensic Science*, **38**, 542–53.
Horai, S., Kondo, R., Murayama, K., Hayashi, S., Koike, H. & Hakai, N. (1991). Phylogenetic affiliation of ancient and contemporary humans inferred by mitochondrial DNA. *Philosophical Transactions of the Royal Society of London Series B*, **333**, 409–17.
Jeffreys, A. J., Allen, M., Hagelberg, E. & Sonnberg, A. (1992). Identification of the skeletal remains of Josef Mengele by DNA analysis. *Forensic Science International*, **56**, 65–76.
Lewin, R. (1987). The unmasking of Mitochondrial Eve. *Science*, **238**, 24–6.
Lum, J. K., Rickards, O., Ching, C. & Cann, R. L. (1994). Polynesian mitochondrial DNAs reveal three deep maternal lineage clusters. *Human Biology*, **66**, 567–90.
Melton, T., Peterson, R., Redd, A. J., Saha, N., Sofro, A. S. M., Martinson, J. J. & Stoneking, M. (1995). Polynesian genetic affinities with Southeast Asian populations as identified by mtDNA analysis. *American Journal of Human Genetics*, **57**, 403–14.
Pääbo, S., Higuchi, R. G. & Wilson, A. C. (1989). Ancient DNA and the polymerase chain reaction. *Journal of Biological Chemistry*, **264**, 9709–12.
Pääbo, S. & Wilson, A. C. (1988). Polymerase chain reaction reveals cloning artefacts. *Nature*, **334**, 387–8.

Saiki, R. K., Scharf, S., Faloona, F., Mullis, K. B., Horn, G. T., Erlich, H. A. & Arnheim, N. (1985). Enzymatic amplification of β-globin genomic sequences and restriction site analysis for diagnosis of sickle cell anemia. *Science*, **230**, 1350–4.

Schurr, T. G., Ballinger, S. W., Gan, Y. -Y., Hodge, J. A., Merriweather, D. A., Lawrence, D. N., Knowler, W. C., Weiss, K. M. & Wallace, D. C. (1990). Amerindian mitochondrial DNAs have rare Asian mutations at high frequencies, suggesting they derived from four primary maternal lineages. *American Journal of Human Genetics*, **46**, 613–23.

Shields, G. F., Hecker, K., Voevoda, M. I. & Reed, J. K. (1992). Absence of the Asian-specific region V mitochondrial marker in native Beringians. *American Journal of Human Genetics*, **50**, 758–65.

Shields, G. F., Schmiechen, A. M., Frazier, B. L., Redd, A., Voevoda, M. I., Reed, J. K. & Ward, R. H. (1993). mtDNA sequences suggest a recent evolutionary divergence for Beringian and northern North American populations. *American Journal of Human Genetics*, **53**, 549–62.

Soodyall, H., Jenkins, T. & Stoneking, M. (1995). 'Polynesian' mtDNA in the Malagasy. *Nature Genetics*, **10**, 377–8.

Stone, A. C. & Stoneking, M. (1993). Ancient DNA from a pre-Columbian Amerind population. *American Journal of Physical Anthropology*, **92**, 463–71.

Stoneking, M. (1993). DNA and recent human evolution. *Evolutionary Anthropology*, **2**, 60–73.

Stoneking, M., Hedgecock, D., Higuchi, R. G., Vigilant, L. & Erlich, H. A (1991). Population variation of human mtDNA control region sequences detected by enzymatic amplification and sequence-specific oligonucleotide probes. *American Journal of Human Genetics*, **48**, 370–82.

Stoneking, M. & Wilson, A. C. (1989). Mitochondrial DNA. In *The Colonization of the Pacific: A Genetic Trail*, ed. A. V. S. Hill and S. W. Serjeantson, pp. 215–45. Oxford: Clarendon Press.

Templeton, A. R. (1992). Human origins and analysis of mitochondrial DNA sequences. *Science*, **255**, 737.

Terrell, J. (1986). *Prehistory in the Pacific Islands*. Cambridge: Cambridge University Press.

Torroni, A., Schurr, T. G., Cabell, M. F., Brown, M. D., Neel, J. V., Larsen, M., Smith, D. G., Vullo, C. M. & Wallace, D. C. (1993*a*). Asian affinities and continental radiation of the four founding native American mtDNAs. *American Journal of Human Genetics*, **53**, 563–90.

Torroni, A., Sukernik, R. I., Schurr, T. G., Starikovskaya, Y. B., Cabell, M. F., Crawford, M. H., Comuzzie, A. G. & Wallace, D. C. (1993*b*). mtDNA variation of aboriginal Siberians reveals distinct genetic affinities with native Americans. *American Journal of Human Genetics*, **53**, 591–608.

Vigilant, L., Stoneking, M., Harpending, H., Hawkes, K. & Wilson, A. C. (1991). African populations and the evolution of human mitochondrial DNA. *Science*, **253**, 1503–7.

Wainscoat, J. (1986). Out of the garden of Eden. *Nature*, **325**, 13.

Wainscoat, J. S., Hill, A. V. S., Boyce, A. J., Flint, J., Hernandez, M., Thein, S. L., Old, J. M., Falusi, A. G., Weatherall, D. J. & Clegg, J. B. (1986). Evolutionary relationships of human populations from an analysis of nuclear DNA polymorphisms. *Nature*, **319**, 491–3.

Wallace, D. C. (1995). Mitochondrial DNA variation in human evolution, degenerative disease, and ageing. *American Journal of Human Genetics*, **57**, 201–23.

Ward, R. H., Redd, A., Valencia, D., Frazier, B. & Pääbo, S. (1993). Genetic and linguistic differentiation in the Americas. *Proceedings of the National Academy of Sciences, USA*, **90**, 10663–7.

Wolpoff, M. R. (1989). Multiregional evolution: the fossil alternative to Eden. In *The Human Revolution*, ed. P. Mellars and C. Stringer. pp. 62–108. Edinburgh: Edinburgh University Press.

Woodward, S. R., Weyand, N. J. & Bunnell, M. (1994). DNA sequence from Cretaceous period bone fragments. *Science*, **265**, 1229–32.

Wrischnik, L. A., Higuchi, R. G., Stoneking, M., Erlich, H. A., Arnheim, N. & Wilson, A. C. (1987). Length mutations in human mitochondrial DNA: direct sequencing of enzymatically amplified DNA. *Nucleic Acids Research*, **15**, 529–42.

2 *Digital DNA typing of human paternal lineages*

M. A. JOBLING, N. BOUZEKRI, N. FRETWELL,
G. A. DOVER AND A. J. JEFFREYS

The Y chromosome

The Y chromosome (for a review see Ellis, 1991) is an unusual part of the human genome; at about 60 million base pairs, it represents around 2% of the total, yet half of the human population survives quite happily without one. The Y is haploid and paternally inherited. The consequence of this haploidy is that, outside of two small pseudoautosomal regions at the tips of the short and long arms of the chromosome, which can recombine with the X (Rouyer *et al.*, 1986; Freije, Helms, Watson & Donis-Keller, 1992), the Y has no chromosomal partner with which to recombine at meiosis. Thus the Y is passed down virtually intact from father to son. This is in contrast to the X chromosome and autosomes, which are continually being reshuffled by recombination, and means that a comparison of Y chromosomes is a direct comparison of individuals. As Y chromosomes pass down paternal lineages from a common paternal ancestor, they accumulate mutations; if we could look at these mutations, in the form of various kinds of DNA polymorphisms, we could firstly try to determine how modern Ys are related to each other and to their ancestors. Secondly, these polymorphisms define Y chromosome types, and if we could measure the frequencies of these types in different populations, we could also attempt to understand some aspects of the histories of these populations (Spurdle & Jenkins, 1992*a*).

Comparison of the Y and mtDNA

In these respects, of course, the Y is analogous to the mitochondrial genome, which is maternally inherited, also escapes from recombination, and can be used in an attempt to construct phylogenies of maternal lineages (Cann, Stoneking & Wilson, 1987). Much of the content of this volume is concerned with mtDNA data on human diversity, but there is little data on the Y. The reasons for this disparity lie in the important differences between the two systems: the mitochondrial genome is a 16.5 kb, circular molecule whose entire DNA sequence is known. It has a very high density of genes, and little 'junk' DNA or simple sequence. It also has a ten-fold higher mutation rate than nuclear DNA, which means that there are a very large number of different mtDNA sequences in the population. These sequence differences can be simply surveyed by restriction digestion of purified mtDNA, or of mtDNA PCR products, or by DNA sequencing. The Y, in contrast, is a large linear molecule of almost entirely unknown sequence. Although about four thousand times the size of the mitochondrial genome, it has fewer known genes (Affara *et al.*, 1994). It has high sequence complexity, with large tracts of non-coding DNA, and many families of dispersed and tandemly arranged repeat sequences. The Y has a low base-substitutional mutation rate compared with mtDNA, and has been found to have a much lower nucleotide diversity than the other nuclear chromosomes, for reasons to do with chromosome population size, the mating structures of human populations, and possibly selective hitch-hiking effects. So, although extensive searches have been made for Y-chromosomal RFLPs (Jakubiczka *et al.*, 1989; Malaspina *et al.*, 1990; Spurdle & Jenkins, 1992*b*), very few have been found, and those that are known are mostly complex and difficult to analyse and interpret.

So, on the one hand, we have a small, simple, highly characterized, highly polymorphic system (mtDNA), and on the other a large, complex, poorly characterized system of apparently low diversity. In consequence, the contribution of Y chromosome studies to issues of human evolution has so far been very poor. However, in a few years' time, through technological advances, this will certainly have changed, and it is likely that the Y will be as informative about paternal lineages as mtDNA is about maternal lineages. The Y actually possesses some inherent advantages over mtDNA, which are to do with the differences in mutation rate and complexity: i) The mutation rate of mtDNA is so high that given point mutations can revert, or occur independently in separate lineages, and ancestral states cannot be determined; this complicates tree construction. On the Y, in contrast, point mutations are so rare that they can be regarded as unique events which are lineage-specific: a group of chromosomes bearing a given point mutation is in all probability monophyletic. In many cases, ancestral states can be determined by the analysis of great ape DNAs. ii) The greater complexity of the Y means that many other classes of polymorphism can occur on this chromosome, associated with large- and small-scale chromosomal rearrangements, such as inversions, insertions and deletions, and with tandem repeat arrays such as satellites, minisatellites, and microsatellites. Thus the

Y carries more polymorphism of more diverse kinds than mtDNA, and should be able to answer more diverse questions.

Many of the tandem repeat loci on the Y chromosome are known to be polymorphic. Hypervariable satellite loci are numerous (Oakey & Tyler-Smith, 1990; Jobling, 1994), and can be used to discriminate between almost any pair of Y chromosomes (Jobling, 1994; Mathias, Bayés & Tyler-Smith, 1994), but are difficult to analyse because of their very large sizes. Several polymorphic microsatellites are known (Roewer *et al.*, 1992; Mathias *et al.*, 1994), but because of allele sharing are of limited informativeness. There is, however, only one known Y-specific minisatellite (Jobling *et al.*, 1994), and this will be the subject of the rest of this chapter.

Minisatellites

Minisatellites (Jeffreys, Wilson & Thein, 1985) are loci composed of tandemly repeated, normally G+C-rich DNA sequences of 10–50 bp, present in many copies in all eukaryotic genomes. Interest in minisatellites since their discovery has been because of their high degree of length polymorphism, which is due to variation in the number of tandem repeats in the array. This has made them useful in many applications, including linkage analysis, individual identification in forensics, paternity testing and immigration casework, transplant screening, tumour clonality analysis, and, to some degree, in studies of human diversity. However, their utility in population studies is limited provided that allele length is the only property considered. Distantly related alleles can share identical lengths by chance, and it therefore becomes necessary to look at several loci and infer relatedness by statistical analysis of allele sharing or allele frequencies (Chakraborty & Jin, 1993).

MVR–PCR

As well as variation in the number of repeats in an array, all minisatellites examined so far have an additional level of variation, in the sequence of the repeat units themselves. Mapping of the positions of different variant repeats along minisatellite arrays can be done simply by Minisatellite Variant Repeat PCR (MVR-PCR; Jeffreys *et al.*, 1991), in which PCR primers which anneal specifically to one particular repeat type give products extending from these repeat types in an array out to a fixed flanking primer (Fig. 2.1). This technique produces digital DNA codes for minisatellite alleles which have enormous diversity. Analysis of the best-characterized autosomal minisatellite, MS32, predicts that there are of the order of 10^8 different alleles worldwide at this locus alone. Apart from

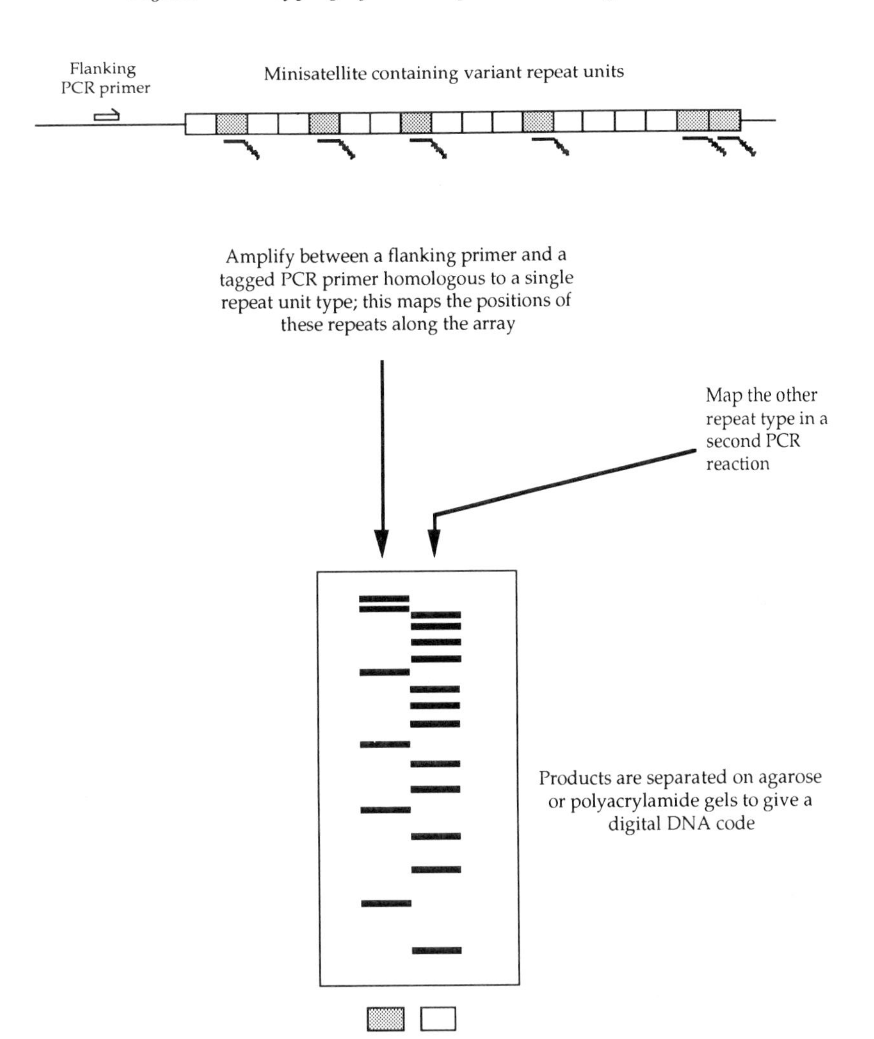

Fig. 2.1. Principle of MVR–PCR. White and grey boxes represent two different repeat types. For a full description of MVR–PCR, including the significance of primer tagging, refer to Jeffreys *et al.* (1991).

its great power for discriminating between alleles, this fine structure mapping by MVR–PCR allows a detailed analysis of mutation events to be done. Mutant (new-length) alleles occur at a frequency of about 1% for MS32. They are detected either by pedigree analysis, or, more readily, by a PCR-based method from sperm DNA (Jeffreys *et al.*, 1994), from which virtually unlimited numbers of mutants can be recovered. Mutants are isolated by electrophoresis, subjected to MVR–PCR analysis, and their structures compared with those of their parental alleles. For autosomal loci such as MS32, these studies have shown that mutations occur predominantly in the germ-line, and mostly involve complex gene-conversion-like events between alleles (Jeffreys *et al.*, 1994).

Thus, minisatellites have very large numbers of alleles which can be unambiguously typed in a simple PCR-based assay, and the mutation rates and processes responsible for their diversity can be directly analysed. A minisatellite on the Y chromosome would clearly be a powerful marker for paternal lineages. It would have the additional advantage that it never passes through female meiosis, so sperm DNA would be a perfect source of mutants – in the case of autosomal loci, it is necessary to analyse female germline events too, which can be identified only in small numbers by extensive pedigree studies.

MSY1

Although many minisatellites are known to lie in the short-arm pseudoautosomal region of the Y, none has previously been described in the Y-specific part of the chromosome. MSY1 (Jobling *et al.*, 1994; Jobling, manuscript in preparation), which was identified fortuitously, is the first such sequence.

MSY1 (*DYF*155*S*1) lies in interval 3 on the short arm of the Y, and consists of an array of 25 bp repeats which, unlike repeats in 'classical' minisatellites, are A + T-rich – on average about 75% A + T. Alleles show length polymorphism of 58–96 repeats (1.7 – 2.7 kb), which is rather limited compared to loci such as MS32, where arrays vary from 12 to 800 or more repeats (Jeffreys, Neumann & Wilson, 1990).

PCR and Southern analysis in the DNAs of chimps, gorillas, and orang-utans show no fragment length polymorphism, and preliminary evidence indicates that the corresponding locus in chimp DNA contains a repeat monomer; this suggests that the MSY1 repeat array has amplified in the human lineage since the human–chimp divergence.

MVR–PCR at MSY1

In order to design an MVR system, it is first necessary to identify variant repeats. Direct sequencing of PCR-amplified alleles from eight individuals of diverse ethnic origins revealed five different repeat types, all 25 bp in length, which differ from each other by base substitutions (Fig. 2.2). All of these repeat types are strongly predicted to form hairpin structures, which has uncertain biological significance, but considerable practical significance: because the repeats are A + T-rich, and are predicted to form

```
Type 1:  CACAATATACATGATGTATATTATA
Type 2:  ..T.........G............
Type 3:  ..C.........C............
Type 4:  ..T.........C............
Type 5:  ..T.........G.......A....
```

```
                           G
                          T-A
                          A-T
                          C-G
                          A-T
                          T-A
      Type 2 repeat:      A-T
                          T-A
                          A-T
                          A-T
                          T-A
                          A-T
                          C*A
```

Fig. 2.2. Repeat types at MSY1. A predicted hairpin structure is shown for a type 2 repeat.

hairpins, the discriminator primers used for MVR–PCR are A + T-rich and may form hairpins (or dimerize) too, making design of an MVR system difficult. One quality which simplifies analysis, however, is that only a single allele exists at this locus, in contrast to the case at the diploid autosomal loci, where it is necessary to ablate one allele so that the other can be mapped in isolation (Monckton *et al.*, 1993).

The MVR–PCR system that is currently used maps the positions of three repeat types, known as type 1, type 3, and type 4, along arrays. Type 2 repeats are confined to a few populations, and are at present refractory to coding. The type 5 repeat is a singleton, lying at the 5′ end of the array in a Biaka Pygmy male (m47).

Three state MVR codes at MSY1

The use of five different discriminator primers, in forward and reverse mapping (Fig. 2.3), gives maps covering whole alleles, which are consistent with sequencing data.

This system has been available for only a short time, so the number of coded alleles is still small. All 47 allele codes obtained so far are different, which is an extremely high level of diversity, particularly when compared with the monotonous results of conventional RFLP searches on the Y (Malaspina *et al.*, 1990). At the same time, some alleles fall into groups,

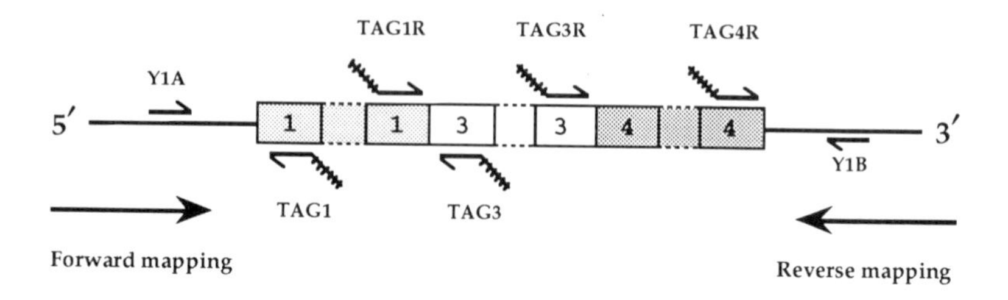

Fig. 2.3. Schematic representation of primers used in MVR–PCR at MSY1. 5′ and 3′ are designated arbitrarily. A detailed description of the system will be published elsewhere (Jobling, manuscript in preparation).

which are consistent with haplotype groups defined by other means. Some examples of allele structures are shown in Fig. 2.4.

What are the properties of these structures? The numbers of repeats in the alleles typed to date vary between 58 and 96. The different repeat types tend to be clustered into blocks, which are often quite large; this in contrast to autosomal loci such as MS32, which have much more richly interspersed patterns of variant repeats, and is probably a reflection of mutation processes such as replication slippage – favoured in A + T-rich DNA – or highly constrained unequal exchange between sister chromatids, and the absence of exchange of sequence information between alleles of different structures (or between distant regions of the same allele). Many alleles have very simple structures, with just two boundaries between these large blocks. The most complex contains nine boundaries between repeat types. Twelve of the forty-seven codes contain repeats which fail to code in the three-state system: these are called null repeats, and will be discussed below.

Allele groups: comparison with chromosome haplotype

Detailed haplotypic information is available for most of these Y chromosomes, derived from the typing of many other polymorphisms, including conventional RFLPs and pulsed-field gel analysis of hypervariable satellite DNA loci (Oakey & Tyler-Smith, 1990; Jobling, 1994; Mathias *et al.*, 1994). This compound haplotyping identifies four major groups of Ys (called Groups 1 to 4) in the chromosomes analysed, as well as a number of chromosomes which do not fall into these groups, and which probably represent further groups in the world population.

A comparison of MSY1 codes with these chromosomal haplotypic groups shows some strong correlations. For example, all Group 1 chromosomes, which are Caucasian-specific and rather homogeneous by other criteria, have very similar tripartite MSY1 structures, with no nulls.

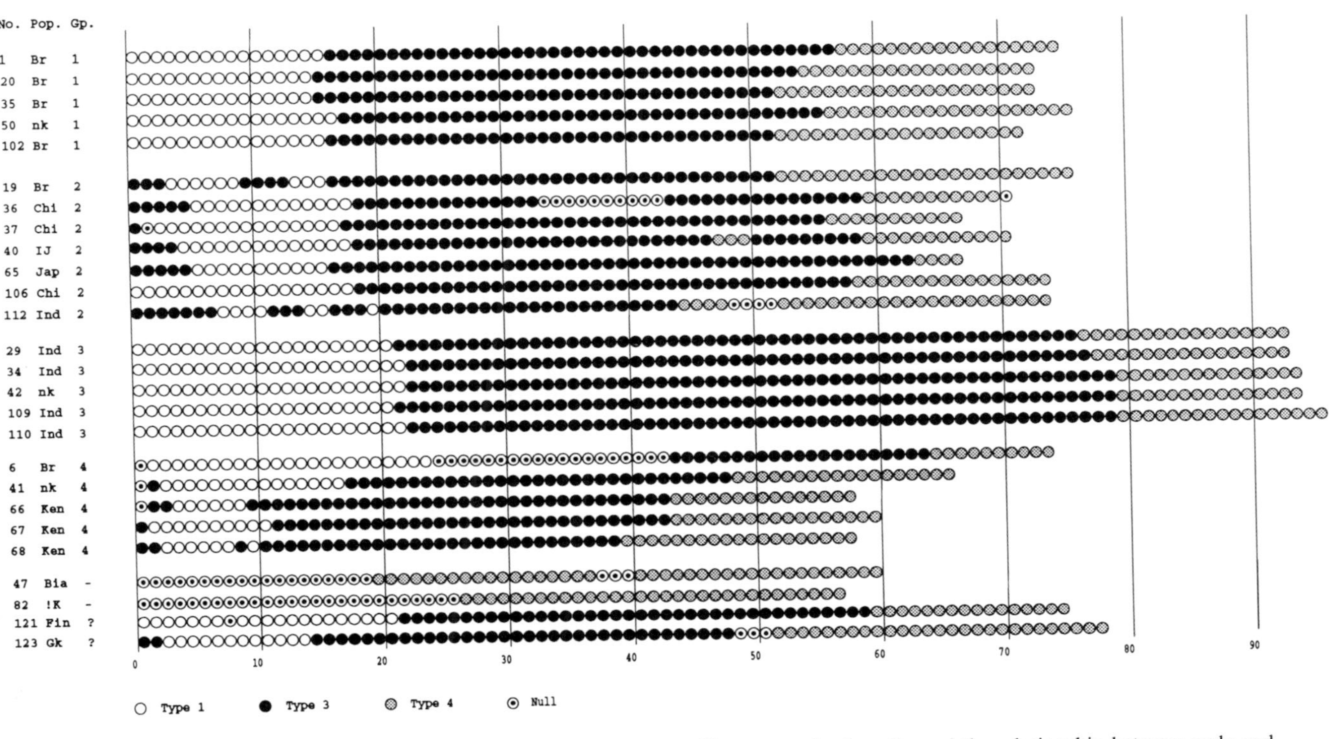

Fig. 2.4. Examples of MSY1 MVR–PCR codes. Twenty-six codes are shown to illustrate code diversity and the relationship between code and haplotypic group (Gp.). All individual numbers (No.) except 121 and 123 are from Jobling (1994) and Mathias, Bayés & Tyler-Smith (1994). Populations (Pop.) are: Br: British; Chi: Chinese; IJ: Iraqi-Jewish (Sirota *et al.*, 1981); Jap: Japanese; Ind: Indian; Ken: Kenyan; Bia: Biaka Pygmy; !K: !Kung; Fin: Finnish; Gr: Greek; nk: not known. Haplotypic groups are as defined by Jobling (1994). '-' indicates non-membership of groups 1–4; '?' indicates group not known.

Similarly, the Group 3 chromosomes, which are probably Indian-subcontinent-specific, have very similar, characteristically larger tripartite structures, again lacking nulls. The four African Group 4 chromosomes have small, more complex alleles, and the Group 2 chromosomes, heterogeneous by other criteria, also have a heterogeneous set of complex MSY1 alleles. Chromosomes which do not fall into any of these groups show additional diversity. If grouping were done blind, at least one of the Group 2 alleles (m106) would be classified as Group 1, indicating that similar allele structures can arise by convergence more than once in different lineages.

Null repeats

The initial sequence analysis of alleles revealed five different repeat types. We now know that there are at least seven more types, from work aimed at the characterization of nulls – repeats which fail to code (as type 1, 3, or 4 repeats) in the existing MVR–PCR system (Bouzekri & Jobling, manuscript in preparation).

Some occasional nulls, like that in the Finnish allele (m121), and those in the Greek (m123), are likely to be useful for identifying specific groups of Y chromosomes within populations. Other nulls constitute major repeat species in some alleles – for example, type 2 repeats constitute large parts of the alleles in a Biaka Pygmy (m47) and a !Kung (m82) male. There is evidence that still other repeat types exist which code normally in the existing three-state system, but contain additional base substitutions. Sequencing of these, and design of specific MVR–PCR systems for their analysis, will improve discrimination at MSY1 and throw light on mutation processes which spread variants within the array. There may be other alleles which have homogenized mutations in all or most of their repeats: examples of this may be alleles in some populations which completely fail to code in the existing MVR–PCR system.

Uses of MSY1

What is MSY1 likely to be useful for? Aside from its uses in population studies, MSY1 will be useful for the basic understanding of mutation processes, and possibly in forensics, as a male-specific identification or exclusion tool. It may also be useful in genealogical studies: males sharing surnames might in some cases be expected to share the same, or significantly similar, MSY1 codes.

MSY1 has many of the criteria of an ideal locus: many alleles, a PCR assay system, and direct access to mutation processes; if mutation rates can be measured, it may be of use in dating the branches of a Y chromosome tree. Rare nulls are likely to be particularly useful, because their population specificity should allow questions of introgression of particular Y chromosome types into different populations to be addressed. However, MSY1 does have the disadvantage that indistinguishable alleles can on occasion arise in different lineages. The isolation of more minisatellites might be a solution to this, but systematic searches for these have proved unsuccessful (unpublished observations). This is where the other kinds of polymorphisms, which have fewer alleles, but little or no chance of occurring independently in separate lineages, are useful. A battery of PCR-typable polymorphisms such as point mutations (Seielstad *et al.*, 1994) and *Alu* element insertions (Hammer, 1994), defining basic, monophyletic Y chromosome groups, together with hypervariable systems such as MSY1 MVR–PCR, will be a powerful combination for paternal lineage analysis.

Future work

Future work at MSY1 will concentrate on population analysis using large samples, characterization of novel repeat types and design of new MVR systems, and the study of mutation processes, both in families and in sperm DNA. The observed level of code diversity suggests a mutation rate to new-length alleles in excess of 1%. An understanding of how mutation works in different allele structures can then be applied to diversity studies.

Analysis of the Y and the mitochondrial genome tell us about different facets of the history of human populations. Trees built using mtDNA data might be very different to those built using Y data, as a result of cultural mating practices, and the different behaviours of males and females in migrations and colonizations. For example, in the construction of the first tree of mtDNA haplotypes (Cann *et al.*, 1987), the use of black Americans to represent African populations was justified by the argument that black females would have mated with white males, and that therefore the maternally inherited mtDNA would be likely to belong to African lineages. The converse would be likely to apply to Y chromosomes.

The information obtained from other nuclear DNA markers is different again to mtDNA- or Y-based data. A serious examination of human population histories, which is the aim of the global and European human genetic diversity projects, will require the combination of all three kinds of analysis.

Acknowledgements

This research was supported by grants from the Medical Research Council and the European Union. We thank John Armour and Chris Tyler-Smith for their contributions to the work.

The Y Chromosome Consortium: The Y Chromosome Consortium is an international collaborative organisation which exists to study genetic variation on the Y. It maintains a cell-line repository and distributes DNAs to members, runs a polymorphism database, and publishes a newsletter. The YCC contacts are Michael Hammer (E-mail: hammer@brahms.biosci.arizona.edu; Fax: +1 602 621 9190) and Nathan Ellis (E-mail: nellis@server.nybc.org; Fax: +1 212 570 3195).

References

Affara, N. A., Lau, Y. -F. C., Briggs, H., Davey, P., Jones, M. H., Khwaja, O., Mitchell, M. & Sargent, C. (1994). Report of the First International Workshop on Y Chromosome Mapping 1994. *Cytogenetics and Cell Genetics*, **67**, 360–87.

Cann, R. L., Stoneking, M. & Wilson, A. C. (1987). Mitochondrial DNA and human evolution. *Nature*, **325**, 31–6.

Chakraborty, R. & Jin, L. (1993). A unified approach to study hypervariable polymorphisms: Statistical considerations of determining relatedness and population distances. In *DNA Fingerprinting: State of the Science*, ed. S. D. J. Pena, R. Chakraborty, J. T. Epplen and A. J. Jeffreys, pp.153–175. Basel: Birkhäuser Verlag.

Ellis, N. A. (1991). The human Y chromosome. *Seminars in Developmental Biology*, **2**, 231–40.

Freije, D., Helms, C., Watson, M. S. & Donis-Keller, H. (1992). Identification of a second pseudoautosomal region near the Xq and Yq telomeres. *Science*, **258**, 1784–7.

Hammer, M. F. (1994). A recent insertion of an *Alu* element on the Y chromosome is a useful marker for human population studies. *Molecular Biology and Evolution*, **11**, 749–61.

Jakubiczka, S., Arnemann, J., Cooke, H. J., Krawczak, M. & Schmidtke, J. (1989). A search for restriction fragment length polymorphism on the human Y chromosome. *Human Genetics*, **84**, 86–8.

Jeffreys, A. J., Wilson, V. & Thein, S. -L. (1985). Hypervariable minisatellite regions in human DNA. *Nature*, **314**, 67–73.

Jeffreys, A. J., Neumann, R. & Wilson, V. (1990). Repeat unit sequence variation in minisatellites: a novel source of DNA polymorphism for studying variation and mutation by single molecule analysis. *Cell*, **60**, 473–85.

Jeffreys, A. J., MacLeod, A., Tamaki, K., Neil, D. L. & Monckton, D. C. (1991). Minisatellite repeat coding as a digital approach to DNA typing. *Nature*, **354**, 204–9.

Jeffreys, A. J., Tamaki, K., MacLeod, A., Monckton, D. G., Neil, D. L. & Armour, J. A. L. (1994). Complex gene conversion events in germline mutation at human minisatellites. *Nature Genetics*, **6**, 136–45.

Jobling, M. A. (1994). A survey of long-range DNA polymorphisms on the human Y chromosome. *Human Molecular Genetics*, **3**, 107–4.

Jobling, M. A., Fretwell, N., Dover, G. A. & Jeffreys, A. J. (1994). Digital coding of human Y chromosomes: MVR–PCR at Y-specific minisatellites. *Cytogenetics and Cell Genetics*, **67**, 390.

Malaspina, P., Persichetti, F., Novelletto, A., Iodice, C., Terrenato., L., Wolfe, J., Ferraro, M. & Prantera, G. (1990). The human Y chromosome shows a low level of DNA polymorphism. *Annals of Human Genetics*, **54**, 297–305.

Mathias, N., Bayés, M. & Tyler-Smith, C. (1994). Highly informative compound haplotypes for the human Y chromosome. *Human Molecular Genetics*, **3**, 115–23.

Monckton, D. G., Tamaki, K., MacLeod, A., Neil, D. L. & Jeffreys, A. J. (1993). Allele-specific MVR-PCR analysis at minisatellite DIS8. *Human Molecular Genetics*, **2**, 513–19.

Oakey, R. & Tyler-Smith, C. (1990). Y chromosome DNA haplotyping suggests that most European and Asian men are descended from one of two males. *Genomics*, **7**, 325–30.

Roewer, L., Arnemann, J., Spurr, N. K., Grzeschik, K. -H. & Epplen, J. T. (1992). Simple repeat sequences on the human Y chromosome are equally polymorphic as their autosomal counterparts. *Human Genetics*, **89**, 389–94.

Rouyer, F., Simmler, M. -C., Johnsson, C., Vergnaud, C., Cooke, H. J. & Weissenbach, J. (1986). A gradient of sex linkage in the pseudoautosomal region of the human sex chromosomes. *Nature*, **319**, 291–5.

Seielstad, M. T., Hebert, J. M., Lin, A. A., Underhill, P. A., Ibrahim, M., Vollrath, D. & Cavalli-Sforza, L. L. (1994). Construction of human Y-chromosomal haplotypes using a new polymorphic A to G transition. *Human Molecular Genetics*, **3**, 2159–61.

Sirota, L., Zlotogora, Y., Shabtai, F., Halbrecht, I. & Elian, E. (1981). 49,XYYYY: a case report. *Clinical Genetics*, **19**, 87–93.

Spurdle, A. M. & Jenkins, T. (1992*a*). The Y chromosome as a tool for studying human evolution. *Current Opinions in Genetics and Development*, **2**, 487–91.

Spurdle, A. M. & Jenkins, T. (1992*b*). The search of Y chromosome polymorphism is extended to negroids. *Human Molecular Genetics*, **1**, 169–70.

3 *Minisatellites as tools for population genetic analysis*

J. FLINT, J. BOND, R. M. HARDING AND J. B. CLEGG

Many multi-allelic loci in the human genome are almost always due to variation in the number of repeats of a short DNA sequence and a number of investigators have demonstrated that the frequency distribution of such loci differs between human ethnic groups, making them a potentially valuable tool in the analysis of population relationships (Balazs *et al.*, 1989; Bowcock *et al.*, 1994; Deka, Chakraborty & Ferrell, 1991; Flint *et al.*, 1989). However, a number of issues have still to be addressed before the full benefit of multi-allelic loci analysis can be felt. These can be divided into questions about the most appropriate ways to analyse allele frequencies, and technical questions about data collection. This chapter is devoted to the latter.

We discuss here one class of multi-allelic locus, the minisatellite or variable number tandem repeat (VNTR) which usually has unit sizes between 10 to 50 bp with alleles varying in size from 500 bp to 30 kb. The issues surrounding the collection of population allele frequencies at minisatellite loci can be divided into two: there are purely technical issues concerning the most effective way of extracting and recording data for analysis, and there are questions about how allele frequencies at one locus compare with those at another, and indeed with other types of DNA, protein and antigen polymorphism. Answers to these questions provide the basis for an understanding of how best to analyse and interpret variation in allele frequencies between different world populations.

Sources of error in the measurement of allele sizes between populations

Currently, most studies of minisatellite allele variation estimate allele frequencies by measuring size differences observed after agarose gel electrophoresis. The resolving power of the method decreases logarithmically in proportion to the size of the allele, so that whereas alleles that differ by 100 bp can be accurately distinguished in the 1 to 2 kilobase size range, with sizes over about 12 kb, differences of less than one kilobase cannot be separated reliably. Errors in the estimate of genotype frequencies arise in four ways. First, allele sizes are subject to measurement error, which is a function of the allele size and is normally distributed with a variance that is a function of the allele size (Devlin & Risch, 1993). Secondly, the resolving power of agarose gel electrophoresis cannot distinguish all variants. Indeed, when an individual has two alleles of similar size, only one band may be visible on the autoradiograph where two bands have amalgamated or coalesced (Devlin, Risch & Roeder, 1990). Thirdly, alleles at some loci may be too small to be detectable (so-called null alleles) so that some individuals are wrongly assumed to be homozygotes for a larger allele (Chakraborty *et al.*, 1994). Finally, even when alleles are accurately measured to be identical in size they may still differ because of internal sequence variation.

Clearly, some sources of error in allele classification are irreducible, particularly the latter two, and simply have to be taken into account in subsequent analyses of gene frequencies (Devlin, Risch & Roeder 1991; Chakraborty *et al.*, 1994). However, there is room for improvement of currently favoured methods for minisatellite allele size measurement for the purposes of population genetic analysis. The largest data sets so far available have been assembled by forensic scientists and from tests of paternity where the emphasis has been on comparisons between individuals (Devlin & Risch 1992; Balazs *et al.*, 1989). Although raw measurements of alleles are continuous, allele sizes are estimated by reference to a set of known size standards, which are used to define a series of 'bins'. Since the same boundaries apply to all data sets, comparisons can be made between data sets, and the method is generally regarded as conservative (Budowle *et al.*, 1991). However, fixed-bin analysis has the shortcoming that the classification is not sensitive to the allele distribution. Thus, in the case where a high frequency allele in one population lies at the extreme of a bin size and a high frequency allele in another population lies at the other extreme, the alleles will be grouped into the same bin. A potentially more accurate approach is to define bins by the allele frequency distribution itself and then to classify alleles by comparison of bins between populations. This is the basis of a floating bin analysis described below.

Semi-automated floating bin measurement of minisatellite alleles

Many analyses of gene frequencies require access to a large sample size; they also require a system of measurement that directly compares allele sizes in different world populations. For high sample throughput we have used a video-based scanning system and software that estimates allele size by comparison with reference ladders at three positions on the autoradiograph. The raw data from the video system are then processed as follows: first, a standard deviation for each size marker on the reference ladder is calculated by repeated measurement of ladder markers. Next, the commonest allele from the raw data is compared with alleles either side. Alleles are assimilated into the same bin if the difference in size is less than the determined measurement imprecision, as defined by the standard deviation of the appropriate size marker. Once all alleles have been compared in this way, the analysis is repeated for the next most common allele, and this process is iterated until all alleles are allocated to bins that lie outside the measurement limits of neighbouring bins. Our initial studies show that there are no serious discrepancies between samples analysed by fixed and floating bin methods, but that the floating bin method allows more flexibility of measurement. It is well suited to robust analysis of many thousands of samples from different world populations.

What is the best strategy to acquire an accurate measure of genetic diversity using minisatellites?

Although the high heterozygosity of some minisatellites appears to provide a useful tool for distinguishing populations by allele frequency, this feature also poses problems of interpretation. First, we need to know what is an appropriate sample size for an accurate assessment of allele frequencies. The distribution of minisatellite alleles is frequently multi-modal, with the majority of allele classes present at low frequencies. When analysis shows that these low frequency alleles provide most of the discrimination between populations it is important that their frequencies are confidently established. Secondly, we need to know whether discrimination between populations is more sensitive to sample size or to the number of loci examined. Where samples are acquired from remote relatively small populations, it may not be possible to collect more than fifty samples; in such cases it will be important to know how many minisatellite loci should be examined to derive a robust estimate of genetic diversity (assuming that we have some way of deciding on the robustness of the estimate). Thirdly, there is little information about the discriminating potential of individual loci. Differences in the heterozygosity of minisatellites suggests that the picture of population relationships may vary according to the loci examined. Whether this is purely a function of heterozygosity or reflects other processes is unknown.

Preliminary results

We have analysed 5000 chromosomes from 18 different world populations using floating bin analysis of 10 minisatellite loci. The large number of chromosomes has allowed us to investigate the effects of sample size on assessments of allele frequencies. We have also been able to investigate whether the pattern of population relationships, as assessed by dendrograms derived from genetic distance estimates, is the same for different minisatellites and to what extent data from additional loci alter this pattern.

Sample size

We find that heterozygosity is relatively robust to sample size. Adequate estimates of heterozygosity are obtained with sample sizes of only 50 chromosomes. However, allele frequency estimates do continue to alter with sample sizes over 200 chromosomes, and for minisatellites with heterozygosities over 80% new allele classes continue to be found even with the largest sample sizes. Provisionally we estimate that at least 400 chromosomes are required for accurate estimates of the number of alleles in a population and of the prevalence of low frequency alleles.

Comparisons between minisatellite loci

We have found striking differences in allele frequencies at some loci. For instance, one locus on 22q has a heterozygosity of over 80% in Caucasian populations but less than 20% in some South Pacific populations. Again using dendrograms as a way of comparing the description of population relationships afforded by different loci we find that there is considerable agreement between loci. Moreover, a stable picture emerges more quickly from using more minisatellite loci than from increasing the number of samples analysed. The dendrogram that emerges from using data from four minisatellites on 50 samples differs little from the picture derived from using 200 samples while increases in sample size to over 500 samples continues to alter the dendrogram when only one locus is examined.

Conclusions

A number of technical problems beset the analysis of minisatellite loci in population genetic research. We have developed a method that we call semi-automated floating bin analysis that permits the acquisition of large amounts of high quality data for the accurate assessment of allele frequency. Using this method we have already accumulated data on about

5000 chromosomes in 18 world populations and have investigated the potential of analysing data from numerous minisatellite loci. We find that large sample sizes (400 chromosomes) are necessary to obtain accurate estimates of low frequency alleles, but that by combining data from 4 minisatellites on samples of only 50 chromosomes, a picture of population relationships emerges that is relatively robust to further increases in sample size. The large database that we have assembled provides the basis for further analysis of population relationships described by minisatellites and will allow us to compare this with data obtained from other polymorphic loci on the same samples.

References

Balazs, I., Baird, M., Clyne, M. & Meade, E. (1989). Human population genetic studies of five hypervariable DNA loci. *American Journal of Human Genetics*, **44**, 182–90.

Bowcock, A. M., Ruiz-Linares, A., Tomfohrde, I., Minch, E., Kidd, J. R. & Cavalli-Sforza, L. L. (1994). High resolution of human evolutionary trees with polymorphic microsatellites. *Nature*, **368**, 455–7.

Budowle, B., Giusti, A. M., Waye, J. S., Baechtel, F. S., Fourney, R. M., Adams, D. E., Presley, L. A., Deadman, H. A. & Monson, K. L. (1991). Fixed-bin analysis for statistical evaluation of continuous distributions of allelic data from VNTR loci, for use in forensic comparisons. *American Journal of Human Genetics*, **48**, 841–55.

Chakraborty, R., Zhong, Y., Jin, L. & Budowle, B. (1994). Nondetectability of restriction fragments and independence of DNA fragment sizes within and between loci in RFLP typing of DNA. *American Journal of Human Genetics*, **55**, 391–401.

Deka, R., Chakraborty, R. & Ferrell R. E. (1991). A population genetic study of six VNTR loci in three ethnically defined populations. *Genomics*, **11**, 83–92.

Devlin, B. & Risch, N. (1992). Ethnic differentiation at VNTR loci, with special reference to forensic applications. *American Journal of Human Genetics*, **51**, 534–48.

Devlin, B. & Risch, N. (1993). Physical properties of VNTR data and their impact on a test of allelic independence. *American Journal of Human Genetics*, **53**, 324–9.

Devlin, B., Risch, N. & Roeder, K. (1990). No excess of homozygosity at loci used for DNA fingerprinting. *Science*, **249**, 1416–20.

Devlin, B., Risch, N. & Roeder, K. (1991). Estimation of allele frequencies for VNTR loci. *American Journal of Human Genetics*, **48**, 662–76.

Flint, J., Boyce, A. J., Martinson, J. J. & Clegg, J. B. (1989). Population bottlenecks in Polynesia revealed by minisatellites. *Human Genetics*, **83**, 257–63.

4 *DNA fingerprinting: development of a technology and its application to the study of human populations*

A. P. RYSKOV, M. I. PROSNYAK, N. S. KUPRIYANOVA,
E. K. KHUSNUTDINOVA, I. M. KHIDIATOVA,
V. V. KALNIN, O. V. KALNINA, K. B. BULAYEVA
AND S. A. LIMBORSKA

Hypervariable sequences and DNA fingerprinting

The most polymorphic loci known in the human genome are tandemly repeated minisatellites (MS) and microsatellites (MCS), hundreds of which have been described and mapped. Jeffreys and his colleagues (Jeffreys, Wilson & Thein, 1985) were the first to show that probes derived from the human myoglobin gene (probes 33.6 and 33.15) could hybridize to several minisatellite loci, producing complex and individual-specific banding patterns called 'DNA fingerprints'. Many other multilocus and single-locus probes have since been described (see Pena & Chakraborty, 1994). Fig. 4.1 shows an example of DNA fingerprints obtained with a multilocus probe, M13 phage DNA, which was independently discovered by Vassart *et al.* in Belgium (Vassart *et al.*, 1987) and by Ryskov *et al.* in Russia (Jincharadze, Ivanov & Ryskov, 1987; Ryskov *et al.*, 1988). This probe contains a minisatellite (M13-ms) which consists of repeating 15 bp sequences GAGGGTGGXGGXTCT. It detects a set of hypervariable regions in human DNA and makes it possible to estimate individual specificity as well as to establish paternity relationships. Ryskov *et al.* (1988) tested M13-ms for hybridization with DNA from many species and demonstrated a wide distribution of this MS in different taxonomic groups including man, animals, plants, bacteria. This peculiar characteristic of M13-ms allows one to apply its analysis for studying genomic diversity in different species.

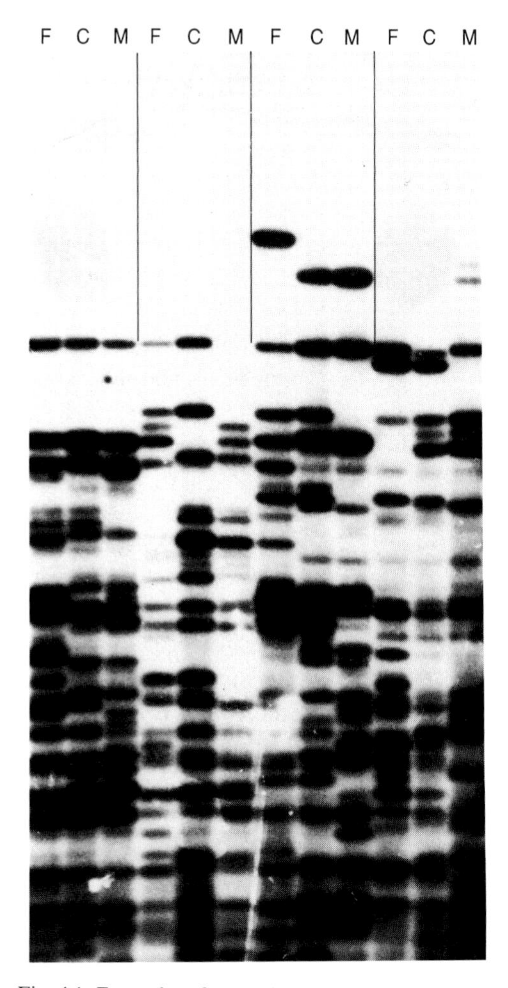

Fig. 4.1. Examples of paternity determination with the multilocus probe M13 phage DNA. The DNA was digested with *Bsp*RI and separated on an 0.8% agarose gel. The blot filter was hybridized with ^{32}P-labelled M13 DNA, washed and autoradiographed without an intensifier screen for 3 days.

Development of new polymorphic DNA probes and DNA fingerprinting

New polymorphic DNA markers isolated from human genomic libraries

To obtain new polymorphic DNA markers, human genomic and chromosome-specific libraries were screened by oligonucleotides homologous to M13-ms and some known MCS (CAC-, TCC-, GACA-motifs). Positive clones were isolated and some of them were analysed by blot hybridization. Fragments containing MS and MCS were recloned in the pUC19 vector, sequenced and checked for genomic variability. Fig. 4.2 shows partial sequences of several MS and MCS containing cloned fragments. The most interesting sequence among them is K25 which represents a new human minisatellite. K25 is composed of an array of 16-25 bp tandem repeats with the copy number equal to 24. It has a partial homology to the G-rich consensus previously detected in repeated units of some minisatellites and to the sequence Chi of *E.coli* (Smith & Stahl, 1985).

Fig. 4.3 shows the application of the above mentioned probes for multilocus DNA fingerprinting. They produce typical individual-specific fingerprints suitable for paternity determination and for studying genetic diversity. One can see that, in the cases of paternity inclusions (a) there is a mutated allele (indicated by an arrowhead) revealed by both the M13 DNA and TCC-mcs probes. It means that the two types of hypervariable clusters are located close to each other in the same locus. Cloned CAC-mcs detects highly informative fingerprints in unrelated individuals (b). Comparison of the M13 DNA and K25-ms probes has revealed differences in the fingerprint patterns of unrelated individuals although common bands can be also observed (c). These common bands may represent loci containing MS clusters of different types or divergent sequences of the same repeat family. The new polymorphic probes isolated from human DNA libraries are also suited as locus-specific markers recognizing only one or two bands of a DNA fingerprint (depending on whether the individual is homozygous or heterozygous at the particular locus).

Construction of multilocus DNA probes of increased sensitivity

One possible way to improve the sensitivity of hybridization in DNA fingerprinting analysis is to use DNA probes modified to increase the stability of their duplexes with complementary DNA sequences compared to the stability of their unmodified analogue. In particular, the substitutions of 2-aminoadenine for adenine and 5-methylcytosine for cytosine in oligo- and polynucleotides are known to increase the thermal stability of the duplexes (Cheong, Tinoco & Chollet, 1988). The improved properties of the doubly modified oligonucleotides in DNA sequencing, PCR amplification and DNA fingerprinting have been recently demonstrated (see Prosnyak *et al.*, 1994). Fig. 4.4 shows the difference between modified $(m^5C\text{-}am^2A\text{-}m^5C)_5$ and unmodified $(CAC)_5$ oligonucleotides as hybridization probes for DNA fingerprinting. The results obtained under different hybridization and washing temperature conditions demonstrate an increased sensitivity of modified analogues. An important point is that the

```
HMM1 Sau3A-Sau3A fragment
gatccttccg catatggaga gtagagacca ttgtgagtgt ggtgggtaat tctttttggg  60
aaggctaagg caccgaaacc attctgaata tgtggagtcc ttccgtgtgg aagagaccat 120
tcgtcgggtg tgggagttcc tcactgaaga agggggagac tattctGAGG GTGGTGGTTC 180
Cttccttata aagggacaga gactattgtg agttagggtg gggcgagtcc ttctttatgt 240
gactagagag ggtccatttt gaggatgggg gtgtctttca ttatgtgcgg aggagaccgt 300
tctgggggct ccctccctaa tggaggtaga gttcattgta cattcgggtg ggagggctca 360
tcgttatatg gtggagaaag aacgttctga gagtgccggg tccttcctta tagcagagag 420
agctcattgt gagtgtgggt agagcaagga aaaattcatt agctagaggg tccgtgtgcg 480
gtggtgggag ggcagaccat gtggagtgca gggaagtatt tcctcatctg gttgagggag 540
ccagaccttt cttggtggga taaccattcc tgatacagga ttcagtccct gggagcatgt 600
gggaatcatt tcttgtattt agtagactat tctggtcgag aggaacccag ttgtcacagg 660
agtgaggggt tggagagacc attctgggta tgaggaggat c
```

```
HMG1 Sau3A-Sau3A fragment
gatctagaca gacaggcagt tgatggatgg aagggcagac aggcaagttg ggcaggtgga  60
agggcagggg aggagtggac tggacagagg aaagggcagg tggttggtgg tggaggtgga 120
gGTGGTGGGT GGAGGTGGTG GAGGTGGTGG GTGGAGGTGG TGGAGGTGGT GGAGGTGGTG 180
GGTGGTGGGT GGTGGAGGTG GTGGGTGGAG GTGGTGGAGG TGGTGGGTGG TGGGTGGTGG 240
GTGGTGGAGG TGGTGGAGGT GGGGTGGCAG TGGTGGGTGG TGGTGGTAGT GGGTGGCAGG 300
GTGgatgaat ggaaaggaga caaggttggc tcaggcaggc tgaggtttgc cctatctctc 360
catcccccca ccatggattg gcagaatgag taggaatctt tctctcctga ctcctcctca 420
gacccctcag cttcctcagg gcacacccta cactggatgg gacccactgg gcacttgcag 480
ccggagatgt aactggcacc aggaacctcc ctccatctta tccgtgtcca cacatcttca 540
cccgtggggt tctttttgcc gactggtctg attactggcg agaacctggt gtttggggat 600
gaaagaagtt caggtctcct ctacacatcc atttcttcc ttcgtccttt ctcaccttct 660
gcctcctcct cctcctccca tgtctttcca gcaaagcaca ggatacccag atggcttggg 720
ccaggttgaa ggcacagcca tagccagcaa ccagccaccc gagccctgcc tgttcggccc 780
catgcccagc cagcgcgggt ccctgctgcc tggtgcccac cccacccacc cggcagctcc 840
gattttcagc ttgggtttga tg
```

```
HMT1 Sau3A-Sau3A fragment
gatctcgctg ggagctgtag accggaggtg ttcctatttg gccatcttgg aagcgactta  60
ctccttcctt ttctgagttt caaaactgtt tatataggtt gggaatttta ttttctaaaa 120
gtctgagtaa cttatttgta aaacatttca aacctggagg ctttggggta gatacttctc 180
tttctcctat ttttccaaag ttactaatcc atgccagttg caacaactag TTCTTCTCCT 240
CCTCGTCCTC CTCCTCCTCC TCCTCCTCCT TCTTCTTCTg agatgtattc tcactctgtc 300
actcaggctg gagtctcagc tcacttcaac ctctgctcct gggttcaaac aattctgccc 360
cagcctccag tgtagctggg actacaggca tgtgacacca ctcctggata ctttttgta 420
tttttactag agacagggtt tcaccatgtt agccaggatg gtctcaatct cctgacctcg 480
```

```
HMT2 Sau3A-Sau3A fragment
gatccttgct tctttcctga aagcctcagg gatggagctg ccacaccagc tgcccgtcag  60
tctgggctct ctttcatcag gactaagttg tagatgctac aaaacaacaa tttcaactac 120
tactTTCTTC_TTCTTCCTCT_TCTTCCTTCC TCCTCCTCCT CCTTTCTCCT CCTCCTCCTT 180
CTTCTTCCTC CTCCTCCTCC TTCTTCTTCC TCCTCCTCCT TCTTCTTCCT CTTCATCCTC 240
TTCTCCTCTT TTTCCTCTTC GTCTTCTTCC TCCTCTTCTT CTTTCTCCTT TCTTCttctt 300
ctttcacctt tctgatattt cagaaaaaaa gcacatagca attttttct gtctaggaca 360
taacaaaaat aataggctta agagatc
```

```
HMK25 Sau3A-Hinf1 fragment
gatCCCTTCA TCCCTCATCC CTCCATTCCC TATCTTCCCA TCCCCCATCC CTCCATCTGC  60
CATCCCCCAA TCCCTCCATC CCATCCTCAA TCCCCCATCC CCCATCCCTT CATCTTCCAG 120
TCCCCCATCC TCCATCCACC ATCCCTCCAT TCTCCCATCC CCCTATCCCT CCATCCCCCA 180
CCCTCCATCC CCTATCCCTC ATATTTCCAT TCCCCATCCC TCTGACCCTC TCACCCTTCC 240
ATCCCCCATC CCCATCTCCT CATCCCCCAT CCCTCCACAC TCCATCCCCA TCCCTCTACT 300
CCATCCCTCC ATCCCTCTAT CCCCCATCCC CCCTCTCCCA CCCCTCCCAT CCCTCCCATC 360
CCCCATCCCT CCATCCCTCC ATCCCCCATT CTCCCATCCC TCCATCCCCC ATCCTCCCAT 420
CCCTCCATCT CTCATCCCCA TCCCTCCATC CCTTCATCCC CATCCCCCAT CTCCTCATCC 480
CCCATCCCTC CATACTCCAT CCCTCCATCT CCCGCATCCC TCTGTCCCTT CATCCTTCCA 540
TCCTTCCATC caaagaacat ttcttcagca caggagaggt gcactcagcc agtttcgggc 600
taggaaaggc ttttaggagg gagaaaccta cgcgagacag gagaatgagt c
```

Fig. 4.2. Partial nucleotide sequences of ms and mcs containing clones isolated from human genome library. Five examples of cloned 'simple' tandem repeats are shown. HMM1 contains only one repeat unit of M13-ms, probably due to loss in the course of cloning. HMK25 represents a new human ms repeat.

patterns observed with both modified and standard probes are identical. Therefore, a stronger binding of modified oligonucleotide does not affect the specificity of binding. Modified oligonucleotides may be advantageous in various applications and especially useful when the stability of duplexes is not sufficient, as in the case of AT-rich sequences or short oligonucleotides.

Another way to increase the sensitivity of a probe in DNA fingerprinting is based on the synthesis of long chains of tandemly repeated sequences. This approach was developed previously (Ijdo *et al.*, 1991; Mariat & Vergnaud, 1992) to obtain probes suitable for *in situ* hybridization and for detection of polymorphic loci in complex genomes. In our case, synthetic tandemly repeated sequences (STRS) were produced by the polymerase chain reaction with two partly complementary oligonucleotides homologous to the M13 core sequence.

Fig. 4.5 shows the results of blot hybridizations of human genomic DNA with the M13 phage DNA (*a*) and M13 STRS probes (*b*), (*c*). The M13 STRS probe hybridized significantly more effectively in comparison with the standard M13 phage DNA probe. The latter needs a ten times longer autoradiographic exposure to obtain hybridization intensities similar to those with the M13 STRS probe. An increased sensitivity of STRS probes was also obtained with other tandemly repeated motifs (CAC, 33.6 and 33.15 core sequences). The high sensitivity of STRS allows their use as ^{33}P-labelled probes in DNA fingerprinting (*c*). In this case, the hybridization pictures are more clear-cut, providing very high band resolution patterns and thus improving the information content of DNA fingerprinting.

Besides having a high band resolution and sensitivity, STRS probes detect many additional polymorphic fragments. This results in an increased individual specificity of DNA fingerprints which is calculated as the probability that all resolved fragments in an average individual A are also present in a second unrelated individual B. Analysis of DNA fingerprints has also revealed fragments simultaneously hybridizing with different types of MS and MCS. It means that some human genome loci are organized as long DNA stretches enriched in different hypervariable clusters. The structural organization and possible functions of such loci are of special interest. One of the ways to search for them is by screening genomic libraries using different polymorphic probes.

Hypervariable clusters and intragenomic polymorphism of human ribosomal RNA genes

We have screened the cosmid library of human chromosome 13 (DH L4/FS 13) obtained from Dr. H. Lerach using oligonucleotide probes $(CAC)_5$,

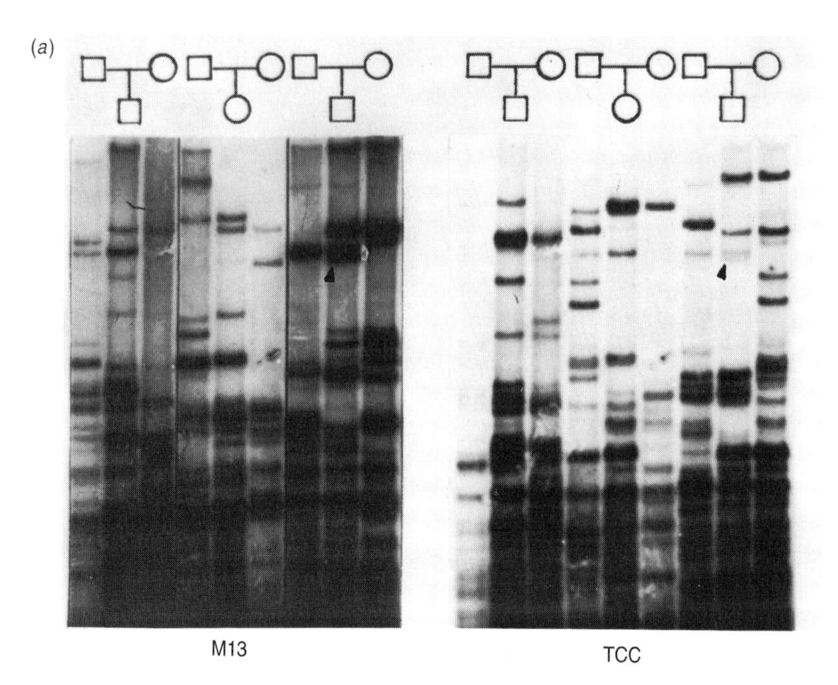
(a)
M13
TCC

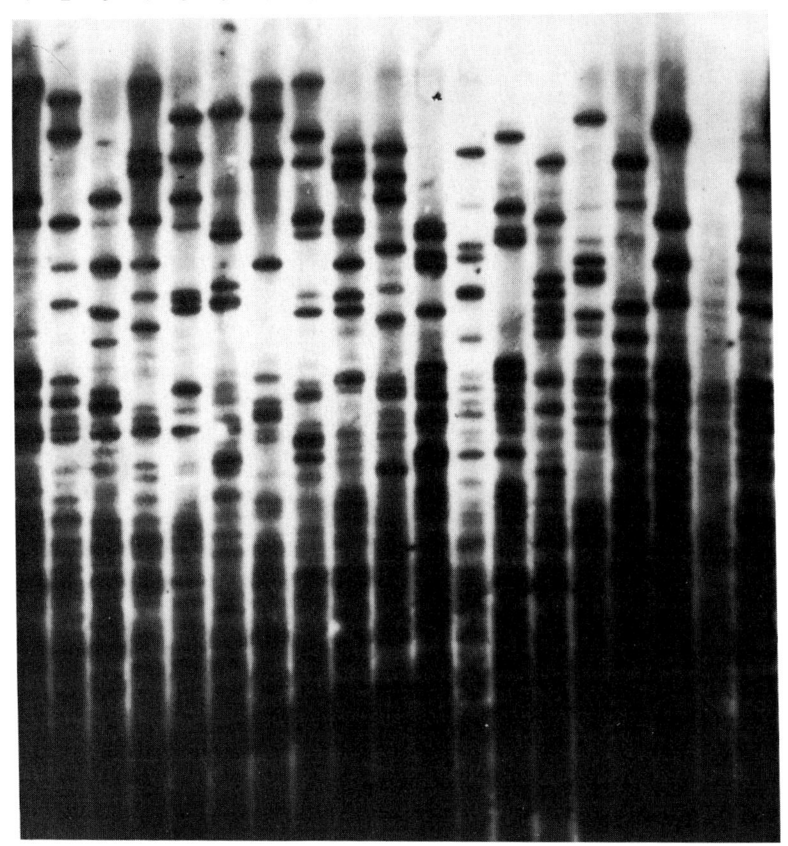
(b)
1 2 3 4 5 6 7 8 9 10 11 12 13 14 15 16 17 18 19

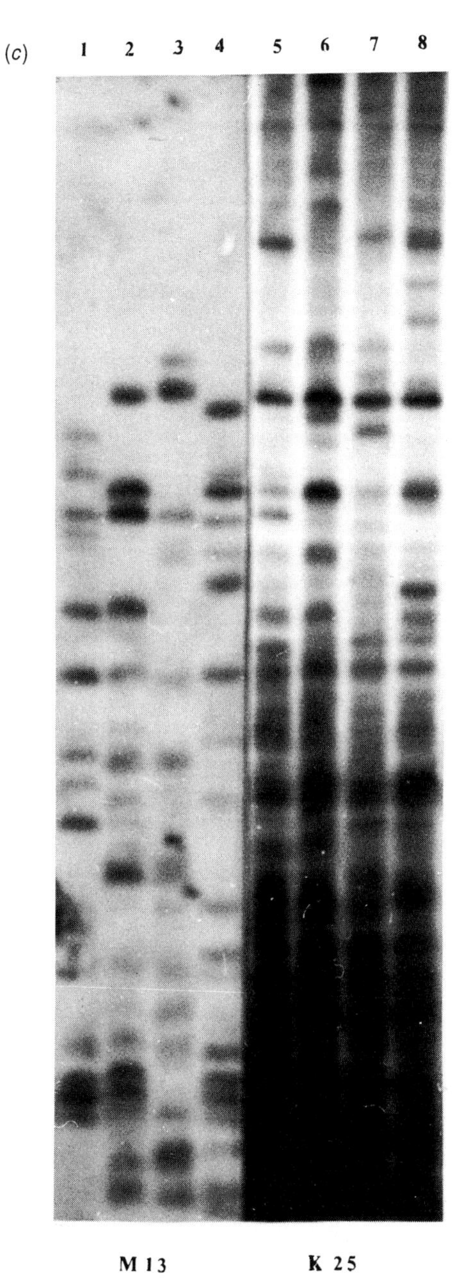

Fig. 4.3. Human DNA fingerprints detected with cloned ms and mcs containing DNA probes. (*a*) paternity testing in three trios with M13 phage DNA (left) and TCC-mcs (clone HMT2, right), arrowhead indicates mutated allele. (*b*) DNA fingerprints of unrelated individuals detected with CAC-mcs (Clone HMgl). (*c*) DNA fingerprints of unrelated individuals detected with M13 phage DNA (1-4) and K25-ms (clone HMK25, 5-8).

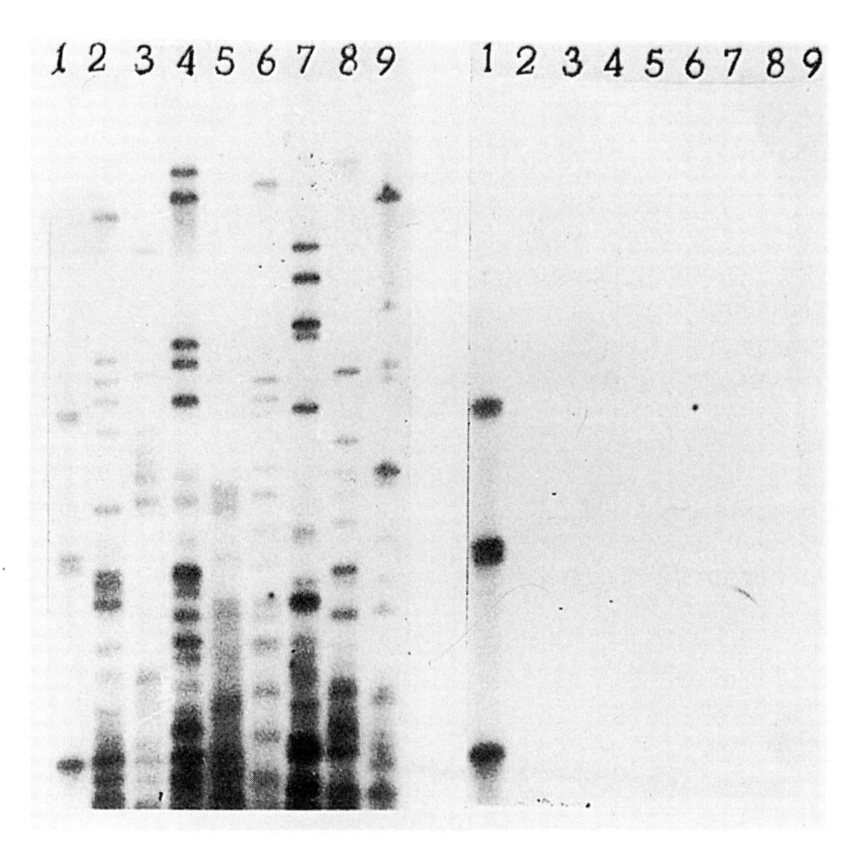

Fig. 4.4. Southern blot hybridization analysis of human genomic DNA, with modified (left) and unmodified (right) $(CAC)_5$ oligonucleotide probes. The DNA was digested with BspRI and separated on a 0.8% agarose gel. Lane 1, the labelled *Hind*III λ phage fragments (from top to bottom 9.4, 6.6 and 4.4 kb). Lanes 2–9, DNA from unrelated individuals.

$(TCC)_5$, $(GACA)_4$, 15 bp core sequence of M13-ms and 16 bp core sequence of K25-ms. Different MS and MCS have a tendency to be concentrated in the same cosmid inserts (Ryskov *et al.*, 1993). Blot-hybridization analysis of a set of such clones shows that identical tandem arrays are often multiply represented in the same clone. Furthermore, a great many MS and MCS containing cosmids can be also hybridized with oligonucleotide probes complementary to the conservative parts of the 18S and 28S rRNA genes (Ryskov *et al.*, 1993). Individual cosmids harbouring ribosomal DNA and different MS and MCS have been isolated and mapped by restriction and blot hybridization methods (Fig. 4.6). These clones represent a collection of individual ribosomal genes because an average size of a cosmid insert ($\approx$35 kb) is comparable to the size of a rDNA repeating unit (45 kb). As a result, a pronounced intragenomic polymorphism of ribosomal operons has been detected. It is manifested in different distribution patterns of MS and MCS and accumulation of point mutations in the conservative and spacer regions of rDNA. Fig. 4.7 shows a schematic representation of the organization of a human ribosomal unit and the examples of its polymorphic variants found in the cosmid library of human chromosome 13. The *Eco* RI restriction map of human rDNA reveals four fragments denoted earlier as A, B, C, D (Sylvester *et al.*, 1986). Chemically synthesized 15- and 16-mer oligonucleotides (RI, RII, RIII, RIY, RY) complementary to definite regions of rRNA genes (indicated by circles) as well as oligos homologous to different MS and MCS were used to map cosmid restriction fragments produced by *Eco* RI and *Hind* III digestion. It is seen that a majority of MS and MCS are located in the ribosomal intergenic spacer (rIGS). Every clone has an individual-specific distribution of these sequences. Only 3 of 12 cosmids have restriction maps typical for human rDNA, the rest show some differences. Considering the transcribing part of the ribosomal gene that comprises 28% (13 kb) of the total rDNA repeating unit length, 3–4 of the 12 cosmid DNAs under study can be expected to be restructured in this region during cloning. In fact, a two times higher level of additional restriction sites is detected resulting in significant alterations in the restriction maps of cloned rDNAs.

An interesting point is that gene variants with a large number of mutations in the transcribed regions are characterized by a high concentration of hypervariable clusters in the same regions. The possibility exists that MS and MCS are hot spot mutations which affect DNA divergence in adjacent regions (for instance, in rRNA genes) providing inter- and intragenomic variability. Our preliminary results have shown that there is a polarity of variations in rDNA units; most of them are concentrated close to the important regulatory parts of rIGS, such as sites of transcription initiation, origins of replication, *top* I and *top* II isomerase association sites (Sylvester, Petersen & Schmickel, 1989; Financsek *et al.*, 1982; Harwood, Tachibana & Meuth, 1991). The aim of further experiments is to check the hypothesis that repetitive elements, such as rRNA genes, can accumulate mutations induced by different factors and therefore be used for monitoring external and internal mutagenic agents as well as for the studying of genetic diversity.

DNA fingerprinting and the study of population differentiation

The population genetics of multilocus DNA fingerprints rests on three main assumptions (Jeffreys, Turner & Debenham, 1991): 1) fingerprint bands are allelic products from different loci and behave independently, 2)

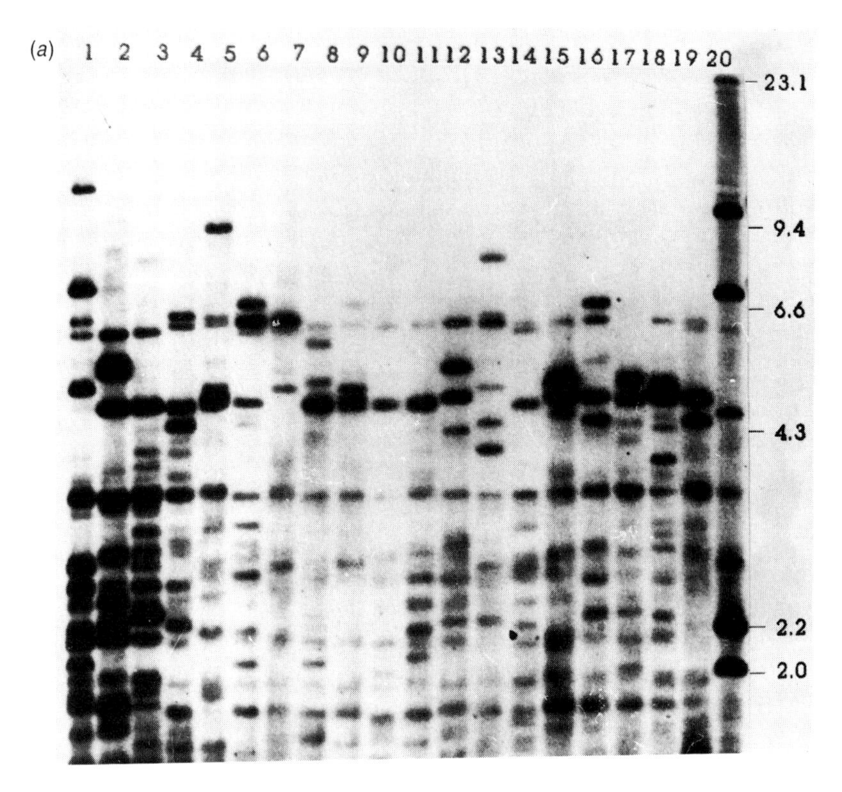
(a)
1 2 3 4 5 6 7 8 9 10 11 12 13 14 15 16 17 18 19 20
23.1
9.4
6.6
4.3
2.2
2.0

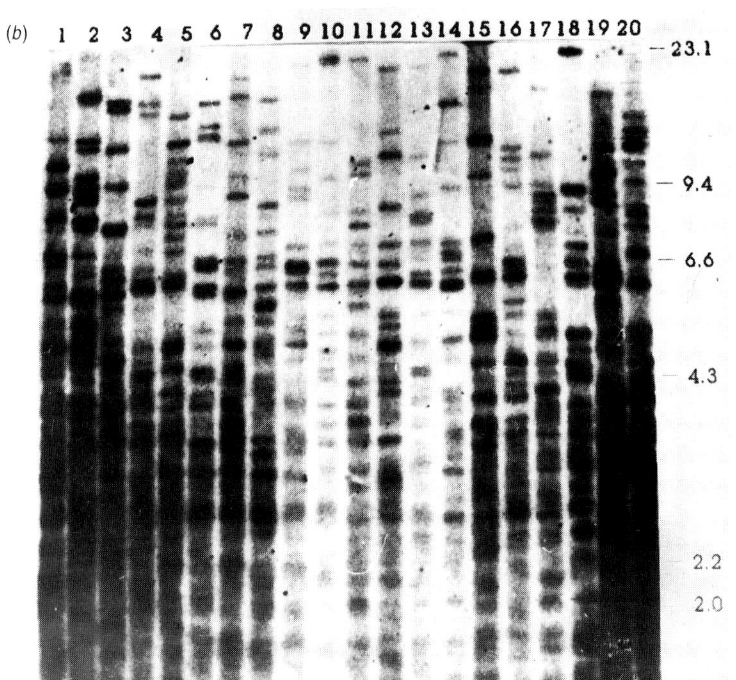
(b)
1 2 3 4 5 6 7 8 9 10 11 12 13 14 15 16 17 18 19 20
23.1
9.4
6.6
4.3
2.2
2.0

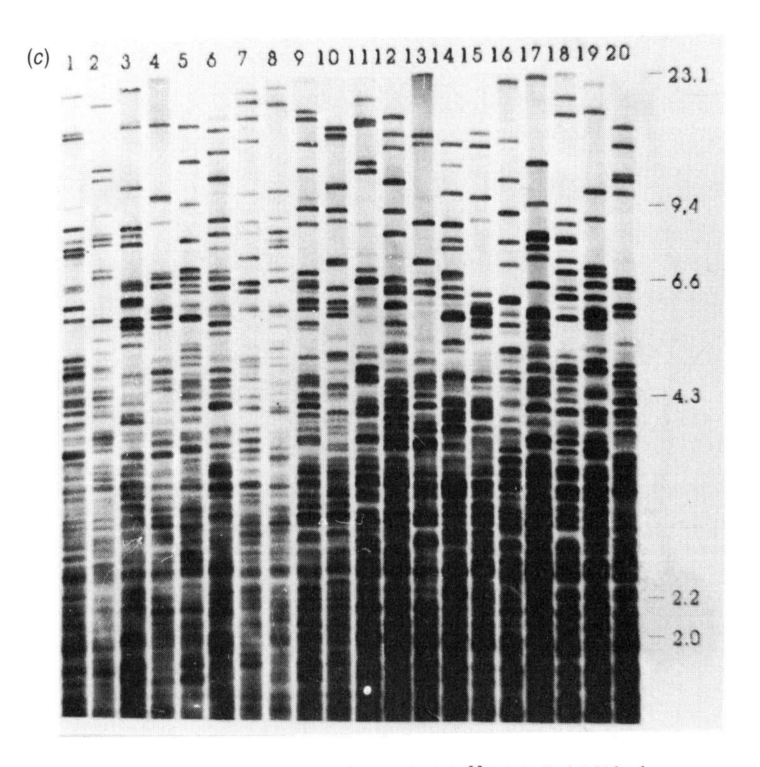

Fig. 4.5. Human DNA fingerprints detected with ^{32}P-labelled M13 phage DNA (*a*), ^{32}P-labelled M13 STRS (*b*) and ^{33}P-labelled M13 STRS (*c*). The blot filters containing equal amounts of BspRI digested human DNA from unrelated individuals were hybridized with M13 DNA and STRS probes. After washing, the filters were autoradiographed for 3 days (*a*), 7 hours (*b*), and 4 days (*c*). *Hind*III phage DNA fragments are used as molecular weight markers.

at each locus, the allelic frequencies are approximately uniformly distributed, and 3) co-migrating bands represent the same allele at the same locus. With these premises, the statistical evaluation of multilocus DNA fingerprints ultimately rests on only one parameter: the average proportion of bands shared between unrelated individuals. This parameter depends more on DNA fingerprinting techniques (hybridization probe, gel band resolution, criteria used to match and score bands) than on the population tested. For instance, for the M13 phage DNA probe it was found to be practically the same in unrelated individuals, husband–wife pairs and several ethnic groups in Russia (Prosnyak *et al.*, 1990; Barysheva *et al.*, 1989; Khusnutdinova *et al.*, 1994 and unpublished results). According to *in situ* hybridization data (Christmann, Lagoda & Zang, 1991), M13 minisatellites are randomly distributed on human chromosomes and

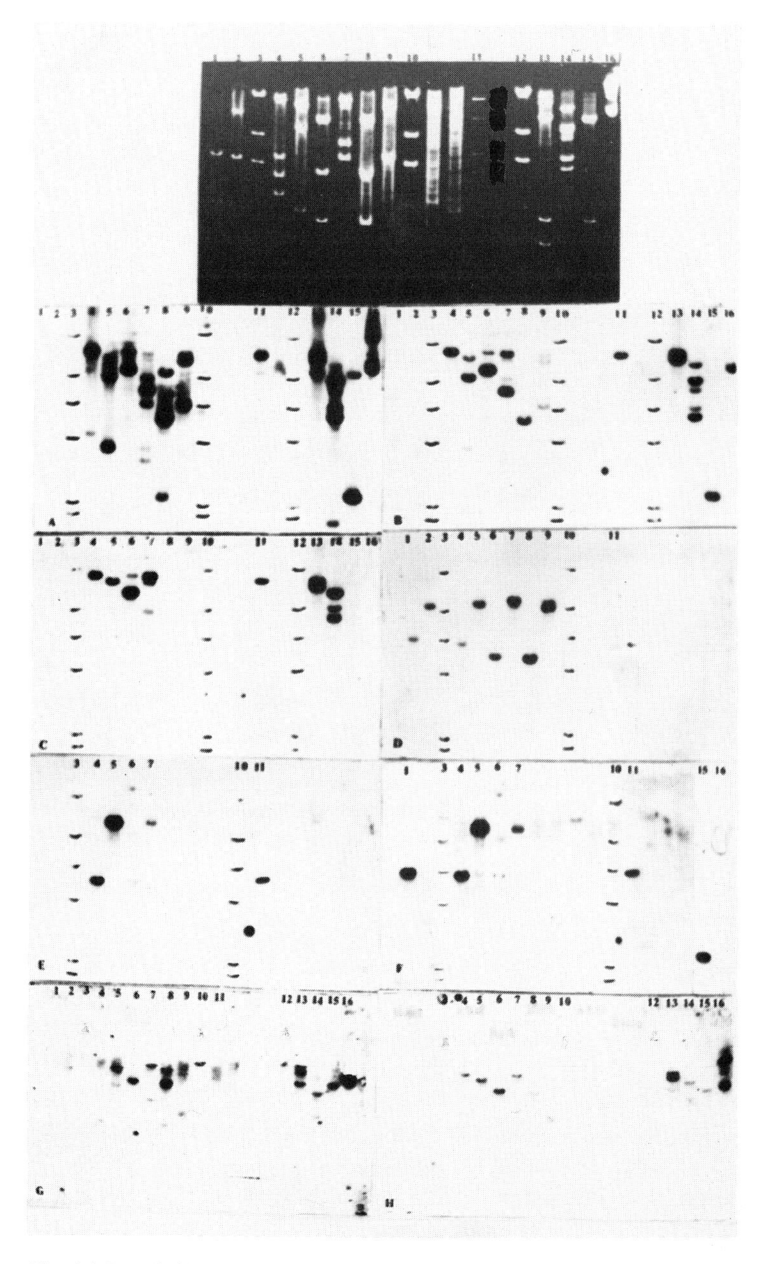

Fig. 4.6. Restriction-hybridization mapping of rDNA containing cosmid clones isolated from the library of human chromosome 13. DNA of individual. clones was digested by *Eco*RI (lanes 1,4,6,8,11,13,15) 'or *Hind*III (lines 2, 5,7,9,14,16) restrictases and electrophoresed in 0.8% agarose. The gel was then dried and subjected to successive hybridizations with different 32P-labelled oligonucleotides: $(GACA)_4$-A, $(CAC)_5$-B, $(TCC)_5$-C; homologues to ribosomal DNA-D,E,F; M13-ms homologue-G; K25-ms homologue-H.Lanes 3,10 and 12 are *Hind*III phage fragments.

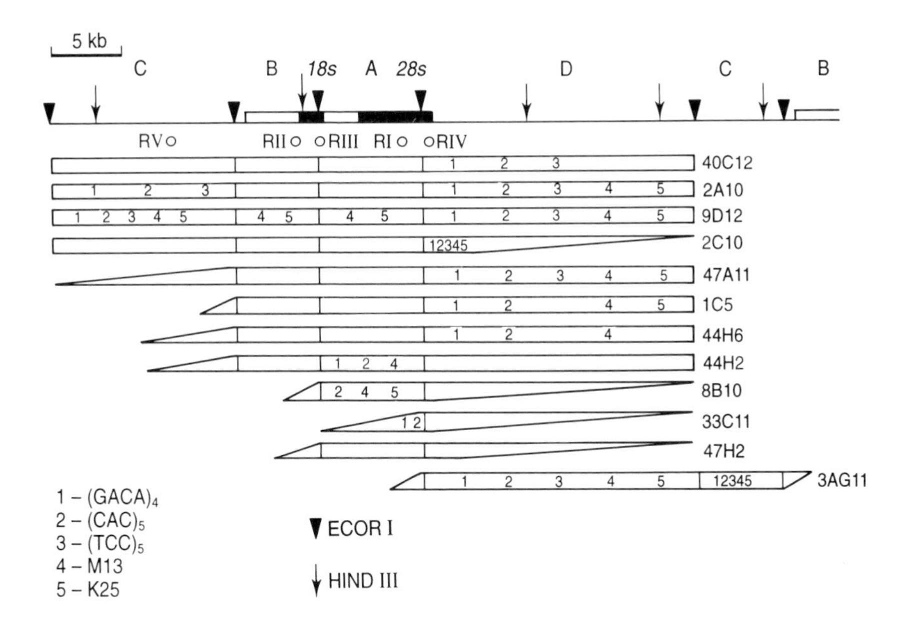

Fig. 4.7. Schematic representation of the organization of human ribosomal RNA genes and examples of their polymorphic variants found in the cosmid library of chromosome 13. Typical *Eco*RI and *Hind*III restriction sites are indicated by arrowheads (*Eco*RI) and by arrows (*Hind*III). Oligonucleotide probes RI, RII, RIII and RIV were used for mapping and identification of *Eco*RI RDNA fragments A, B, C and D. Locations of ms and mcs are indicated inside every cloned RDNA units. Split rectangles refer to cosmids having non-typical RDNA restriction maps.

represent an ubiquitous family of polymorphic markers. DNA fingerprinting with M13 DNA combined with factor correspondence analysis (FCA) of data (Lebart, Morinea & Warwick, 1985) shows a good correlation between population position and genetic proximity (Tokarskaya *et al.*, 1994; Kalnin *et al.*, 1995).

Five human populations differing in the degree of ethnogenetic relationship and in their geographical location have been analysed (Fig. 4.8). One of them, the Talish population lives in Southern Azerbaijani and belongs to the Indo-Iranian group. It is an ethnic minority, the total number of the Talish is about 100 000. Their language (Talish) is closely related to the Indo-Iranian linguistic group. They have inhabited this area (Southern Caucasus) since before the Christian era. All the Talish are Muslim. The Muslim religion and geographical isolation led to strong endogamy and inbreeding in the Talish population as in other Caucasian ethnic popula-

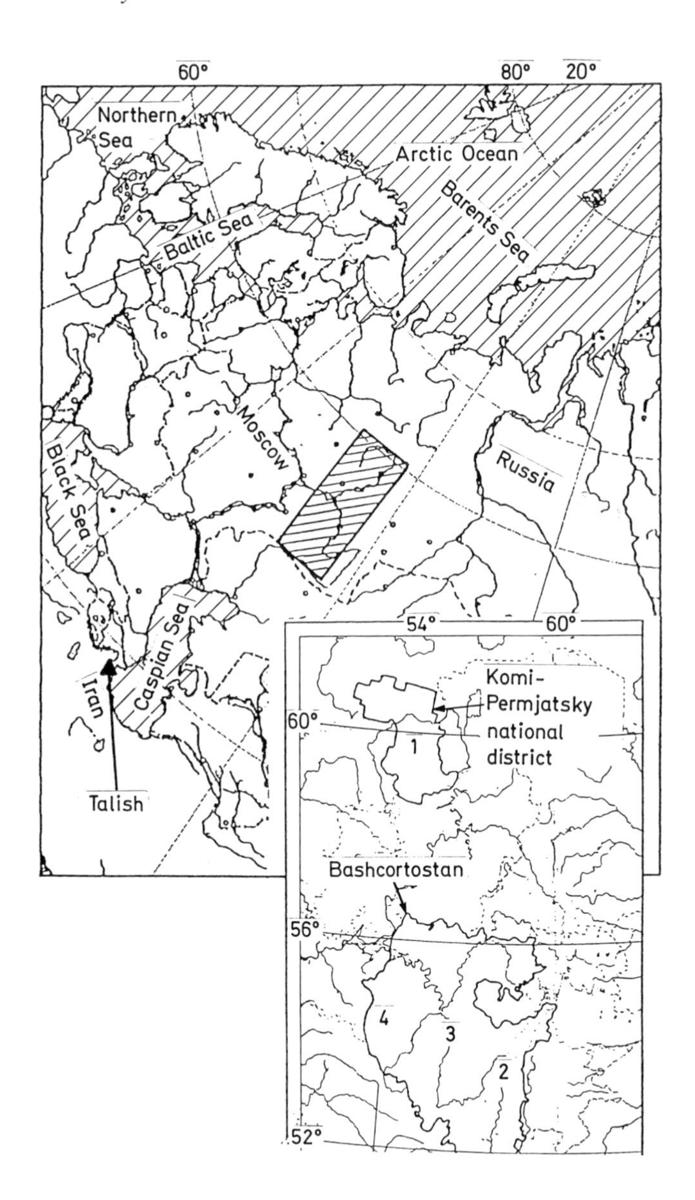

Fig. 4.8. Geographical locations of populations investigated. The arrow designates the location of the Talish population (47 individuals studied). The insert shows the Ural region, designated on the map by hatching: The locations of the Komi (1, 42 individuals analysed) and the Bashkir populations. The Abzelilovsky (2, 63), Arkhangelsky (3, 70) and Ilishevsky (4, 56). The Talish sample is derived from an isolated subpopulation with a high degree of inbreeding.

tions (Bulayeva & Pavlova, 1993). The representatives of the four other populations live in central Russia (the Urals). The first of these, the Komi subpopulation, represents the Finno-Ugric ethnic group. The total number of ethnic Komi in Russia is about 300 000. The Komi population studied inhabits the flat lowlands to the west of the Urals. All the Komi are Christians. Three others are the Bashkirs, representatives of a mixed origin group (Turkic and Finno-Ugric). The Bashkir population samples were taken from three regions referred to different tribal subdivisions on the basis of archaeological, ethnographical and anthropological evidence. The total number of Bashkirs is about one million individuals. All the Bashkirs have been Muslim since the tenth century. The economic activities and lifestyle in Komi and Bashkir populations are same: both ethnic groups had a migrant lifestyle for many centuries, their economy activities are related to agriculture, cattle- and horse-breeding as well as with gas- and oil-industries in the twentieth century.

In total, DNA fingerprints of 278 individuals were obtained and analysed. The analysis involved: 1) identification of the positions of fragments on autoradiographs, 2) identification of the position of a corresponding fragment on different autoradiographs, 3) correct integration of information from different blots. The DNA fingerprinting techniques are not amenable to complete standardization of all parameters and conditions, which results in differences between experiments. Therefore there may be a risk that differences revealed between two groups might be due simply to interblot variations. To overcome these difficulties, external and internal molecular weight markers were used, and DNAs of some individuals were repeatedly analysed on different gels and in different combinations with other samples.

As a result, 150 band positions were identified (the number of different bands scored on any one blot varied from 60 to 70). The mean numbers of bands scored per individual were 11.7, 14.1, 15.3, 17.4 and 13.8 for the Talish, Komi, Abzelilovsky, Arkhangelsky, and Ilishevsky populations, respectively. The Talish population has the lowest number; this may be due to the high degree of inbreeding characterizing this group, and also of the low degree of heterozygosity, which is related to several factors, including neurophysiological sensitivity to the numerous environments (Bulayeva *et al.*, 1993).

The restriction fragment frequency profiles for the molecular weight in each population are shown in Fig. 4.9. Restriction fragments of 7.5–14 kb are characterized by relatively low band frequencies (mean number 0.012). The frequencies of smaller fragments (below 7.5 kb) are significantly higher. One band (fragment 99, 3.4 kb) was always (Fig. 4.9(*a*), (*c*), (*d*)) or almost always (Fig. 4.9(*b*), (*e*)) present in all tested populations (this fragment was

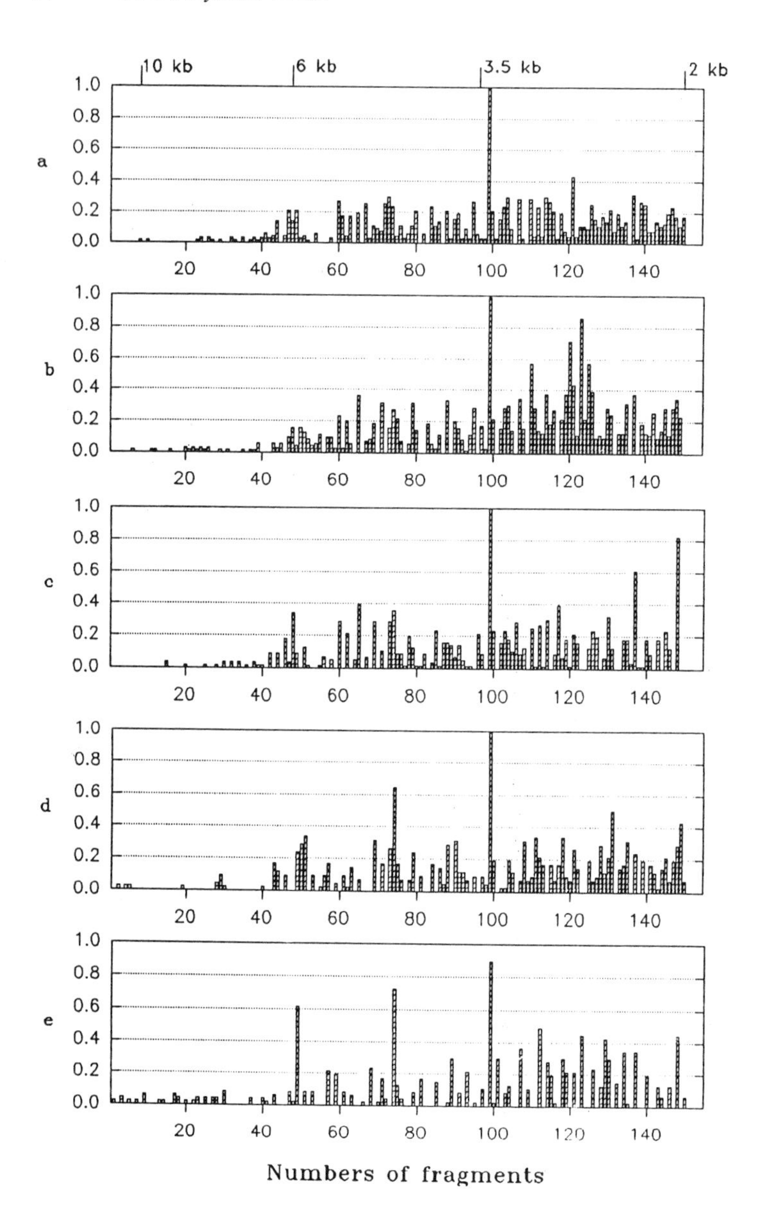

Fig. 4.9. Restriction fragment frequency profiles of populations investigated. Frequencies are given for each of 150 fragments numbered in order of decreasing size, ranging from 2 to 14 kb. a, Abzelilovsky; b, Arkhangelsky; c, Ilishevsky; d, Komi; e, Talish.

used as an internal marker to verify the positions of other bands). Obviously, these histograms are not informative enough for comparison of populations.

DNA fingerprinting data tabulated in the form of a binary data matrix were analysed by the Factor Correspondence Analysis (FCA) method (using the SYSTAT program). The distribution of individuals in the coordinate system provided by the first two factors obtained by the FCA method is presented in Fig. 4.10(*a*), where each point corresponds to a single individual. Different symbols indicate different populations. One of the distinctive features of this method is that it allows the analysis of individuals rather than traits as in traditional population studies. It can be seen that the Talish population is clearly distinguished from the others (Ural group) by the first factor, reflecting the fact that the Talish population is historically and geographically remote from the others. Fig. 4.10(*a*) also shows that the four Ural populations overlap in this factor space. Fig. 4.10(*b*) shows the distribution of the Ural individuals (without the Talish population) in the coordinate system provided by the second and third factors. The Ural populations can be distinguished. One of the Bashkir subpopulations, Abzelilovsky, overlaps significantly with the Komi population. This was expected because the Bashkir nation originates from the Finno-Ugric group. Another Bashkir population, Arkhangelsky, lies in the same overlapping region and is intermediate in geographical location between the Komi and Abzelilovsky populations. The third Bashkir population, Ilishevsky, lives near Tatarstan and shows little overlap with the others. These data may be interpreted as being the result of interethnic crosses. This process seems to play a key role in the formation of the Bashkir nation (Kuseev, 1979). The hypothesis is supported by the fact that the Arkhangelsky population is characterized by the highest mean number of bands scored per individual.

The total data matrix was also subjected to a discriminant analysis (using the SPSS program), allowing the most pronounced distinctions between the groups to be revealed. The results of the analysis revealed that the populations studied were differentiated mostly by 71 restriction fragments out of 150. In the analysis four statistically significant discriminant functions were obtained. The variances of the functions were from 38.29% to 13.62% (for each function respectively: $\chi^2 = 972.00$; 655.94; 408.35; 168.81; df=264; 195; 128; 63; P = 0.000).

The results, expressed both in the form of a territorial map, i.e. the characterization of the populations based on DNA fingerprint bands (considered as traits, Fig. 4.11 (*a*)) and in the form of all-groups Scatterplot, i.e. the classification of the population members (Fig. 4.11 (*b*)), are very similar.

One can see that in the space of two discriminant functions the

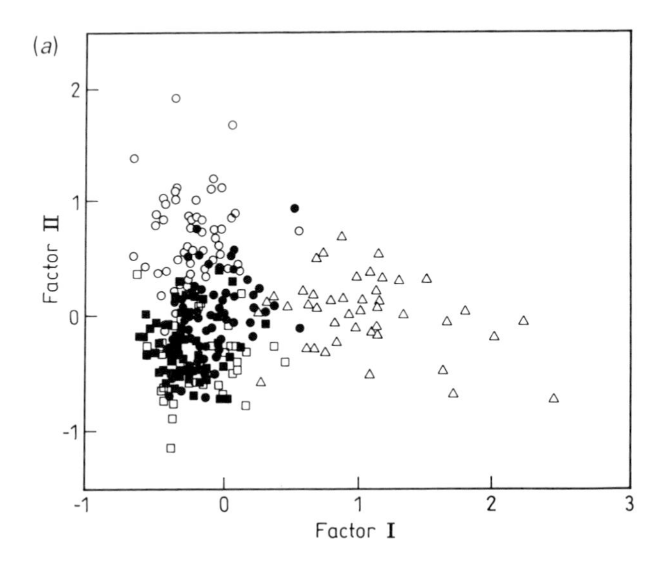

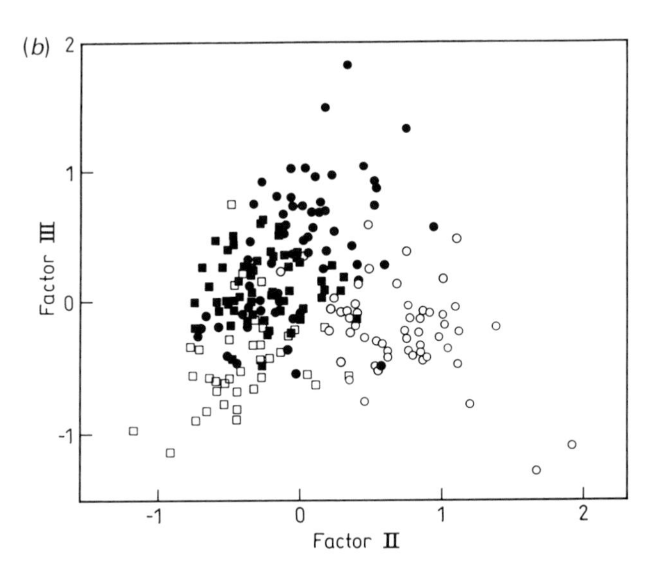

Fig. 4.10. The disposition of five human populations studied in the space defined by the first three factors obtained by means of factor correspondence analysis. (*a*) All populations studied are distributed in the space defined by first and second factors; each point corresponds to one individual. Different symbols designate different populations: △ designate individuals from the Talish population; □, Komi; ●, Abzelilovsky; ■, Arkhangelsky; ○, Ilishevsky. (*b*) Ural populations are disposed in the space of second and third factors.

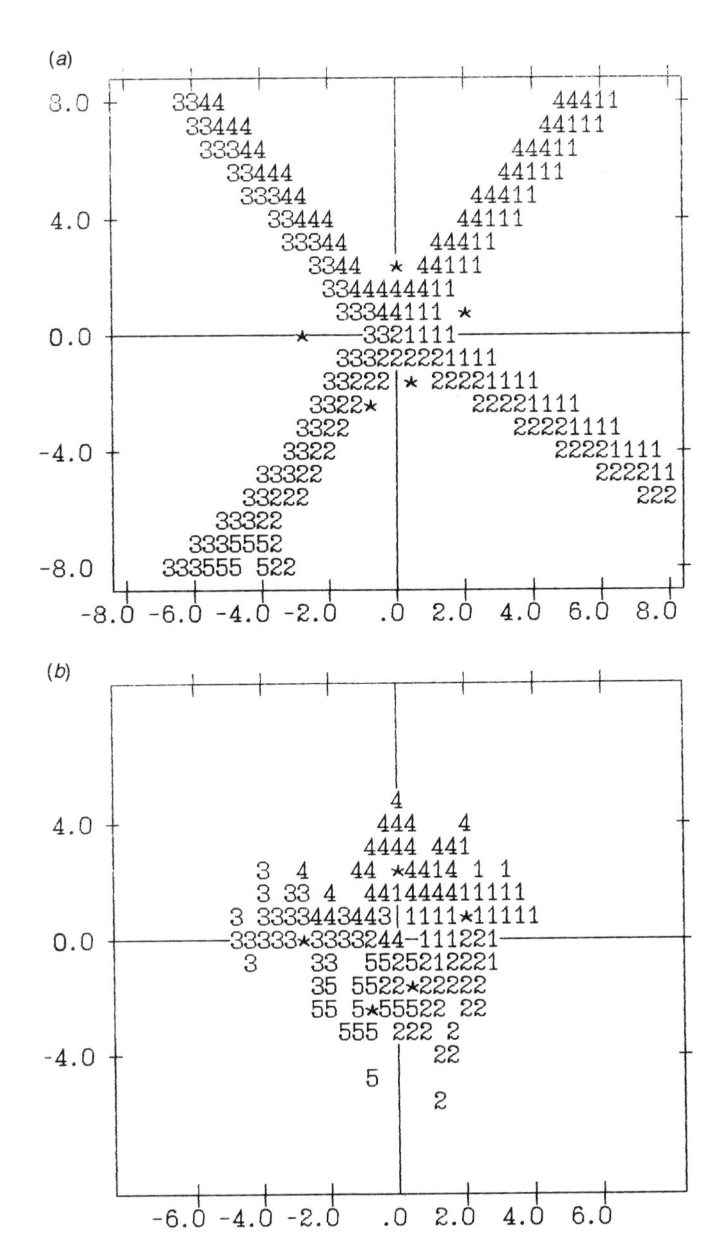

Fig. 4.11. Discriminant analysis of the populations studied.
(*a*) Characterization of the populations based on DNA fingerprint bands (territorial map). (*b*) Discrimination of the population members (all groups Scatterplot). The asterisk denote group centroids (mean values). The positions of the symbols (populations 1,2,3,4,5) correspond to every individual from five different populations: Bashkir: l- Arkhangelsky, 2- Abselilovsky, 3- Ilishevsky; 4- Komi; 5- Talish.

Table 4.1. *Population membership classification by discriminant analysis of DNA fingerprinting data*

Actual group	Number of cases	Predicted group membership 1	2	3	4	5
1. Arkhangelsky	70	69 98.6%	0 0%	0 0%	1 1.4%	0 0%
2. Abselilovsky	63	4 6.3%	57 90.5%	1 1.6%	0 0%	1 1.6%
3. Ilishevsky	56	1 1.8%	0 0%	55 98.2%	0 0%	0 0%
4. Komi	42	1 2.4%	0 0%	0 0%	41 97.6%	0 0%
5. Talish	20	0 0%	0 0%	0 0%	0 0%	20 100%

Percentage of 'grouped' cases correctly classified: 96.41%.

populations are clearly distinguished from each other.

The population membership classification obtained by the discriminant analysis of DNA fingerprinting data is shown in Table 4.1. For the most part, the members of a given population are assigned to their own group. Only a few individuals are ascribed to other groups. The level of cases correctly classified is 96.4%.

Thus these data demonstrate new possibilities for using multilocus DNA fingerprint probes in population studies as well as in ethno-historical investigations. Theoretically, it is also possible to solve the complementary problem – to identify individuals belonging to a given, previously characterized population. We hope that further development of the approach will make it applicable to different fields of population biology, such as the evaluation of the genetic diversity of a small ethnic group; the genetic epidemiology in the different ethnic populations and so on.

Acknowledgements

This work was partly supported by the Russian State Programmes: Fundamental Research Foundation, Biodiversity, Frontiers in Genetics and Human Genome.

References

Barysheva, E. V., Prosnyak, M. I., Vlasov, M. S. Limborska, S.A. and Ginter, E. K. (1989). The use of wild-type M13 DNA for analysis of interindividual human DNA polymorphisms demonstrated in a population study in Krasnodar. *Genetika*, **25**, 2079–82.

Bulayeva, K. B. & Pavlova, T. A. (1993). Behavior genetic differences within and between defined human populations. *Behavior Genetics*, **23**, 433–41.

Bulayeva, K. B., Pavlova, T. A., Dubinin, N. P., Hay, D. A. & Foley, D. (1993). Phenotypic and genetic affinities among ethnic populations in Daghestan (Caucasus, Russia): a comparison of polymorphic, physical, neurophysiological and psychological traits. *Annals of Human Biology*, **20**, 455–67.

Cheong, C., Tinoco, I., Jr. & Chollet, A. (1988). Thermodynamic studies of base pairing involving 2,6-diaminopurine. *Nucleic Acids Research*, **16**, 5115–22.

Christmann, A., Lagoda, P. J. L. & Zang, K. D. (1991). Non-radioactive *in situ* hybridization pattern of the M13 minisatellite sequences on human metaphase chromosomes. *Human Genetics*, **86**, 487–90.

Financsek, I., Mizumoto, K., Mishima, J. & Muramatsu, M. (1982). Human ribosomal RNA gene: Nucleotide sequence of the transcription initiation region and comparison of three mammalian genes. *Proceedings of the National Academy of Sciences of the USA*, **79**, 3092–6.

Harwood, I. J., Tachibana, A. & Meuth, M. (1991). Multiple dispersed spontaneous mutations: a novel pathway of mutagenesis in a malignant human cell line. *Molecular and Cell Biology*, **11**, 3163–70.

Ijdo, J. W., Wells, R. A., Baldini, A. & Reeds, S. T. (1991). Improved telomere detection using a telomere repeat probe $(TTAGGG)_n$ generated by PCR. *Nucleic Acids Research*, **19**, 4780.

Jeffreys, A. J., Wilson, V. & Thein, S. L. (1985). Hypervariable 'minisatellite' regions in human DNA. *Nature*, **314**, 67–73.

Jeffreys, A. J., Turner, M. & Debenham, P. (1991). The efficiency of multilocus DNA fingerprint probes for individualization and establishment of family relationships determined from extensive casework. *American Journal of Human Genetics*, **48**, 824–40.

Jincharadze, A. G., Ivanov, P. L. & Ryskov A. P. (1987). Genomic fingerprinting. Characterization of cloned human DNA sequence JIN600. *Dokl.Acad.Nauk (USSR)*, **295**, 230–3.

Kalnin, V. V., Kalnina, O. V., Prosnyak, M. I., Khidiatova, I. M., Khusnutdinova, E. K., Raphicov, K. S. & Limborska, S. A. (1995). Use of DNA fingerprinting for human population genetic studies. *Molecular and General Genetics*, **247**, 488–93.

Khusnutdinova, E. K., Khidiatova, I. M., Prosnyak, M. I., Raphicov, K. S. & Limborska, S. A. (1994). Analysis of DNA polymorphism revealed by the method of DNA fingerprinting on the basis of M13 phage DNA in Bashkir and Komi populations. *Genetika*, **30**, 1621–5.

Kuseev, R. G. (1974). The origin of the Bashkir people. In *Nauka, Moscow*, ed. T. A. Zhdanko, 6–112.

Lebart, L., Morinea, A. & Warwick, K. M. (1985). *Correspondence Analysis and Related Techniques for Large Matrices*. New York: Wiley.

Mariat, D. & Vergnaud, G. (1992). Detection of polymorphic loci in complex genomes with synthetic tandem repeats. *Genomics*, **12**, 454–8.

Pena, S. D. J. & Chakraborty, R. (1994). Paternity testing in the DNA era. *Trends in Genetics*, **10**, 204–8.

Prosnyak, M. I., Kartel, N. A., Perebityuk, S. A., Limborska, S. A. & Ryskov, A. P. (1990). Non-isotopic method for DNA fingerprinting based on the M13 phage DNA in human parentage studies. *Genetika*, **26**, 134–7.

Prosnyak, M. I., Veselovskava, S. I., Myasnikov, V. A., Efremova, E. J., Potapov, V. K., Limborska, S. A. & Sverdlov, E. D. (1994). Substitution of 2-aminoadenine and 5-methylcytosine for adenine and cytosine in hybridization probes increases the sensitivity of DNA fingerprinting. *Genomics*, **21**, 490–4.

Ryskov, A. P., Jincharadze, A. G., Prosnyak, M. I., Ivanov, P. L. & Limborska, S. A. (1988). M13 phage DNA as a universal marker for DNA fingerprinting of animals, plants and microorganisms. *FEBS Letters*, **233**, 388–92.

Ryskov, A. P., Kupriyanova, N. S., Kapanadze, B. I., Netchvolodov, K. K., Pozmogova, G. E., Prosnyak, M. I. & Yankovsky, N. K. (1993). The frequency of mini- and microsatellites in the human 13th chromosome. *Genetika*, **29**, 1750–4.

Smith, G. R. & Stahl, F. W. (1985). Homologous recombination promoted by Chi sites and Rec BC enzyme of *Escherichia coli*. *Bioessays*, **2**, 244–9.

Sylvester, J. E., Whiteman, D. A., Podolsky, R., Pozsgay, J. M., Respess, J. & Schmiskel, R. D. (1986). The human ribosomal RNA genes: structure and organization of the complete repeating unit. *Human Genetics*, **73**, 193–8.

Sylvester, J. E., Petersen, R. & Schmickel, R. D. (1989). Human ribosomal DNA: novel sequence organization in a 4.5-kb region upstream from the promoter. *Gene*, **84**, 193–6.

Tokarskaya, O. N., Kalnin,V. V., Panchenko, V. G. & Ryskov, A. P. (1994). Genetic differentiation in a captive population of the endangered crane (*Grus leucogeranus* Pall.). *Molecular and General Genetics*, **245**, 658–60.

Vassart, G., Georges, M., Monsieur, R., Brocas, H., Lequarre, A-S. & Christophe, D. (1987). A sequence in M13 phage detects hypervariable minisatellites in human and animal DNA. *Science*, **235**, 683–4.

5 *Kinship, inbreeding, and matching probabilities*

N. E. MORTON AND J. W. TEAGUE

To a geneticist, population structure signifies the relation between gene frequencies and genotype frequencies of individuals and pairs of individuals. The key concept is *kinship*, the probability that two alleles sampled in a specified way are identical by descent. If the alleles are in the same individual, kinship is called *inbreeding*, which is the same as kinship between the parents of that individual. Together with other coefficients that measure identity by descent in pairs of individuals, kinship and inbreeding determine the *matching probability* that two individuals sampled in a specified way have the same genotype. Kinship leads to inferences about the similarity and therefore (with less assurance) about the phylogeny of different populations and the optimal strategy for conserving variability in a species that may be endangered. Inbreeding leads to identifiable risks and inferences about the mutational load. Matching probabilities are the basis for DNA forensic science. The relationship of these concepts is the *pons asinorum* of population genetics. Much of the literature is marred by use of genetic distance and other surrogates for kinship, inefficient estimation, selection of extreme populations, and failure to allow for biases due to sample size and incompleteness of genealogies.

Kinship

It is possible to predict kinship from genealogies or migration, but the ultimate test is kinship bioassay. Much of the literature is based on bioassay from homozygosity under local panmixia. If Q_k is the frequency of the kth allele G_k in an array of interrelated populations and q_{ik}, q_{jk} are the corresponding frequencies in populations i and j, the homozygosity estimate of kinship is

$$Y_{ij}(H) = \frac{\sum q_{ik}q_{jk} - \sum Q_k^2}{1 - \sum Q_k^2} \tag{5.1}$$

We shall see that this estimate, which underlies the work of Nei (1972) and Cockerham (1969), is inefficient.

Kinship based on a sample of N_i disomic individuals from the ith population has bias $1/2N_i$. An unbiased estimate is

$$\phi_{ij} = \begin{cases} \dfrac{Y_{ii} - 1/2N_i}{1 - 1/2N_i} & \text{for } i = j \\ Y_{ij} & \text{for } i \neq j \end{cases} \tag{5.2}$$

The mean of ϕ_{ij} is

$$\bar{\phi} = \sum_i \sum_j N_i N_j \phi_{ij} \bigg/ \sum_i \sum_j N_i N_j = 0 \tag{5.3}$$

since two genes drawn at random from an array of populations with gene frequencies $\{Q\}$ generate Hardy–Weinberg genotype frequencies. The weighted mean of the diagonal elements is

$$F_{ST} = \sum_i w_i \phi_{ii} \bigg/ \sum_i w_i \tag{5.4}$$

where w_i is the size of the ith population. The single parameter F_{ST} of Wright (1943) summarizes the $n(n+1)/2$ distinct elements of the symmetrical kinship matrix of Malecot (1950) in the same sense that a mean summarises a distribution. This is not very useful unless the populations have the same effective size and migrational pattern, which is seldom true. Sometimes F_{ST} is used incorrectly as a synonym for specific kinship ϕ_{ii}.

Genealogy, migration, and isolation by distance provide an estimate of random kinship Φ_R relative to founders, which is obviously not the same as mean kinship $\bar{\phi}$. Expressed in other terms, the kinship matrix $\{\phi\}$ from bioassay is conditional on current gene frequencies in the array of populations, whereas the absolute kinship matrix $\{\Phi\}$ is relative to gene frequencies in founders, or

$$\phi_{ij} = \frac{\Phi_{ij} - \Phi_R}{1 - \Phi_R} \tag{5.5}$$

Evolutionary genetics deals with absolute kinship $\Phi(t)$ where t is the number of generations from the founders. Under simple models $\Phi(t)$ goes from 0 to some equilibrium value $\Phi(\infty) < 1$ at which stochastic variation determined by effective population size is balanced by systematic pressure (migration, mutation and perhaps selection). Since founders are usually arbitrary and their gene frequencies unknown, applications to forensic and other real populations deal with conditional kinship.

Isolation by distance summarizes the kinship matrix in terms of three parameters (a,b,L). Kinship of two random individuals or populations at distance d is approximately

$$\Phi(d) = ae^{-bd}$$
$$\phi(d) = (1 - L)ae^{-bd} + L \tag{5.6}$$

where L is the kinship at large distances. Substituting Φ_R for ae^{-bd} and setting $\Phi(d) = 0$ according to equation (5.3), we obtain

$$\Phi_R = -\frac{L}{1 - L} \tag{5.7}$$

corresponding to random kinship within the examined array of populations, relative to a total that includes the surrounding array. For example, French populations may be used to estimate random kinship in France relative to Western Europe, blurring the distinction between F_{ST} and Φ_R since France may be considered a random strain S within a total T (Morton, 1992).

If we let $\sum Q_k^2$ approach zero in equation (5.1), we obtain $y_{ii} = \sum_k q_{ik}^2$ as limiting kinship. This is the 'infinite alleles model' about which much nonsense has been written. It has bias $1/2N_i$, is highly variable among loci, cannot be used to estimate genotype frequencies or matching probabilities, and corresponds to an array of populations in which any contemporary species is only a small element. Like any other measure of similarity it may be used to construct dendrograms and make evolutionary inferences which are not much worse than from kinship relative to contemporary gene frequencies $\{Q\}$, which has much richer applications.

There are two sources of variation in kinship bioassay, corresponding to fixed sampling of $2N_i$ gametes from the contemporary ith population and the accumulated variation of replicate loci or populations. The latter is usually unknown and is the same for different estimates of kinship. Morton, Collins and Balazs (1993) showed by simulation that the sampling variance of an efficient estimate for a codominant system as $\phi_{ii} \rightarrow 0$ is $1/2N_i(2N_i - 1)(K - 1)$, where K is the number of alleles in the array of populations. For example, Pearsonian χ^2 testing homogeneity of the $\{q\}$ with $\{Q\}$ is $\chi^2 = 2n_i\{\sum(q_{ik}^2/Q_k) - 1\}$, which corresponds to assigning weight $1/Q_k$ to the expression $E(q_{ik}^2) = Q_k^2 + y_{ii}Q_k(1 - Q_k)$ and summing to give $y_{ii}(C) = \sum(q_{ik}^2/Q_k) - 1\}/(K - 1)$ where C denotes Pearsonian χ^2 and so $\chi^2 = 2N_i(K - 1)y_{ii}$. The variance of central χ^2 is $2(K - 1)$, and therefore the result of Morton *et al.* (1993) is demonstrated. The asymptotic equivalence of Pearsonian χ^2 and likelihood ratio theory gives $y_{ii}(S) = 2\{\sum q_{ik} \ln (q_{ik}/Q_k)\}/(K - 1)$. Finally, maximum likelihood (ML) theory for $\phi_{ii} = 0$ scores $G_k G_k$ as $(1 - Q_k)/Q_k$ and $G_k G_{\bar{k}}$ as -1, the difference of $1/Q_k$ motivating χ^2 and giving information $n(K - 1)$ for a sample of n individuals as $\phi_{ii} \to 0$. If $2N_i$ genes are paired without replacement, there are $N_i(2N_i - 1)$ pairs of genes, and so the amount of information is $N_i(2N_i - 1)(K - 1)$ as above. Whereas the other formulae pair alleles with replacement to give $\sum q_{ik}^2$ and so require the bias correction of equation (5.2), the ML estimate is unbiased (Morton *et al.*, 1993). These results show the close connection between kinship and correspondence analysis (Benzecri, 1973).

If gene frequencies are $Q_k = 1/K$, then $E(q_{ik}^2) = [1 + y_{ii}(K - 1)]/K^2$. Summing gives equation (5.1), showing that the homozygosity estimate is fully efficient in this special case. However, in general the sampling variance of equation (5.1) as $\phi_{ii} \to 0$ is $\sum Q_k^2/N_i(2N_i - 1)(1 - \sum Q_k^2)$, and so the relative efficiency is

$$R = (1 - \sum Q_k^2)/(K - 1)\sum Q_k^2 \tag{5.8}$$

The worst possible situation is a very common gene and $K - 1$ rare ones, like the transferrin (TF) electromorphs. In that case R approaches 0. This is not serious in phylogenetic studies at the species level, because sampling variance is negligible by comparison with accumulated variation among loci. However, for human populations (especially the large, poorly differentiated populations of forensic importance), inefficient estimates based on homozygosity should be avoided. Pooling asymptotically efficient estimates over loci gives

$$\bar{\phi}_{ii} = \frac{\sum\{N_i(2N_i - 1)(K - 1)\}\phi_{ii}}{\sum\{N_i(2N_i - 1)(K - 1)\}} \tag{5.9}$$

Equations (5.8) and (5.9) were derived on the assumption that $\phi_{ii} = 0$. The general case is being examined.

Inbreeding

Bioassay of inbreeding is subject to large errors, because each individual contributes only one scorable pair of genes and so the amount of information for a codominant system is only $N_i(K - 1)$, or $1/(2N_i - 1)$ as much as for kinship (Morton, 1992). Furthermore typing errors and binning of alleles generate artefacts that are confounded with inbreeding. The ML theory (Yasuda, 1968*a*) provides a foundation for kinship as well as inbreeding.

Most estimates of inbreeding are derived from genealogies, migration or marital isonymy. Methods are available to extrapolate from a few generations to total inbreeding, assuming that effective population size and migration remain constant (Morton, 1982). The estimate from migration assumes that only one member of each marital pair is an immigrant. Then the frequency of endogamy is $2p_{ii} - 1$ where p_{ii} is the probability that a parent of a child born in i was also born in i. Therefore inbreeding is

$$\alpha_i = 2\sum_h p_{hi}\Phi_{hi} - \Phi_{ii} \tag{5.10}$$

where p_{hi} is the probability that a parent of a child born in i was born in h, and both h and i are in the array of populations being modelled. Migration theory neglects avoidance or preference for consanguineous marriage.

A few years ago inbreeding was a bone of contention in the forensic use of DNA. Controversy was resolved by showing that inbreeding initially decreases the probability that two individuals drawn at random from an array of populations have the same genotype (Li & Chakravarti, 1994; Collins & Morton, 1994). Only if α exceeds a quantity that is approximately twice the homozygosity does the mean matching probability exceed its value under panmixia ($\alpha = 0$). An expert system for forensic identification must provide the critical value of α for each locus.

Table 5.1 gives estimates of kinship and inbreeding in several populations that were studied because inbreeding was relatively high. The difference from kinship is too small to be consequential, and so estimates of local kinship and inbreeding are usually taken to be equivalent (Morton, 1992; Weir, 1994).

Matching probabilities

Forensic use of DNA depends on samples contributed by the suspect S and culprit C. The population of the latter and relationship between them are usually unknown. Let H_0 be a null hypothesis about their relationship and H_1 be an alternative hypothesis specifying a closer relationship. The general likelihood ratio for information contributed by the hth locus is

$$\lambda_h = \frac{P(E_c^h|E_s^h, H_1)}{P(E_c^h|E_s^h, H_0)} \tag{5.11}$$

evaluated under some assumption about the population of C. The λ_h are independent over multiple unlinked loci, and so the total likelihood ratio is $\Lambda = \prod_h \lambda_h$. In the *exclusion test*, H_0 specifies two random individuals from the population and H_1 corresponds to replicate samples from the same individual. Choice of population has some effect on the distribution under H_0, but kinship need not be considered. With VNTR loci allelic bins are heterogeneous and their width is arbitrary. It is conventional to take the

Table 5.1. *Kinship ϕ and inbreeding α in various populations*

Source	Population	Kinship (ϕ)	Inbreeding (α)	Reference
Genealogy	Artas village, generation 3	0.0194	0.0200	Morton (1982)
	Artas village, generation 4	0.0220	0.0224	Morton (1982)
	Saas commune	0.0056	0.0067	Morton *et al.* (1973)
	Bergkirchen	0.0025	0.0043	Morton *et al.* (1973)
	Namu atoll	0.0178	0.0249	Morton *et al.* (1973)
	Pingelap atoll	0.0756	0.0698	Morton *et al.* (1973)
	Mokil atoll	0.0404	0.0338	Morton *et al.* (1973)
Migration	Artas village	0.0213	0.0196	Morton (1982)
	Kallerwan village M	0.0155	0.0154	Morton (1982)
	Kallerwan village N	0.0311	0.0310	Morton (1982)
	Barra island	0.0077	0.0075	Morton (1982)
	Switzerland	0.0026	0.0021	Morton *et al.* (1973)
	Alpine isolates	0.0071	0.0060	Morton *et al.* (1973)
Mean		0.0207	0.0203	

logarithm of fragment size and to define bin width in terms of the radius of coalescence that gives the same value of $\sum Q_k^2$ as the frequency of single band phenotypes (Collins & Morton, 1994). For microsatellites, bins are specified by antimodes, logarithmic transformation is not helpful and bins are assigned to 'molecular units' corresponding to estimated base pairs (bp). It is convenient to express matching probability in terms of the sum and difference of corresponding fragment lengths which are nearly orthogonal. Under H_0 (and under H_1 for microsatellites) they are not normal, and so the empirical distribution of λ is useful.

The remaining tests (coincidence and kinship) depend on bin assignment rather than estimated fragment length and are of interest only if exclusion cannot be made. If coincidence and kinship are both rejected, the only alternative is identity. Choice of population affects bin frequencies under H_0, whereas the probability under H_1 is $1 - \varepsilon$, where ε is the frequency with which replicates are assigned to different bins, or effectively 1 if binning is accurate. Then λ is the reciprocal of the matching probability under H_0. The test can be made nearly independent of population by taking the mean matching probability, which entails a tolerable loss of information (Collins & Morton, 1994). In the coincidence test H_0 denotes random sampling from an array of populations. We have discussed robustness of the coincidence test with respect to inbreeding.

Rejection of coincidence leads to consideration of kinship, which has a greater and much more variable effect than inbreeding without kinship. For the kinship test, the suspect and culprit are drawn from the same

Table 5.2. *Coefficients of identity for regular relatives*

Relationship	Kinship (ϕ)	Identity coefficients			Degree of kinship (k)
		c_2	c_1	c_0	
Identical twins	1/2	1	0	0	0
Sibs	1/4	1/4	1/2	1/4	1
Double first cousins	1/8	1/16	6/16	9/16	2
Parent–child	1/4	0	1	0	1
Grandparent–grandchild (= uncle–niece = half sibs)	1/8	0	1/2	1/2	2
First cousins (= great grandparent-great grandchild = great uncle–niece)	1/16	0	1/4	3/4	3
First cousins once removed	1/32	0	1/8	7/8	4
Second cousins	1/64	0	1/16	15/16	5
Equal bilineal	$(1/2)^{k+1}$	$4\phi^2$	$4\phi(1-2\phi)$	$(1-2\phi)^2$	k
Unilineal	$(1/2)^{k+1}$	0	4ϕ	$1-4\phi$	k

population rather than at random from the array of populations. Collins and Morton (1994) give solutions for regular relatives (i.e. neither inbred), including robust means of matching probabilities. The *affinal model* considers relatives related as closely as spouses. Following Yasuda (1968*b*) solutions were given linear in α, but more general cases have recently been considered (Weir, 1994). Following Wright (1931) and Malecot (1950), Balding and Nichols (1994) examined a model in which each population is randomly drawn from a Dirichlet distribution generated by equilibrium between drift and linear systematic pressure. Their approach can be extended to relatives who share a defined close relationship with probabilities c_0, c_1, c_2 of 0, 1, 2 genes identical by descent in a population where there is remote relationship measured by inbreeding α (or kinship ϕ) relative to an array of populations with gene frequencies Q_k (Table 5.2). The relatives are regular if $\alpha = 0$ and irregular otherwise. It is not assumed that α is the same for every population, simply that α is independent of gene frequencies: in the set of populations with inbreeding α, gene frequencies are on the average the same as in the array.

Let $P(R^nS^m)$ denote the probability of drawing n alleles G_r, and m alleles G_s in a sample of $n + m$ from the same population. Table 5.3 gives the matching probability for two close relatives in such a population and Table 5.4 gives examples. The results of Weir (1994, Table 2) derived from a more complex model, satisfy the Dirichlet distribution, and numerical errors in his Tables 4–6 have been corrected (personal communication). This illustrates that forensic use of DNA is now normal science, where advances can be made from established principles and without trauma.

Table 5.3. *Probabilities in a Dirichlet distribution with parameter* α *(Balding and Nichols, 1994)*

$P(R) = Q_s$	$P(S) = Q_s$
$P(R^2)$	$= Q_r[\alpha + (1-\alpha)Q_r]$
$P(RS)$	$= Q_rQ_s(1-\alpha)$
$P(r^4)$	$= P(R^2)\ \{2\alpha + (1-\alpha)Q_r\}\ \{3\alpha + (1-\alpha)Q_r\} / (1+\alpha)(1+2\alpha)$
$P(s^4)$	$= P(S^2)\ \{2\alpha + (1-\alpha)Q_s\}\ \{3\alpha + (1-\alpha)Q_s\} / (1+\alpha)(1+2\alpha)$
$P(R^2S^2)$	$= P(RS)\ \{\alpha + (1-\alpha)Q_r\}\ \{\alpha + (1-\alpha)Q_s\} / (1+\alpha)(1+2\alpha)$
$P(R^3)$	$= P(R^2)\ \{2\alpha + (1-\alpha)Q_r\} / (1+\alpha),\ P(S^3) = P(s^2)\ \{2\alpha + (1-\alpha)Q_s\} / (1+\alpha)$
$P(R^2S)$	$= P(RS)\ \{\alpha + (1-\alpha)Q_r\} / (1+\alpha),\ P(RS^2) = P(RS)\ \{\alpha + (1-\alpha)Q_s\} / (1+\alpha)$
$P(R^2 \mid R^2)$	$= \{c_0P(R^4) + c_1P(R^3) + c_2P(R^2)\} / P(R^2)$
$P(RS \mid RS)$	$= \{2c_0P(R^2S^2) + c_1[P(R^2S) + P(RS^2)]/2 + c_2P(RS)\} / P(RS)$

Table 5.4. *Likelihood ratio (forensic index) under combined effects of kinship and inbreeding*

Q	Inbreeding (α)	Unrelated homozygote	Unrelated heterozygote	First cousin homozygote	First cousin heterozygote
0.01	0	10000	5000	388	377
	0.001	6440	4152	321	342
	0.010	864	1301	121	182
	0.100	20	56	12	26
0.10	0	100	50	31	25
	0.001	96	49	30	25
	0.010	67	43	25	23
	0.100	12	18	8	12

We have argued that equating ϕ and α is not restrictive: the boundaries of populations are not precise, and the difference between α and ϕ is small (Table 5.1). Forensic practice does and should err on the side of the defendant by overestimating α when there is doubt, in this way avoiding fudged gene frequencies. However, the expert witness is ignorant of the culprit's identity, and therefore of his population and relationship to the suspect. The court must decide what hypothesis to favour and therefore how conservative to be within the domain of scientifically admissible tests. In exceptional circumstances a value of α as high as 0.01 may be justified in Caucasian communities and 0.05 or more in strongly endogamous populations. In many circumstances close relationship may reasonably be excluded by other evidence, circumstantial or molecular. However, the general solution is neat and will be useful in some cases. The prosecution must be prepared to type enough loci so that even a skilful appeal to kinship will carry little weight, and the computer program that calculates

Table 5.5. *Three approaches to population structure (Morton, 1994)*

	Tectonic Wright Malecot	Halieutic Lewontin/Hartl	Icarian Lander/NRC
Kinship explicit	yes	no	no
Gene frequencies	estimated	mongrelized	fabricated
Frequencies depend on population	yes	yes	no
Allowance for sampling bias	yes	no	no
Misuse of confidence interval	no	no	yes
Population	forensic	singular	imaginary
Used in population genetics	yes	no	no
Supported by migration, genealogy, etc	yes	no	no
Results are probabilities	yes	yes	no
Likelihood ratios applicable	yes	yes	no

likelihood ratio must be subjected to as much quality control as laboratory data and chain of custody.

Forensic DNA typing is now ten years old. The tectonic approach favoured by nearly all population geneticists for the past 75 years and used here has been attacked from two directions (Table 5.5). The halieutic approach of Lewontin and Hartl (1991) insists that the culprit be assigned to the same subpopulation as the suspect, although the latter is formally irrelevant unless the culprit is a relative (which the halieutic approach does not allow for). Usually there is little information about the culprit's subpopulation, and even the assignment of the suspect to a narrowly defined group may be poorly supported: DNA evidence should not be dismissed because the suspect claims to be the last of the Mohicans. This problem is easily handled by the tectonic approach through an appropriate choice of kinship.

The Icarian approach of Lander (1991) and the Committee on DNA Technology in Forensic Science (1992) replaces gene frequencies with numbers specified by an arbitrary rule (the 'ceiling principle') and neglects kinship. The results are not probabilities, and the 'ceiling principle' is neither a ceiling nor principle. Endorsement by a Committee of the National Research Council places the National Academy on the side of authority against science. It is to be hoped that a second committee will attempt to inform rather than to legislate and will not make a fraudulent compromise between genetics and the 'ceiling principle'. In all likelihood the O.J.Simpson trial has discredited the 'ceiling principle' as the least scientific and therefore most vulnerable way in which the prosecution can present DNA evidence. There is every reason to expect that American courts will be reconciled to science, and that future debate will be confined to methods that do not violate population genetics and statistics. Then calculations presented as evidence will be subjected to as much quality control as laboratory performance, and the enormously expensive databases now being constructed for forensics and the study of human diversity will provide efficient measures of population structure and the operating characteristics of alternative matching probabilities.

References

Balding, D. J. & Nichols, R. A. (1994). DNA profile match probability calculation: how to allow for population stratification, relatedness, database selection and single bands. *Forensic Science International*, **64**, 125–40.

Benzecri, J. P. (1973). *L'analyse des données, l'analyse des correspondences*. Paris: Dunod.

Cockerham, C. C. (1969). Variance of gene frequencies. *Evolution*, **23**, 72–84.

Collins, A. & Morton N. E. (1994). Likelihood ratios for DNA identification. *Proceedings of the National Academy of Sciences, USA*, **91**, 6007–11.

National Committee on DNA Technology in Forensic Science, National Research Council (1992). *DNA Technology in Forensic Science*. Washington: National Academy Press.

Lander, E. S. (1991). Research on DNA typing catching up with courtroom application. *American Journal of Human Genetics*, **48**, 819–23.

Lewontin, R. C. & Hartl, D. L. (1991). Population genetics in forensic DNA typing. *Science*, **254**, 1745–50.

Li, C. C. & Chakravarti, A. (1994). DNA profile similarity in a subdivided population. *Human Heredity*, **44**, 100–9.

Malecot, G. (1950). Quelques schemas probabilistes sur la variabilité des populations naturelles. *Annales Université Lyon* A, **13**, 37–60.

Morton, N. E. (1982). Kinship and inbreeding in populations of Middle Eastern origin and controls. In *Current Developments in Anthropological Genetics*, ed. M. H.Crawford and J. H. Mielke, *Vol* 2, pp. 449–66. New York: Plenum Press.

Morton, N. E. (1992). Genetic structure of forensic populations. *Proceedings of the National Academy of Sciences, USA*, **89**, 2556–60.

Morton, N. E. (1994). Genetic structure of forensic populations. *American Journal of Human Genetics*, **55**, 587–8.

Morton, N. E., Klein, D., Hussels, J.E., Dodinval, P., Todorov, A., Lew, R. & Yee, S. (1973) Genitic structure of Switzerland. *American Journal of Human Genetics*, **25**, 347–61.

Morton, N. E., Collins, A. and Balazs, I. (1993). Kinship bioassay on hypervariable loci in Blacks and Caucasians. *Proceedings of the National Academy of Sciences, USA*, **90**, 1892–6.

Nei, M. (1972). Genetic distance between populations. *American Naturalist*, **106**, 283–92.

Weir, B. S. (1994). The effects of inbreeding on forensic calculations. *Annual Review of Genetics*, **28**, 597–621.

Wright, S. (1931). Evolution of mendelian populations. *Genetics*, **16**, 97–159.

Wright, S. (1943). Isolation by distance. *Genetics*, **28**, 114–38.

Yasuda, N. (1968*a*). Estimation of the inbreeding coefficient from phenotype frequencies by a method of maximum likelihood scoring. *Biometrics*, **24**, 915–35.
Yasuda, N. (1968*b*). An extension of Wahlund principle to evaluate mating type frequency. *American Journal of Human Genetics*, **20**, 1–23.

Appendix

Variance of kinship ϕ

1. Maximum likelihood

A sample of $2N_i$ genes gives $n_k(n_k - 1)/2$ pairs G_kG_k with expected frequency $Q_k[Q_k + \phi(1 - Q_k)]$.
The frequency of heterozygous pairs is

$$\binom{2N}{2} - \sum n_k(n_k - 1)/2 = 2N_i^2 - \sum n_k^2/2.$$

Define

$$w_k = \begin{cases} n_k(n_k - 1)/2 & \text{for } G_kG_k \\ 2N_i^2 - \sum n_k^2/2 & \text{for heterozygotes} \end{cases}$$

$$u_k = \frac{\partial \ln L}{\partial \phi} = \begin{cases} \dfrac{1 - Q_k}{Q_k + \phi(1 - Q_k)} & \text{for } G_kG_k \\ -1/(1 - \phi) & \text{for heterozygotes} \end{cases}$$

The maximum likelihood score is $U = \sum w_k u_k$ with information $\sum w_k u_k^2$, and the variance of ϕ is the reciprocal.

2. Chi-square

The expected value of χ^2 is

$$\begin{aligned} \chi^2 &= 2N_i(K - 1)y_{ii} \\ &= (K - 1)[1 + (2N_i - 1)\phi] \end{aligned}$$

The variance of is χ^2 is $2(K - 1) + 4(2N_i - 1)(K - 1)\phi$. Therefore the variance of ϕ is $[2 + 4(2N_i - 1)\phi]/(K-1)(2N_i - 1)^2$. If $\phi < 1/(2N_i - 1)$, Equations 5.8 and 5.9 apply to a good approximation. If $\phi \gg 1/2N_i - 1)$ and constant among loci, the information is proportional to $(2N_i - 1)(K - 1)$. If ϕ varies among loci, the information becomes approximately $(2N_i - 1)(K - 1)/4\phi$.

3. Homozygosity

In a sample of $N_i(2N_i - 1)$ pairs without replacement, the expected frequency of homozygosity is $\sum Q_k^2 + (1 - \sum Q_k^2)\phi$. Therefore the variance of kinship estimated from homozygosity is

$$\frac{(1 - \phi)[\sum Q_k^2 + (1 - \sum Q_k^2)\phi]}{N_i(2N_i - 1)(- \sum Q_k^2)}$$

which approaches $\phi(1 - \phi)/N_i(2N_i - 1)$ for $\phi \gg \sum Q_k^2$.

6 *Using the coalescent to interpret gene trees*

R. M. HARDING

Human genetic diversity: is it old?

In the last decade, molecular biology has provided sophisticated technology for assaying DNA polymorphisms, and considerable research effort has been devoted to characterizing human variability at the DNA level. Many geneticists are concentrating on the DNA variation that explains human diversity at the phenotypic level, with particular attention being given to genetic diseases. Most of the disease alleles so far identified have restricted geographic distributions, are relatively rare, and were probably generated by recent mutation events. Cystic fibrosis is an example of a genetic disease prevalent in Europe and estimated to have arisen perhaps 53 000 years ago (Morral *et al.*, 1993). A few disease alleles with similarly restricted geographical distributions, are much more common, clearly as the result of powerful selection. Such are the haemoglobinopathies which are subject to malarial selection. Allelic variants occur at both the α- and β-globin genes, causing α- and β-thalassaemias, sickle-cell anaemia and other traits and diseases. It is difficult to judge whether they arose very recently, possibly even within the last 10 000 years, or are older, say 50 000–100 000 years (Flint *et al.*, 1993). Studies of variation not accounting for obvious functional differences have been taken up by biological anthropologists. Much of the diversity they study has an ancestry that is distributed world-wide and is probably old. Just how old is a hotly debated topic. It is not an easy question to answer. Although there are clues in the patterns of DNA variation, interpreting them calls for the expertise of population genetics.

There is a long-standing debate in population genetics about the nature of genetic diversity and whether or not it results from the maintenance of specific alleles at selective equilibrium or from a dynamic turnover of transient neutral (or nearly neutral) variation (Kimura, 1983). Determining the age of alleles subject to selection so powerful that it establishes equilibrium frequencies requires much more information than is required for ageing neutral alleles. The high frequency of the β^S variant of β-globin, which as a homozygous genotype causes sickle-cell anaemia, but as a heterozygous genotype reduces the risk of severe malarial infection, provides a classic example of a genetic polymorphism maintained by strong selection. To determine the age of such a polymorphism requires information about when malaria began to make a selective impact. That information should be consistent with the minimum time needed for an allele to attain its observed frequency in a population, but at selective equilibria current frequencies do not indicate age. For neutral polymorphisms, conversely, there is a continual turnover of functionally equivalent allelic variants as loss by genetic drift balances new mutational input, and frequencies do provide information about age. In the simplest population genetics models for the turnover of neutral polymorphism, diversity statistics based on allelic frequencies are time-scaled by mutation rate, assuming a constant molecular clock and constant population size. In more powerful models for time-scaling neutral diversity, population size is a variable and describes demographic history.

Human genetic diversity: is it neutral?

There is considerable evidence for many loci that genetic drift is a more powerful process than selection, and some suggestion that human demographic history provides a reasonable explanation even for allelic variation known to code for functional variation. Consider cystic fibrosis. There are several cystic fibrosis alleles having a combined frequency in European populations that seems very high, given their selective disadvantage in an individual homozygous for these alleles. A selective advantage for carriers of these alleles under specific historical circumstances has been proposed to explain the prevalence of this disease (Morral *et al.*, 1993; Sereth *et al.*, 1993). However, an alternative explanation is that, in demographic expansions happening for reasons unrelated to fitness contributions by cystic fibrosis alleles, a number of the founding families were, by chance, carriers. How many genetic diseases were unleashed by human population expansions in the last 50 000 years? As studies of different loci accumulate, it should become possible to discern common patterns due to human demographic history, from distinctive patterns due to selection.

Models have been constructed which suggest time-scalings for the mutations generating genetic diversity, based on the assumption that allelic frequencies are determined by selection and that demographic history is irrelevant. They can suggest controversial time-scalings (see Hill, Chapter 7). That different models are being used to provide time-scalings for genetic diversity is evidence of a widening interest in the problem. Models are instructive not only for studies of human evolution. A better understanding of the age and evolutionary turnover of DNA sequence variation provides the context for analyses of the genetic diversity implicated in disease. Ideally, population genetic models incorporating selection should be used. At present, however, the most powerful model available, the coalescent, requires the assumption of neutrality. Options do exist. Selection can be introduced into branching process models of genetic drift. However, while these models are appropriate for simulation studies, they are neither powerful, nor efficient, for estimation problems.

Is it appropriate to model the evolution of genetic diversity as a stochastic process rather than as a fitness-determined process? After all, if the progeny who reproduce most successfully are characterized by particular genetic variants, adaptation by natural selection occurs: there is abundant evidence for this process. However, this is not to say that all differential reproductive success is determined by genotype. Many chance elements, such as buses, falling meteorites, wars and famines, intervene. Furthermore, the relative contributions to the fitness of an individual by alleles at a single locus may be negligibly small. Putting aside examples such as the maintenance of β^{S} globin variants, many polymorphisms, including most of those found at the β-globin locus, (Fullerton *et al.*, 1994) confer no measurable advantages in reproductive success. So far, the different levels of polymorphism observed in mtDNA, most autosomal nuclear loci (except HLA) and in Y chromosomes, can be explained by the four-fold difference in effective population size for mtDNA and Y variation compared with nuclear DNA, and higher mutation rates in mtDNA. A neutral model of DNA diversity therefore seems a satisfactory place to start.

The apparent neutrality of much allelic variation, even when contributions to functional differences may be suspected, is due to the fact that selection is probably too weak to have much impact in the face of larger random forces. Like the outcome of a coin toss, the persistence of alleles carried by small numbers of individuals in a population depends on a multitude of events happening for complex reasons, that are best accepted as the luck of the draw. In small populations the numbers of copies of all alleles are low by definition. Whether or not populations viewed from the perspective of evolutionary time are small or large was first debated by Wright and Fisher and remains unresolved. For humans, the size of the population for evolutionary analysis seems most appropriately determined by a world-wide count, following both Fisher's recommendation of enumerating population size at the species level, and Wright's recommendation of enumerating population size at the neighbourhood level.

Neighbourhoods comprise randomly mating groups extended by inclusion of other groups with which there is substantial gene exchange over evolutionary time. Counting the world-wide distribution of humans suggests a very large population size. However, the current census count of several billions is the result of expansion that has occurred in the blink of an evolutionary eye. The size of the population over the greatest part of human evolutionary history has been much smaller. How small?

Surveys of nucleotide diversity in the human genome of individuals living now (Li & Sadler, 1991) suggest an effective human population size of 10 000. This estimate is substantially smaller than the current world population size. An effective population size is a theoretical value representing a constant, panmictic population comprising individuals that have equal probabilities of reproductive success over some fixed time parameter. The relationship of an effective size to a realistic population size that is neither constant nor panmictic is not well defined. However, for the human population a long-term Pleistocene census size of between 10 000 and 100 000 has been proposed (Takahata, 1993) with expansion to current numbers too recent to be reflected by single-site nuclear polymorphisms, which have mutation rates on the order of 10^{-6} per kilobase. This small effective population size implies average rates of genetic drift of allelic variation on the order of 10^{-4} per locus, assuming linkage is maintained in human populations over typical genic intervals. Unless rates of selection maintaining allelic variation are greater, there will be a stochastic turnover of diversity subject to the influence of a population's demographic history.

Modelling genealogy

Genetic variability is generated by mutation and recombination. Genetic drift, selection and gene flow acting on this mutational source shape the patterns of genetic diversity at each locus for individuals in a population. If all progeny survived and reproduced, a population would continue expanding infinitely and genetic diversity would increase infinitely. Not all progeny survive. From generation to generation the progeny who survive and reproduce are related by descent. If the numbers of reproductively successful progeny vary, a few ancestors after many generations have large numbers of descendants. Most of the contemporaries of these successful ancestors have, after many generations, no descendants. Contributing to reproductive success are differences in relative fitness for individuals in the environments in which they live, but random factors probably play a major role.

The genealogical ancestry of a population traced backwards in time for one DNA locus represents the genetic drift, selection and gene flow operating forwards in time at that locus. The coalescent is a stochastic model for the genealogical ancestry of a sample of DNA sequences traced back to their most recent common ancestor, assuming that the variance in reproductive success is attributed only to random factors. It was introduced by Kingman (1982*a*) as a continuous-time approximation, obtained in the limit of large population size, to this ancestral process. Kingman (1982*b*) also provided a very useful invariance principle showing that essentially all the exchangeable reproductive models can be approximated by this coalescent. In fact, the coalescent can approximate an even wider range of models (Donnelly & Kurtz, 1995). Among the range of demographic histories that can be represented by the coalescent are population constancy, fluctuation, expansion and subdivision. This is one of the strengths of the coalescent compared with established population genetic models.

The coalescent is the most powerful and efficient model available for time-scaling DNA sequence diversity, taking demographic history into account (Donnelly & Tavaré, 1995). The coalescent is an efficient model because the reproductive contribution of individuals, other than those ancestral to a sample of interest from the current generation, is ignored, enabling the simulation with a desktop computer of large populations for a range of demographic histories. The coalescent is powerful because it makes best use of all the information available in DNA sequence data for inferences on genealogical processes. This information includes the allelic frequencies of all the DNA sequences in a sample and the nucleotide differences between these sequences along branches connecting them in a gene tree (Griffiths & Tavaré, 1994*a*). The time-scaling of gene trees makes them very informative about the evolution of modern humans. Gene trees can also reveal the tell-tale evidence of demographic history. The coalescent model permits the distinction of sequence diversity generated during rapid population expansion, from diversity generated over a longer time period in a stable population. This is important because, assuming neutrality, different time-scalings of DNA polymorphism are required for different demographic histories.

Stochastic models indicate large evolutionary variances

The coalescent is a model for genetic drift and mutation. Using this model the genealogical process can be separated from the mutation process (Hudson, 1990). Conceptually, the coalescent works backwards in time to construct a coalescent tree, Ancestral lines going backwards in time coalesce when they share an ancestor, i.e. a parent, a grandparent or a (great-)^n grandparent. Coalescences occur between pairs of individual lineages and in each generation in which a coalescence occurs, the number of ancestral lines is reduced by one, generating a bifurcating coalescent tree. A coalescent tree represents one run of evolution by genetic drift. A set of six scaled coalescent trees is shown in a review of coalescent models by Donnelly & Tavaré (1995). Having simulated a coalescent tree back to the most recent common ancestor (MRCA), the mutation process can be followed forwards in time from the MRCA along the branches of the tree. A diploid population with size, N, of 10 000 has 20 000 ($2N$) DNA lineages for each autosomal nuclear locus. The time taken for these lineages in the current generation to coalesce until only two are left is $2N$ generations, on average. This is 400 000 years if average generations are 20 years. After another $2N$ generations, again on average, the last two lineages coalesce as the single common ancestor. The MRCA defines the root. The expected total coalescence time is $4N$ generations, approximately 800 000 years. However, any single coalescent tree may be shorter or longer and there is a large evolutionary variance in total coalescence time. Coalescent trees can also be constructed when population size has not remained constant in time. The rate at which lineages coalesce in these trees is faster or slower than in a constant-sized population, depending on the demographic history (Donnelly & Tavaré, 1995).

A gene tree reconstructs the history of mutation events, describing some of the genealogical structure that occurred in the coalescent tree, in particular providing details of the order in which pairs of lineages coalesced. However, time-scaling which is present in the coalescent tree, is lost from the gene tree. Given a sample of DNA sequences, a coalescence model can be used to estimate the time-scale of a gene tree. Both genealogy and mutation are stochastic processes. Simulating the coalescent a thousand times with the same starting parameters, generates a thousand gene trees, astounding in their variety of topology and time-depth. There is no clearer demonstration of the magnitude of the evolutionary variance inherent in neutral models than the graphical output of a coalescence model.

A set of ten gene trees each for 50 DNA sequences simulated for a constant population size model are presented in Fig. 6.1. All of these trees were generated using a mutation-drift parameter, θ, of 3, where $\theta = 4N\mu$. N is the diploid population size and μ is the mutation rate. For N of 10 000, θ of 3 is given by $\mu = 7.5 \times 10^{-5}$ per allele per generation. Assuming a 2.5 kilobase DNA sequence and 20 years per generation, μ is 1.5×10^{-9} per nucleotide per year, which is a realistic mutation rate for nuclear DNA (Nei, 1987). The root of each gene tree is shown by a bold circle. The numbers in each circle are the frequencies of that DNA sequence in the sample. Note that the sequence at the root is not always observed in the

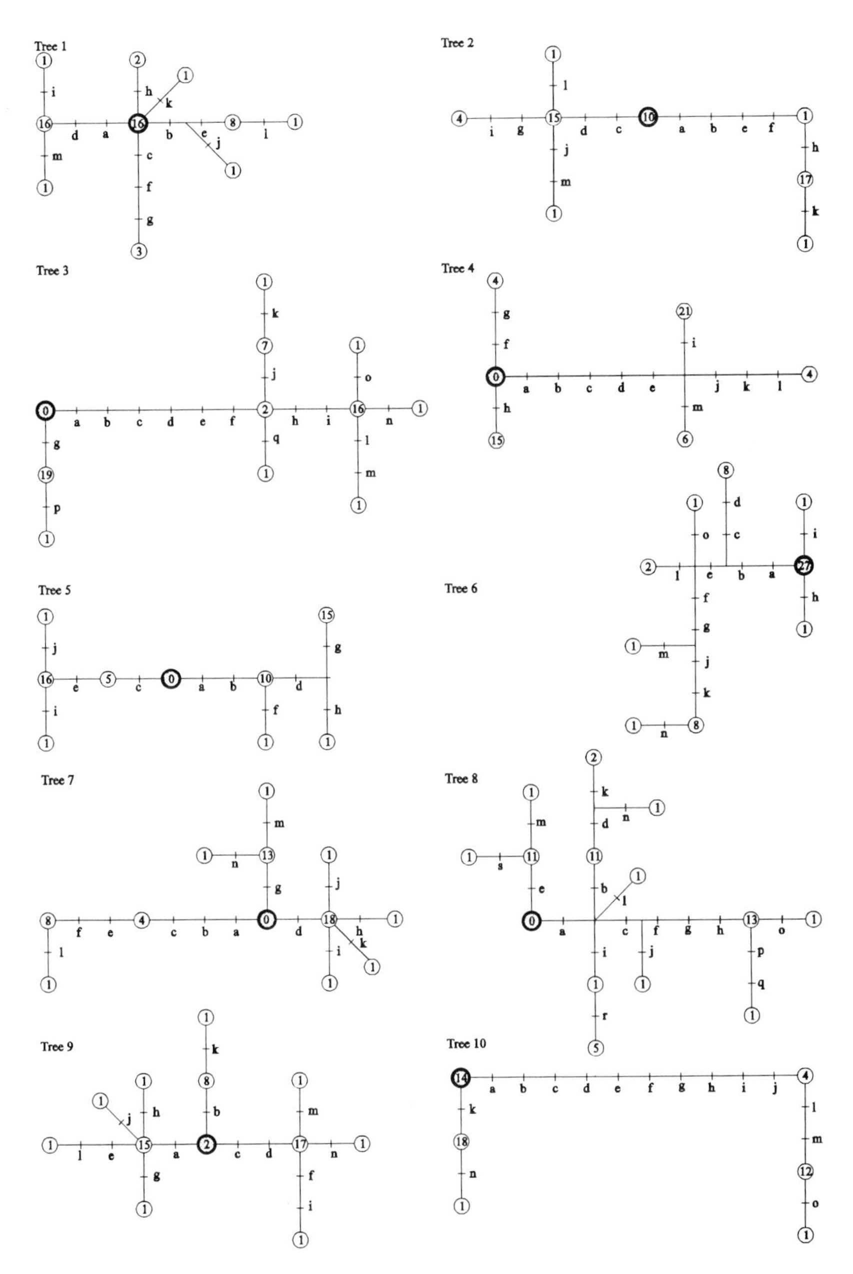

Fig. 6.1. Simulated gene trees for a constant-sized population, with current values of $\theta = 3$.

sample. Letters indicate mutation events, with order represented alphabetically.

For gene trees constructed from actual DNA sequences, the information to determine the root and the order of mutation events is usually limited. The connectivity in the gene tree is fully resolved, however, provided that the data are consistent with an infinite-sites mutation process (Griffiths & Tavaré, 1994*a*). The infinite-sites assumption requires that each mutation occurs at a different nucleotide site, so that the mutations in the gene tree fully represent the mutation history of the sample since the MRCA. Genetic data which indicate recombination or recent parallel mutations due to either recurrent mutation or gene conversion are not infinite-sites compatible. For a set of infinite-sites-compatible DNA sequences, there is only one unrooted gene tree which is easy to construct. If there are sites that have been subject to recombination, recurrent mutation or gene conversion, it may be reasonable to remove sites or sequences in order to resolve a tree consistent with the majority of the data. Alternatively, there are methods being developed to analyse data assuming recurrent mutation and recombination (Griffiths & Tavaré, 1994*b*).

An unrooted gene tree together with the frequencies of each of the DNA sequences in the sample, contains all the information in the sample needed for estimating time-scales and demographic history. It is not a lot to go on, but it is complete. Distributions of pairwise sequence difference are an alternative and popular way of graphically representing DNA sequence variation. They contain less information because numbers of nucleotide differences between sequences taken in pairs is only a subset of the connectivity relationships. Furthermore, pairwise difference distributions distort the allele frequency information.

The evolutionary variance among gene trees for the same mutation-drift parameter θ, of 3 shown graphically in Fig. 6.1, is demonstrated numerically in Table 6.1 by times to the MRCA given in the first column adjacent to the tree number. These TMRCA values were computed during the simulation. Estimates of θ given the tree, of the TMRCA given θ and the tree, and the standard deviations in TMRCA are given in the next three columns. These estimates can be compared with alternative estimates of θ as π, the average pairwise difference, and as s, the standardized number of segregating sites, and with estimates of coalescence times given the maximal pairwise difference, k, following Tajima (1983). Tajima (1983) formulated the expected times back to their respective common ancestors for pairs of sequences as a function of the distribution of k, the number of pairwise site differences: $E(T|k) = 2N[(1 + k)/(1 + S)]$. Here, $S = s/(\sum 1/i)$, for $i = 1, \ldots, n - 1$, where s is the number of segregating sites in the sample of n sequences. The times are given in units of $2N$ in Table 6.1. Given

Table 6.1. *Coalescence times in units of* $2N$ $(\theta = 3)$

Tree	TMRCA	$E(\theta)$	E(TMRCA)	s.d.	π	S	$E(T\|k)$	s.d.$(T\|k)$
1	2.007	3.40	1.248	0.401	2.19	2.90	1.794	0.678
2	1.421	2.90	1.848	0.556	3.85	2.90	2.819	0.850
3	1.606	3.55	2.465	0.658	4.94	3.80	2.711	0.752
4	3.022	2.35	2.673	0.756	4.30	2.90	2.819	0.850
5	1.242	2.20	2.056	0.666	3.02	2.23	2.475	0.875
6	3.075	3.70	1.590	0.456	3.47	3.35	2.300	0.727
7	2.222	3.55	1.483	0.444	2.99	3.13	2.182	0.727
8	1.906	5.30	1.164	0.325	3.80	4.24	1.908	0.603
9	0.860	4.05	1.096	0.344	2.16	3.13	1.939	0.686
10	3.045	2.40	3.846	1.013	5.93	3.35	3.679	0.920
Av.	2.041	3.34	1.947		3.67	3.19	2.463	

$2N = \theta/2\mu$, estimates of N and of coalescence times in years are listed in Table 6.2. Tables 6.1 and 6.2 show that single gene trees are poor estimators of population parameters and that estimates based on maximum pairwise difference are worse. For this constant population size example, estimates averaged over the set of ten simulated trees are very good. It does not follow that estimates on actual DNA sequences for ten loci will be equally as good, because of violations in the assumptions of neutrality, infinite-sites mutation, a molecular clock and population size constancy. However, sets of ten loci will certainly provide improvements over estimates based on a single locus, such as mtDNA.

The consequences of population size expansion can be further investigated by simulation of a coalescence model. In Fig. 6.2 are ten gene trees simulated for a mutation-drift parameter, $\theta = 7$, in the current generation, and exponential expansion in the time since the MRCA with rate, $\beta = 4$. This is roughly an order of magnitude change in population size. A general feature of these trees are the shorter average branch lengths, counted in numbers of mutations, compared with those for trees in Fig. 6.1. Tree 3 in Fig. 6.2 has a star-shaped topology, and a number of other trees in this set combine star-shaped topology with longer branches. The rest of the trees in Fig. 6.2 are not distinguishable from trees in Fig. 6.1 simulated under the assumption of constant population size.

In Table 6.3 are estimates of θ given a gene tree and of TMRCA given θ and a tree, made using the information that the rate of population expansion, β is 4. This parameter can be estimated together with θ by maximum-likelihood techniques from the gene trees, but presentation of such analysis is beyond the scope of this chapter. In the first column adjacent to the tree numbers are the times to the MRCA generated by the

Table 6.2. *Coalescence times in years* ($\theta = 3$)

Tree	TMRCA	$N=\theta/4\mu$	E(TMRCA)	s.d.	$N=\pi/4\mu$	$E(T\mid k)$	s.d.$(T\mid k)$	$N=S/4\mu$	$E(T\mid k)$	s.d.$(T\mid k)$
1	802800	11333	565760	181781	7306	524233	198141	9674	694206	262359
2	568400	9666	714560	214972	12827	1446320	436082	9674	1090840	328916
3	642400	11833	1166734	311445	16482	1787254	495695	12651	1371874	380542
4	1208800	8333	891000	251990	14321	1614760	486868	9674	1090840	328916
5	496800	7333	603093	195351	10079	997746	352756	7442	736758	260470
6	1230000	12333	784379	224954	11565	1063705	336373	11163	1026996	324620
7	888800	11833	701934	210154	9973	870236	290079	10419	909370	302985
8	762400	17667	822560	229671	12675	967208	305858	14139	1079088	341033
9	344000	13500	591840	185760	7205	558893	197598	10419	808098	285897
10	1218000	8000	1230720	324160	19752	2906894	726723	11163	1642747	410798
av.	816240	11183	807258		12219	1273725		10642	1045082	

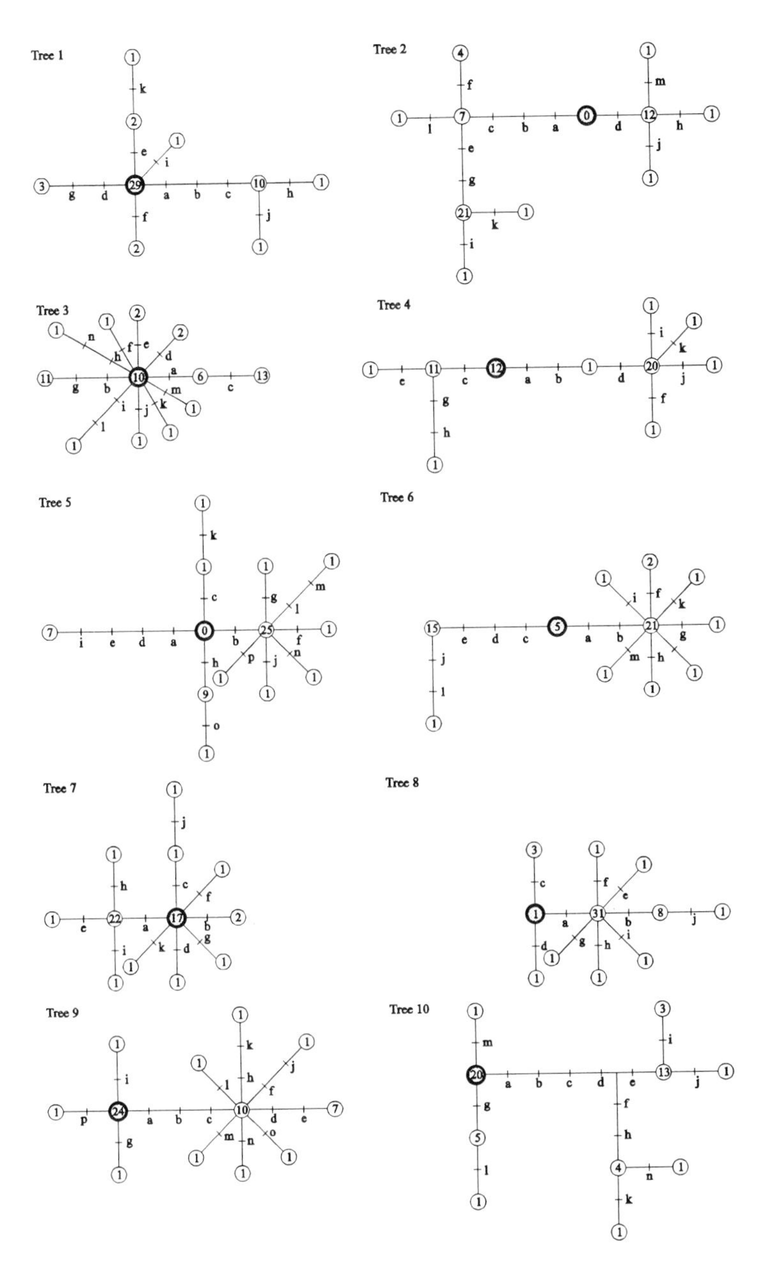

Fig. 6.2. Simulated gene trees for a population that has expanded exponentially with rate $\beta = 4$ to current values of $\theta = 7$.

Table 6.3. *Coalescence times in units of* $2N$ ($\theta = 7.00$)

Times given population expansion ($\beta = 4.0$)						Times assuming constant population size					
Tree	TMRCA	exp(−TNRCA.β)	$E(\theta)$	E(TMRCA)	s.d.	$E(\theta)$	E(TMRCA)	s.d.S	$E(T\|k)$	s.d.$(T\|k)$	
1	0.651	0.074	5.20	0.525	0.101	2.85	1.356	0.496	2.46	2.026	0.766
2	0.749	0.050	5.95	0.627	0.098	3.00	1.975	0.600	2.90	2.306	0.769
3	0.368	0.229	6.70	0.488	0.090	4.10	0.991	0.322	3.13	1.212	0.542
4	0.484	0.144	5.10	0.542	0.103	2.90	1.379	0.462	2.46	2.315	0.819
5	0.443	0.170	7.40	0.529	0.095	4.40	1.202	0.364	3.57	1/750	0.619
6	0.562	0.106	6.55	0.557	0.096	3.70	1.379	0.419	3.13	2.182	0.727
7	0.510	0.130	5.15	0.477	0.102	3.45	0.928	0.326	2.46	1.447	0.647
8	0.467	0.154	4.65	0.525	0.109	2.90	1.137	0.401	2.23	1.547	0.692
9	0.552	0.110	7.30	0.549	0.109	4.55	1.255	0.394	3.57	1.531	0.579
10	0.641	0.077	6.35	0.612	0.107	3.20	1.963	0.588	3.13	2.424	0.767
Av.	0.543	0.124	6.04	0.543		3.51	1.357		2.90	1.874	

simulation. In the next column are the population size contractions for the TMRCA intervals, going backwards in time, $e^{-TMRCA.\beta}$. Estimates of θ and coalescence times, given the demographic history, are compared with two sets of estimates assuming population constancy. Of the latter, estimates of θ and TMRCA based on the trees, are compared with estimates based on maximal pairwise difference using Tajima's (1983) formula. The estimates of θ vary substantially from 7. Even with the information about the population expansion, the average estimate of θ from the gene trees is 6.04. Without this information the average estimates of θ, given the gene trees (3.51), and given only the number of segregating sites (2.9), are not substantially different from the estimates of θ for the constant population trees (Fig. 6.1, Table 6.1). Yet, although there is an overlap in the types of trees presented, Fig. 6.1 and 6.2 constitute very different sets. It is clearly important to study a number of independent loci for inferences on population parameters and to consider the effect of population expansion.

Population sizes and coalescence times in years are reported in Table 6.4. With $\theta = 7$ and the same mutation rates as for the constant population size simulations, N in the current generation is 23 333. Adjacent to the column of times to the MRCA generated by the simulations are the population sizes at the time of the MRCA. The next three columns give estimates of N in the current generation, TMRCA and standard deviation in years, given the gene trees and population expansion parameters. Compare these values with the next three columns for estimates of N, TMRCA and standard deviation given the gene trees, assuming population constancy, and the last three columns for estimates of N based on s, and coalescence times given maximum pairwise differences. Coalescence times are overestimated if population constancy is wrongly assumed. If there has been population expansion, the problem of estimating coalescence time is much harder, and the estimates over ten gene trees are not as good compared with those over ten gene trees for the constant population size simulations. Although the evolutionary variance around the coalescence times for different gene trees representing the same population history is smaller for exponentially expanding populations than for constant-sized populations, error attributed to sampling variance increases with increasing θ. It follows that the larger the population size in the generation being sampled, the greater the number of independent gene trees needed for robust estimates of demographic history.

Unimodal pairwise difference distributions, which are generated by gene trees with star-shaped topologies, have been studied by Rogers & Harpending (1992), who find the parameters for the expansion of modern humans by simulating exponential population growth and fitting the average pairwise difference distribution to mtDNA data. They suggest

Table 6.4. *Coalescence times in years* ($\theta = 7.00$)

Times given population expansion ($\beta = 4.0$)						Times assuming constant population size					
Tree	TMRCA	$N(T=0)$	$N=\theta/4\mu$	E(TMRCA)	s.d.	$N=\theta/4\mu$	E(TMRCA)	s.d.	$N=S/4\mu$	$E(T\vert k)$	s.d.$(T\vert k)$
1	607591	1282	17333	364000	70027	9500	515280	188480	8187	663447	250708
2	699057	991	19833	497420	77747	10000	790000	240000	9667	891653	297347
3	343462	5125	22333	435947	80400	13667	541747	176027	10417	505000	225833
4	451727	2453	17000	368560	70040	9667	533213	178640	8187	758085	268195
5	413461	4193	24667	521947	93733	14667	705173	213547	11907	833467	194809
6	524526	2306	21833	486447	83840	12333	680307	206707	10420	909458	303014
7	475993	2232	17167	327540	70040	11500	426880	149960	8187	473844	211871
8	435860	1294	1500	325500	67580	9667	439640	155053	7443	460593	206031
9	515193	2675	24333	534360	106093	15167	761367	239027	11907	729164	275758
10	598258	1630	21167	518160	90593	10667	837547	250880	10420	1010323	319686
Av.	506513	2528	20117	437988		11683	623115		9674	723504	

exponential expansions of populations from size 800 (or 1600) to 137 000 (or 274 000) females, for estimates of μ per sequence of 1.5×10^{-3} (or 7.5×10^{-4}) . There are some problems with these estimates. First, it is inappropriate to judge gene trees or pairwise difference distributions for single loci by comparison with the averages of a large simulated set (Marjoram & Donnelly, 1994). Secondly, the controversial inference made from these analyses is not that mtDNA diversity reflects recent population expansion, but that it was squeezed by a bottleneck in human demographic history 60 000 to 120 000 years ago. Whereas population expansions in general generate star-shaped gene trees among a range of other topologies, decreasing the size for the founding population in particular, rather than increasing the final population size, increases the probability of star-shaped trees (Marjoram & Donnelly, 1994). A bottleneck serves the purpose of decreasing the evolutionary variance in the simulations so that the average outcome matches the mtDNA data. However, until gene trees at more loci are analysed, their evolutionary variance is unknown.

Can we turn to the nuclear genome for multiple loci?

Surveys of genetic variability have been conducted at the DNA level for more than a decade, and the use of techniques from molecular biology to score polymorphisms is routine. A more complex task is to score a set of polymorphisms linked in a sequence in order to provide data for constructing gene trees. Haploid genomes such as mitochondrial DNA (mtDNA) are considerably easier to survey than diploid nuclear DNA because there are no ambiguous linkage relationships arising from multiple nucleotide sites that vary within an individual. Although heteroplasmy for mtDNA types has been found, it is rare compared with the extent of heterozygosity in the diploid genome. This is one of several reasons why population genetic studies have favoured mtDNA analyses although nuclear sequence data for large samples of individuals are becoming available (Fullerton *et al.*, 1994).

Another reason for favouring the mtDNA locus is that sequence variants are generated only by mutation events and not by recombination. In the nuclear genome DNA sequences may recombine between homologous chromosomes at high rates erasing linkage relationships between polymorphic sites. However, recombination does not occur with uniform probability throughout the human genome. It is becoming apparent that recombination rates are elevated at hotspots and that in their neighbourhood, linkage disequilibrium persists over large intervals (Jorde *et al.*, 1993). If the rate of recombination in such regions is lower than the rate of genetic drift, say by an order of magnitude or more, it is quite likely that recombination events among a sample of DNA sequences will be rare in the time since their divergence from a common ancestor. This is because, if the rate of genetic drift is high, the coalescence time for a sample of DNA sequences will be short. High rates of genetic drift occur when the population size is small. Against rates of drift on the order of 10^{-4} at nuclear loci, recombination rates lower than 10^{-5} per base are unlikely to maintain linkage equilibrium over intervals of less than 100 kilobases. Furthermore, recombination events in the turnover time of shorter DNA sequences are likely to be rare. It is, therefore, most likely that linkage in nuclear loci will permit the construction and analysis of DNA sequences as gene trees, in accordance with the assumption of infinite-sites mutation. There may even be fewer problems compared with analyses of mtDNA, for which rates of recurrent mutation are high. This is important because for evolutionary studies, analyses of multiple loci are needed.

Conclusions

If analyses of multiple loci support an explanation for the limited genetic variability observed for modern humans based on human demographic history, rather than genome-wide selective constraint, it follows that genetic drift has been a major force shaping patterns of functional genetic variation tracing back into the Pleistocene. Accordingly, we should not be surprised if different loci show very different gene trees. Stochastic population genetic models are necessary to handle this evolutionary variability in analyses, not only of human evolution, but also in those of the dynamics of disease genes and other functionally important genetic variation.

As the modern human population has expanded, it is not unlikely that the relative importance of selection on patterns of diversity has been increasing within the last 50 000 years. Perhaps this is the case for cystic fibrosis, the haemoglobinopathies, mitochondrial DNA and many other genetic variants. Selection and population expansion may be confounding factors, distorting time scales estimated under the assumptions of neutrality and population size constancy. However, detecting selection on recent allelic variants against a background of older polymorphism remains a difficult task unless selection has been a powerful force for a long time. If selection cannot be detected, time-scaling on the assumption of neutrality seems reasonable.

To summarize in brief: (1) it is reasonable to begin with a neutrality assumption in population genetics models; (2) genealogical models, such as the coalescent, are neutral models for demographic history which is genetic drift run backwards; (3) the time-scaling of gene trees for DNA sequence polymorphism depends on demographic history; and (4) stochastic models imply large evolutionary variances and therefore studies of multiple loci are essential for inferences on the evolution of modem humans.

References

Donnelly, P. & Kurtz, T. G. (1995). A countable representation of the Fleming–Viot measure-valued diffusion. *Annals of Probability* (in press).

Donnelly, P. & Tavaré, S. (1995). Coalescents and genealogical structure under neutrality. *Annual Review of Genetics* **29**, 401–21.

Flint, J., Harding, R. M., Clegg, J. B. & Boyce, A. J. (1993). Why are some genetic diseases common? Distinguishing selection from other processes by molecular analysis of globin gene variants. *Human Genetics*, **91**, 91–117.

Fullerton, S. M., Harding, R. M., Boyce, A. J. & Clegg, J. B. (1994). Molecular and population genetic analysis of allelic sequence diversity at the human β-globin locus. *Proceedings of the National Academy of Sciences, USA*, **91**, 1805–9.

Griffiths, R. C. & Tavaré, S. (1994*a*). Ancestral inference in population genetics. *Statistical Science*, **9**, 307–19.

Griffiths, R. C. & Tavaré, S. (1994*b*). Simulating probability distributions in the coalescent. *Theoretical Population Biology*, **46**, 131–59.

Hudson, R. R. (1990). Gene genealogies and the coalescent process. *Oxford Surveys in Evolutionary Biology*, **7**, 1–44.

Jorde, L. B., Watkins, W. S., Viskochil, D., O'Connell, P. & Ward, K. (1993). Linkage disequilibrium in the Neurofibromatosis I (NFI) region: implications for gene mapping. *American Journal of Human Genetics*, **53**, 1038–50.

Kimura, M. (1983). *The Neutral Theory of Molecular Evolution*. Cambridge: Cambridge University Press.

Kingman, J. F. C. (1982*a*). On the genealogy of large populations. *Journal of Applied Probability*, **19A**, 27–43.

Kingman, J. F. C. (1982*b*). Exchangeability and the evolution of large populations. In *Exchangeability in Probability and Statistics*, ed. G. Koch and F. Spizzichino, pp. 97–112. Amsterdam: North-Holland.

Li, W. -H. & Sadler, L. A. (1991). Low nucleotide diversity in Man. *Genetics*, **129**, 513–23.

Marjoram, P. & Donnelly, P. (1994). Pairwise comparisons of mitochondrial DNA sequences in subdivided populations and implications for early human evolution. *Genetics*, **136**, 673–83.

Morral, N., Nunes, V. Casals, T., Chillón, M., Giménez, J., Bertranpetit, J. & Estivill, X. (1993). Microsatellite haplotypes for cystic fibrosis: mutation frameworks and evolutionary tracers. *Human Molecular Genetics*, **2**, 1015–22.

Nei, M. (1987). *Molecular Evolutionary Genetics*. New York: Columbia University Press.

Rogers, A. R. & Harpending, H. (1992). Population growth makes waves in the distribution of pairwise genetic differences. *Molecular Biology and Evolution*, **9**, 552–69.

Sereth, H., Shoshani, T., Bashan, N. & Kerem, B. (1993). Extended haplotype analysis of cystic fibrosis mutations and its implications for the selective advantage hypothesis. *Human Genetics*, **92**, 289–95.

Tajima, F. (1983). Evolutionary relationship of DNA sequences in finite populations. *Genetics*, **105**, 437–60.
Takahata, N. (1993). Allelic genealogy and human evolution. *Molecular Biology and Evolution*, **10**, 2–22.

7 *Some attempts at measuring natural selection by malaria*

A. V. S. HILL

Introduction

The longevity of the neutralist–selectionist controversy in population genetics reflects the difficulty of measuring the extent of natural selection in both field and laboratory studies. The advent of powerful molecular tools for genetic analysis has advanced but not resolved this issue. Although there is extensive indirect evidence suggesting a major role for selection in molecular evolution, direct measures of these selection pressures are difficult. Indeed, most selective advantages are probably too small to be measurable in practical terms. However, much may be learned about the likely nature of very weak selection pressures, and of their effects, by analysing less typical large selection pressures that are within range of current studies.

Our extensive and ever-increasing knowledge of human molecular genetic diversity makes ours a species particularly amenable to observational studies of natural selection. This knowledge combined with epidemiological and clinical surveillance offers an alternative approach to studies of experimental animals with more convenient generation times and manipulable environments. Thus, the classical example of heterozygote advantage operating in natural populations remains the resistance of human heterozygotes for haemoglobin S to severe *Plasmodium falciparum* malaria (Allison, 1964). Other major infectious diseases of humans have been less studied, with the exception of investigations of the major histocompatibility complex where associations have been documented for diseases caused by several major pathogens (Todd, West & McDonald, 1990; Hill *et al.*, 1991; Brahmajothi *et al.*, 1991; Thursz *et al.*, 1995). However, there is now evidence that as many as 12 well-studied genetic loci affect susceptibility to malaria in human populations (Hill, 1992; McGuire *et al.*, 1994) and I shall review some recent studies that have provided provisional estimates of the magnitude of the selection pressures exerted by malaria on particular alleles.

The intention is to serve as a reminder that natural selection by infectious disease remains as an important influence on the genome in most parts of the world. Global mortality from malaria, HIV, tuberculosis and other major pathogens is increasing and genetic analysis of susceptibility may offer clues to pathogenetic mechanisms and new control options. In this volume, these examples will hopefully serve to provide some balance to the widespread assumption of neutral evolution employed in many evolutionary models (e.g. Harding, Chapter 6). An inability to measure small selection pressures directly does not negate their importance in evolutionary processes, and an assumption of no selection is just as arbitrary as that of an estimated selection pressure. Mathematical tractability places considerable constraints on the incorporation of both constant and, particularly, fluctuating selection pressures in evolutionary models. So, there is a natural tendency to wish that such neutrality reflects reality. However, once these simpler models have been worked through, incorporating plausible estimates of selection pressures will be the next step.

Haemoglobinopathies

In 1964 Allison reviewed early work on the protection of haemoglobin S against malaria. The degree of controversy that the evidence engendered was clear. In retrospect, this was caused in part by different clinical measures of malaria being used in different studies. The strongest and clearest protection was against death from malaria with little protection against being parasitized, particularly in older children. Measuring protection against severe malaria remains the best surrogate of measuring protection against death from malaria. Early studies suggested almost complete protection of haemoglobin S heterozygotes and a recent large study in The Gambia found a protective efficacy of about 90% (Hill *et al.*, 1991). Without very large studies the 95% confidence intervals on these estimates can be very wide.

Much less information is available from case-control studies of the protection afforded by β thalassaemia. This has only been demonstrated directly in Liberia by Willcox *et al.*, (1983) whose data suggest about 50% protection afforded to heterozygotes. The β thalassaemia found in this population is phenotypically very mild and it would be useful to have an estimate of the protective efficacy of more typical β thalassaemia alleles. Although Hb C is sufficiently prevalent in northern Ghana and surrounding countries to allow estimation of its value in malaria resistance no such studies have been reported. For haemoglobin E, the commonest haemoglobin variant in Asians, there is again no good measurement of the protection this allele may provide against malaria. Similarly, for α thalassaemia no published studies have reported case-control data on whether this disorder can protect against either *P. falciparum* or *P. vivax* malaria. However, preliminary data from studies in The Gambia and Kilifi, Kenya suggest that any protection afforded, particularly to heterozygotes, must be small, and much less than that afforded by haemoglobin S. Homozygotes for the mild α^+ type of α thalassaemia were less frequent amongst children with severe malaria, suggesting a 25% reduction in risk associated with this genotype (Yates *et al.*, 1993; Yates, 1995).

The available data allow some crude but useful estimates of the time required for the selection of such variants to be made. Fortunately, direct estimates of malaria mortality in Africa today are available from epidemiological studies (e.g. Greenwood *et al.*, 1987) and an indirect estimate of recent malaria mortality can be obtained from the current prevalence of haemoglobin S in a population by assuming that this allele is at equilibrium. These estimates are of the order of 5–15% mortality from malaria in sub-Saharan Africa. Assuming that the haemoglobinopathies affect only survival in childhood, and do not affect fertility, it is evident that haemoglobin S reaches present-day frequencies very quickly, in perhaps a thousand years (Cavalli-Sforza & Bodmer, 1971). A major determinant of the time required is the assumed starting frequency of the mutant allele. A frequency of 0.001, representing one carrier in a village of 500 people, is often used and may not be unreasonable. If β thalassaemia is associated with 50–100% protection against malaria it too should reach observed frequencies in less than a few thousand years.

The evolutionary dynamics of a thalassaemia alleles may be different. Deletional forms of α thalassaemia may arise at a much higher rate than β thalassaemia alleles. This may account in part for the high frequency of triplicated α-globin genes, that are probably generated during such unequal crossovers at meiosis and are found in most human populations at allele frequencies approaching 0.01. Even allowing for a high (0.01) starting frequency of $-\alpha$ alleles, the time required for selection of such alleles to currently observed frequencies in sub-Saharan Africa may be very long, of the order of some tens of thousands of years. This results from a selective advantage only to homozygotes so that the frequency of the $-\alpha$ allele rises very slowly for many thousands of years, even in the absence of any disadvantage associated with the homozygous $-\alpha$ genotype. Thus, these data suggest that the time depth of malarial selection, at least in sub-Saharan Africa, may be rather long, of the order of tens of thousands of years rather than just 5000–7000 years (Yates *et al.*, 1993).

These very simple estimations of allele frequency dynamics under constant selection pressures make several assumptions that may be unrealistic simplifications, particularly in relation to population structure.

Often it is possible to deduce the direction of the error that may be introduced, but more difficult to estimate its magnitude. For example, an expanding population will speed the rise in frequency of protective alleles. More work is required on the extent to which such concerns affect the dynamics of selection, and the extent to which comparisons between different alleles selected in the same population can allow these population processes to be estimated and allowed for.

G6PD deficiency

It has been clear for many years that the high frequencies of glucose-6-phosphate dehydrogenase deficiency observed in most tropical and subtropical populations must relate to natural selection by malaria. However, until recently no useful estimate of the degree of protection afforded was available. Also, for this X-linked disorder it has been uncertain, and controversial, whether only female heterozygotes or all deficient genotypes were protected. Analysis of the common African deficiency allele, G6PD A-, in both The Gambia and Kenya has recently provided data for this relatively mild variant. Both male hemizygotes and female heterozygotes showed a 46–58% reduction in risk of severe malaria (Ruwende *et al.*, 1995). Too few homozygotes were seen to provide an estimate of their degree of resistance, but this may be similar to that of the equally enzyme-deficient hemizygote.

Estimates of the time required for selection have again allowed interesting inferences to be made. With a malarial mortality of 8% the G6PD A-allele should reach the frequency observed today in just a few thousand years and should reach fixation in less than 10 000 years (Fig. 7.1). Because G6PD deficiency allele frequencies are less than 0.50 in almost all seven malaria-endemic areas, Ruwende *et al.* (1995) infer that there must be a counterbalancing selective pressure slowing the rise of the G6PD A-allele in Africa. This may be an increased mortality of enzyme-deficient children from other infectious pathogens. It is unclear whether the allele frequencies observed today are at or near equilibrium.

Human leukocyte antigens

The pattern of selection of HLA variants by malaria may differ from that of the variants described above. Early studies of HLA associations with other infectious diseases, particularly leprosy, suggested several associations but these often differed from population to population. This engendered scepticism as to the correctness of these associations (Klein, 1987). Often the studies undertaken were small, the typing methodology imprecise and the number of comparisons performed was large. However, more recently larger studies have supported both the existence of associations with leprosy (Todd *et al.*, 1990) and tuberculosis (Brahmajothi *et al.*, 1991; Khomenko *et al.*, 1990) and the finding of heterogeneity in associations between populations.

In West Africa we found a clear association between particular HLA class I and II alleles (HLA-B53 and HLA-DRB1*1302) and resistance to severe malaria in a very large case control study of Gambian children (Hill *et al.*, 1991). A more recent study of severe malaria in Kenyan children has again shown a significant influence of HLA-DR antigens but the particular resistance alleles observed in The Gambia and Kenya are different (Yates, 1995). Hence, as with leprosy and tuberculosis, there appears to be population heterogeneity in HLA associations that may relate to observed allele frequency differences in some polymorphic parasite antigens between East and West Africa.

This evidence for geographical variation in selection pressures by malaria between East and West Africa urges caution in calculation of selection times using HLA variants. If selection pressures on HLA types reflect the composition of the local parasite population, substantial geographical variation in selection may be observed. Moreover, if parasites evolve to avoid the most prevalent host HLA molecules, as suggested in some models of frequency-dependent selection, substantial temporal as well as spatial variation in selection pressures is to be expected. Hence, estimates of the time depth of malarial selection are probably better made with red cell variants than with HLA data.

However, direct estimates of the magnitudes of HLA association with major infectious diseases are nonetheless of importance for understanding the evolutionary pressures exerted on HLA polymorphism. Indeed, it is only recently that there has been a general acceptance that pathogen-driven selection plays an important role in the maintenance of MHC polymorphism. Further data are required to distinguish between the two most favoured mechanisms that may underlie such selection. These are over-dominant selection and frequency-dependent selection with rare allele advantage (Takahata & Nei, 1990). Additionally, fluctuations in selection pressures may, under certain conditions, contribute to the maintenance of polymorphism (Hill *et al.*, 1994). Although the contribution from each of these factors remains to be established, it seems likely that all play a role in maintaining MHC diversity.

Two groups of malaria resistance genes

Analysis of ABO blood group data in the Gambian severe malaria study (unpublished data) suggests that there is a significant association between blood groups O and resistance to severe malaria. However, comparison

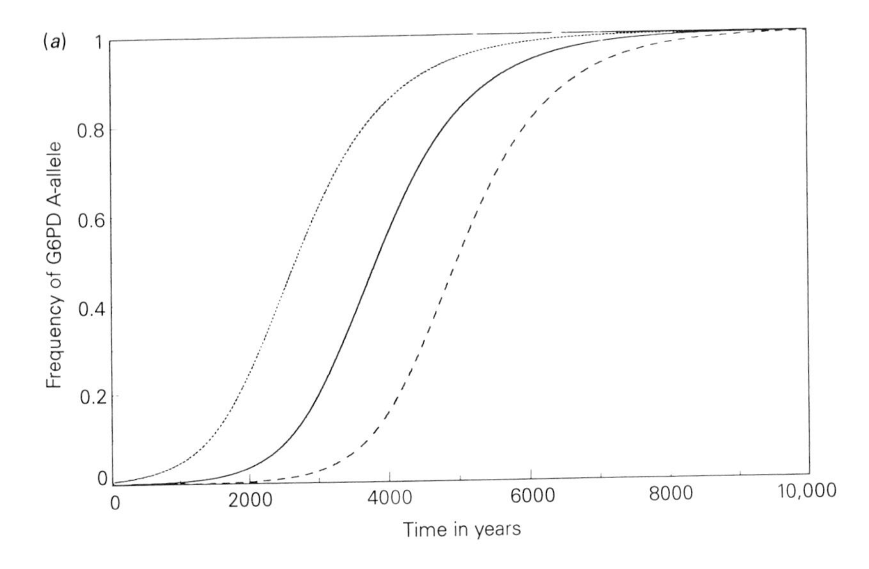

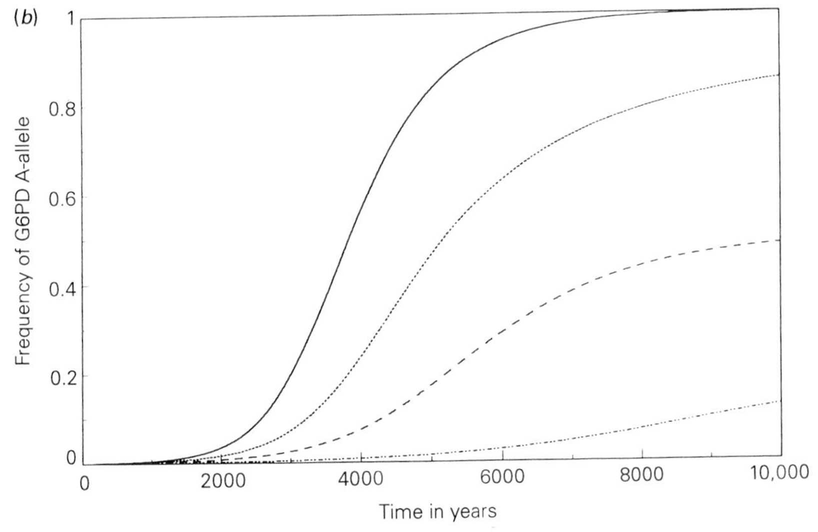

Fig. 7.1. Selection of a common glucose-6-phosphate dehydrogenase deficiency allele in Africa. (*a*) Rise in frequency of the G6PD A-allele in males and females (the lines are superimposed) produced by a constant malarial selection pressure with varying initial frequencies of the A-allele of 1% (dotted line), 0.1% (continuous) and 0.01% (dashed). With a starting frequency of 0.1% (continuous line) the allele reaches the current frequency in The Gambia of 5.9% in 2280 years and nears fixation after 7000 years. (*b*) As malaria appears to have been a major selective force in Africa for at least 5000 to 7000 years, alterations in genotype fitness that reduce the rate of increase in frequency of the G6PD A-allele are shown. The continuous line is as in (*a*). If the selective advantage to the hemi- and homozygote that is produced by malaria resistance is exactly counterbalanced by their increased mortality from other causes, so

with the results from other studies indicates that again there appears to be geographical heterogeneity in this association. Recently, it has been found that strains of malaria parasite differ in their abilities to undergo rosetting *in vitro* and that this is in part dependent on the ABO blood group of the red cells that they are grown in (Carlson & Wahlgren, 1992). Because rosetting parasites have been associated with severe malaria in some studies (Carlson *et al.*, 1990), it is possible that variable susceptibility to severe malaria associated with ABO blood groups might relate to the rosetting characteristics of the local parasite population.

It appears therefore that a preliminary classification of malaria resistance alleles into two groups may be proposed. In group 1 (see Table 7.1) I have listed loci that are unlikely to show substantial variation in their protective efficacies from one location to another. Here are included the globin variants and G6PD deficiency, which probably protect against all strains of *P. falciparum*. In group 2 are the variants that appear to show geographical variation in the degree of resistance afforded. In this group are the HLA variants and, more tentatively, ABO blood group variants. In general, we might expect that resistance genes that mediate protection through an interaction with a variable part of the malaria parasite are likely to fall into group 2. However, there are several other reasons why protective efficacies of resistance alleles may vary, relating both to local epidemiological conditions and to the relative importance of particular immune defence mechanisms in areas with different transmission conditions. Hence, it remains to be seen whether variants such as cytokine gene polymorphisms (McGuire *et al.*, 1994) fall into group 1 or 2. Cytokines, such as TNF, probably act against all parasite strains but are involved in many different types of immune defence, and probably also in disease pathogenesis.

Selection and population affinities

A major application of the study of genetic variation in humans is to the measurement of population affinities, often with a view to illuminating

Caption for fig. 7.1 (*cont.*)

that their overall fitness is equal to that of individuals with a normal G6PD genotype, the rate of rise of allele frequency is reduced and the allele eventually fixes at 50% frequency (dashed line). The dotted line represents a counterbalancing selection pressure corresponding to half the size of the fitness advantage conferred by malaria, and the lowest line shows an overall reduction in fitness of hemizygotes and homozygotes equivalent to the increase in their fitness conferred by malaria resistance, i.e. the selective disadvantage is double the size of the selective advantage. Reproduced from Ruwende *et al.* (1995) who conclude that some counterbalancing selective disadvantage to G6PD deficient individuals has been slowing the rate of rise of the G6PD A-allele in Africa.

Table 7.1. *Malaria resistance genes*

Group 1	Group 2
β-Globin	HLA-B
α-Globin	HLA-DR
G-6-P Dehydrogenase	Blood group O

Division of malaria resistance genes into two groups. In group 1 the resistance alleles appear to show little variation in their protective efficacies from one population to another. In group 2 the protective alleles show significant geographical variation in their protective efficacies.

human prehistory and, in particular, putative prehistoric migration routes. Frequently dendrograms are used to display genetic distances between populations to facilitate inferences as to affinities. There are several problems in the use of this approach and I shall discuss just one of them. This is the general assumption that no selection is acting to affect the genetic differentiation of the populations studied.

However, most of the population markers that have been used to measure affinities between human populations (Cavalli-Sforza, Menozzi & Piazza, 1994) are likely to be affected by selection. Few would now dispute this for many of the phenotypic markers which were used in early studies: blood groups, HLA types, haemoglobin variants, immunoglobulin allotypes and electrophoretic variants of major serum proteins such as group-specific component. Moving to DNA polymorphisms has not solved this problem. Many restriction fragment length polymorphism haplotypes, although composed of changes only in DNA flanking genes or in introns, show population distributions suggestive of selection (Maeda, Bliska & Smithies, 1983). It remains to be determined whether microsatellite (Bowcock *et al.*, 1994) and minisatellite DNA polymorphisms (Flint *et al.*, this volume) can completely overcome this problem, but they are likely to be better.

From the comparison of selected with more neutral markers, such as mitochondrial DNA and simple sequence repeat polymorphisms, a perhaps surprising picture is beginning to emerge. These very different sets of markers often provide rather similar pictures of population histories and affinities. Indeed, it is the polymorphisms most clearly subject to selection, such as HLA and immunoglobulin allotype variants, that often appear to display population affinities most clearly and informatively (e.g. Serjeantson, Ryan & Thompson, 1982; Zhao & Lee, 1989; Allsopp *et al.*, 1992). This is fortunate, as the earliest data on population comparisons relied heavily on selected markers. However, it remains uncertain whether this apparently satisfactory classification of populations is actually correct. The usual criteria for such judgements are congruence with linguistic and archaeological data and other sources of information on population history and prehistory. None the less, the fit with more neutral markers is often good, and the question then arises as to why selection has apparently served to magnify rather than obscure population relationships? The answer is unclear, but geographical variation is selection pressures, such as suggested above for HLA and infectious pathogens, may be important. Also, the higher frequencies of many alleles subject to ongoing selection pressures might preserve evidence of population history more faithfully than very low frequency variants.

Prospects

There are still some major questions to be answered about malarial selection of known variants using straightforward case-controls studies. A few of these were mentioned above, such as selection of β^0 thalassaemia and haemoglobins C and E. Also, a study in Melanesia could address the extent of protection provided by heterozygosity for the band 3 variant causing ovalocytosis and determine the phenotype of homozygotes for this genotype. The latter genotype may not be viable. Additionally, very little is known about whether any of the variants discussed provides resistance to *Plasmodium vivax* malaria. Although the absence of the Duffy blood group in Africans provides complete resistance to *P. vivax* malaria (Miller *et al.*, 1976) the issue of how such a currently non-lethal malaria could have selected the Duffy negative allele(s) to near-fixation remains unresolved.

The most exciting prospects for studying the genetics of resistance to malaria and other major infectious diseases come from recent progress in mapping the human genome. Using family studies it should now be possible to map and subsequently identify the major loci influencing susceptibility to many such multifactorial diseases (Lander & Schork, 1994). This may well uncover yet more loci for malaria resistance and should also indicate whether the most important loci have already been found. Judging by the results for malaria, resistance genes for other infectious diseases will be many and varied and the relative importance of particular loci is likely to vary from one population to another.

These gene mapping linkage studies will lead on to further association studies to identify the relevant loci and alleles associated with susceptibility and resistance. These studies will need to be conducted on large sample sizes to show convincing results, particularly if evidence of interaction between loci is to be sought. Towards this goal, the major advance is likely to be the introduction of automated techniques for typing allelic variants with the minimization or elimination of sample processing after PCR amplification. Progress in this direction is already being made (Livak, Marmaro & Todd, 1995).

It seems likely that a huge assortment of allelic diversity of relevance to infectious disease resistance may be about to be uncovered over the next ten years or so, that should lead to better estimates of some of the larger selection pressures operating during recent human evolution. This may be a tough time to be a neutralist.

Acknowledgements

I thank my numerous colleagues in Oxford, London, The Gambia and Kenya who participated in the studies reviewed here and generated many of the ideas discussed. The author is a Wellcome Senior Research Fellow in Clinical Science.

References

Allison, A. C. (1964). Polymorphism and natural selection in human populations. *Cold Spring Harbor Symposium on Quantitative Biology*, **29**, 137–149.

Allsopp, C. E. M., Harding, R. M., Taylor, C., Bunce, M., Kwiatkowski, D., Anstey, N., Brewster, D. , McMichael, A. J., Greenwood, B. M. & Hill A. V. S. (1992). Interethnic genetic differentiation in Africa: HLA class I antigens in The Gambia. *American Journal of Human Genetics*, **50**, 411–21.

Bowcock, A. M., Ruiz-Linares, A., Tomfohrde, J., Minch, E., Kidd, J. R. & Cavalli-Sforza, L. L. (1994). High resolution of human evolutionary trees with polymorphic microsatellites. *Nature*, **368**, 455–57.

Brahmajothi, V., Pitchappan, R. M., Kakkanaiah, V. N., Sashidhar, M., Rajaram, K., Ranu, S., Palanimurugan, K., Paramasivan, C. N. & Prabhakar, R. (1991). Association of tuberculosis and HLA in South India. *Tubercle*, **72**, 123–32.

Carlson, J., Helmby, H., Hill, A. V. S., Brewster, D., Greenwood, B. M., & Wahlgren, M. (1990). Human cerebral malaria: association with erythrocyte rosetting and the lack of antirosetting antibodies. *Lancet*, **336**, 1457–60.

Carlson, J. & Wahlgren, M. (1992). *Plasmodium falciparum* erythrocyte rosetting is mediated by promiscuous lectin-like interactions. *Journal of Experimental Medicine*, **176**, 1311–7.

Cavalli-Sforza, L. L & Bodmer, W. F (1971). *The Genetics of Human Populations*. San Francisco: W. H. Freeman and Co.

Cavalli-Sforza, L. L., Menozzi, P. & Piazza, A. (1994). *The History and Geography of Human Genes*. Princeton: Princeton University Press.

Greenwood, B. M., Bradley, A. K., Greenwood, A. M., Byass, P., Jammeh, K., Marsh, K., Tulloch, S., Oldfield, F. S. & Hayes, R. (1987). Mortality and morbidity from malaria among children in a rural area of The Gambia, West Africa. *Transactions of the Royal Society of Tropical Medicine and Hygiene*, **81**, 476–86.

Hill, A. V. S. (1991). HLA associations with malaria in Africa: some implications for

MHC evolution. In *Molecular Evolution of the Major Histocompatibility Complex*, ed. J. Klein and D. Klein, pp. 403–20. Berlin: Springer.

Hill, A. V. S. (1992). Malaria resistance genes: a natural selection. *Transactions of the Royal Society of Tropical Medicine and Hygiene*, **86**, 225–6.

Hill, A. V. S., Allsopp, C. E. M., Kwiatkowski, D., Anstey, N. M., Twumasi, P., Rowe, P. A., Bennett, S., Brewster, D., McMichael, A. J. & Greenwood, B. M. (1991). Common West African HLA antigens are associated with protection from severe malaria. *Nature*, **352**, 595–600.

Hill, A. V. S., Yates, S. N. R., Allsopp, C. E. M., Gupta, S., Gilbert, S. C., Lalvani, A., Aidoo, M., Davenport, M. & Plebanski, M. (1994). Human leukocyte antigens and natural selection by malaria. *Philosophical Transactions of the Royal Society of London B*, **346**, 379–85.

Khomenko, A. G., Litvinov, V. I., Chukanova, V. P & Pospelov, L. E. (1990). Tuberculosis in patients with various HLA phenotypes. *Tubercle*, **71**, 187–92.

Klein, J. (1987). Origin of major histocompatibility complex polymorphism: the trans-species hypothesis. *Human Immunology*, **19**, 155–62.

Lander, E. S. & Schork, N. J. (1994). Genetic dissection of complex traits. *Science*, **265**, 2037–48.

Livak, K. J., Marmaro, J. & Todd, J. A. (1995). Towards fully automated genome-wide polymorphism screening. *Nature Genetics*, **9**, 341–2.

Maeda, N., Bliska, J. B. & Smithies, O. (1983). Recombination and balanced chromosome polymorphism suggested by DNA sequences 5′ to the human delta-globin gene. *Proceedings of the National Academy of Sciences, USA*, **80**, 5012–16.

McGuire, W., Hill, A. V. S., Allsopp, C. E. M., Greenwood, B. M. & Kwiatkowski, D. (1994). Variation in the TNF-A promotor region associated with susceptibility to cerebral malaria. *Nature*, **371**, 508–11.

Miller, L. H., Mason, S. J., Clyde, D. F. & McGinniss, M. H. (1976). The resistance factor to *Plasmodium vivax* in blacks. The Duffy-blood-group genotype, FyFy. *New England Journal of Medicine*, **295**, 302–4.

Ruwende, C., Khoo, S. C., Snow, A. W., Yates, S. N. R., Kwiatkowski, D., Gupta, S., Warn, P., Allsopp, C. E. M., Gilbert, S. C., Peschu, N., Newbold, C. I., Greenwood, B. M., Marsh, K. & Hill, A. V. S. (1995). Natural selection of hemizygotes and heterozygotes for glucose-6-phosphate dehydrogenase deficiency by resistance to severe malaria. *Nature*, **376**, 246–9.

Serjeantson, S. W., Ryan, D. P. & Thompson, A. R. (1982). The colonization of the Pacific: the story according to human leukocyte antigens. *American Journal of Human Genetics*, **34**, 904–18.

Takahata, N. & Nei, M. (1990). Allelic genealogy under overdominant and frequency dependent selection and polymorphism of the major histocompatibility complex. *Genetics*, **124**, 967–78.

Thursz, M. R., Kwiatkowski, D., Allsopp, C. E. M., Greenwood, B. M., Thomas, H. C. & Hill, A. V. S. (1995). Association of an HLA class II allele with clearance of hepatitis B virus infection in The Gambia. *New England Journal of Medicine*, **332**, 1065–9.

Todd, J. R., West, B. C. & McDonald, J. C. (1990). Human leukocyte antigens and leprosy: study in Northern Louisiana and review. *Reviews of Infectious Diseases*, **12**, 63–74.

Willcox, M., Bjorkman, A., Brohult, J., Pehrson, P. O., Rombo, L. & Bengtsson, E.

(1983). A case-control study in northern Liberia of *Plasmodium falciparum* malaria in haemoglobin S and beta-thalassaemia traits. *Annals of Tropical Medicine and Parasitology*, **77**, 239–46.

Yates S. N. R (1995). Human Genetic Diversity and Selection by Malaria in Africa. Unpublished DPhil thesis. University of Oxford.

Yates, S. N. R., Snow, R. W., Allsopp, C. E. M., Newton, C. R. J. C., Kwiatkowski, D., Palmer, D., Peschu, N., Greenwood, B. M., Marsh, K., Newbold, C. I. & Hill, A. V. S. (1993). Resistance of homozygotes but not heterozygotes for α^+ thalassaemia to severe malaria: implications for the time depth of natural selection. *Proceedings of the British Society for Parasitology*, 5th Malaria Meeting, pp. 13–14 (abstract).

Zhao, T. M. & Lee, T. D. (1989). Gm and Km allotypes in 74 Chinese populations: a hypothesis of the origin of the Chinese nation. *Human Genetics*, **83**, 101–10.

8 *AIDA: Geographical patterns of DNA diversity investigated by autocorrelation statistics*

G. BARBUJANI AND G. BERTORELLE

To Andrea Marconato (1957–1995) and to Mariella

Introduction

Geographical patterns of genetic diversity contain information on the evolutionary processes affecting individuals and populations. For example, random patterns, clines, and patchy distributions may result from different evolutionary phenomena (like panmixia, response to a selective gradient, and isolation by distance, respectively), which one may be willing to reconstruct. In some cases, a simple visual inspection of data is sufficient. Figure 8.1 shows two maps of human allele frequencies in Europe based on several tens of population data (Sokal, Harding & Oden, 1989). These maps were obtained by interpolation, and different shades of grey indicate different classes of allele frequencies. The upper distribution refers to an allele of the HLA-B locus, HLA*BW15. There is no need of sophisticated statistical tests to see that its frequencies are distributed in a gradient, with maxima in Scandinavia and Finland, and an area of low values around the Mediterranean. Other cases are, however, ambiguous. One cannot easily tell if the lower part of Fig. 8.1 (allele 1 of the Group-specific component) is a cline extending from the Southwest to the Northeast on which random variation has been superimposed, if it should be regarded as the overlapping of two clinal patterns, or even if it may simply reflect drift and local gene flow, i.e., isolation by distance. It is in cases like this that quantitative approaches are necessary.

Spatial autocorrelation is defined as the dependence of one variable upon its values at other localities. Spatial autocorrelation statistics (Sokal & Oden, 1978) summarize geographical patterns in an easy-to-understand manner. The sampling localities are compared pairwise at different, arbitrary spatial distances, e.g. between 0 and 50 km, between 51 and 100, and so on. For each distance class, a coefficient of genetic similarity is

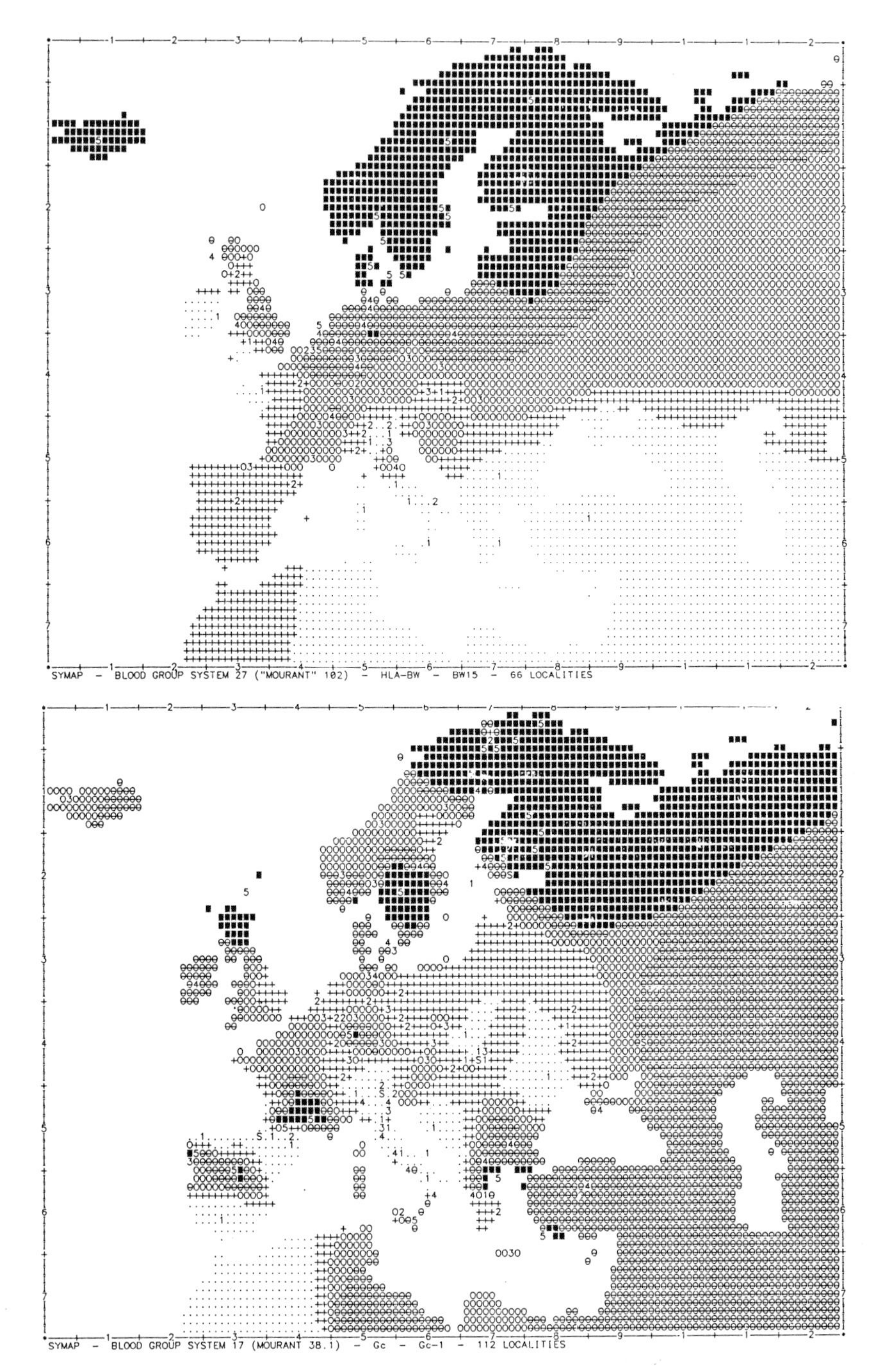

Fig. 8.1. Interpolated distributions of the alleles HLA*BW15 (top) and GC*1 (bottom in Europe. Different shades of grey represent increasing allele frequencies, from lighter to darker shades. The numbers 1–5 indicate sampling localities.

calculated. In this chapter we shall focus on Moran's I and on its molecular equivalent, called II, but other coefficients are available as well. Both I and II are distributed between -1 and $+1$, with positive values indicating genetic similarity and negative values indicating dissimilarity. For both statistics, the expected value is very close to 0, under a randomization hypothesis. Therefore, I and II can be interpreted as correlation coefficients, even though here one is really comparing the values of one variable at different locations, and not two different variables. The set of autocorrelation coefficients thus calculated, or correlogram, has been called the 'signature' (Sokal, 1979) of the locus considered.

The first part of this chapter describes an autocorrelation statistic specifically designed for the analysis of molecular data; we call it II. In the second part we apply II to a dataset of mitochondrial DNA types in twelve Italian populations. The pattern of mtDNA thus described, along with the results obtained using other approaches based on genetic and non-genetic information, will suggest a probable evolutionary scenario and a time-frame for it.

AIDAs: autocorrelation indices for DNA analysis

Two populations are said to resemble each other genetically when their allele frequencies are more similar than pairs of allele frequencies sampled at random locations. When information on DNA sequences is available (either from direct sequencing, or from RFLP analysis), a second factor becomes relevant, namely the sequence itself. Two localities are considered similar if (1) the same haplotypes occur at similar frequencies in the two localities, and (2) the various haplotypes present differ by few substitutions. Whether or not a certain number of substitutions are 'few' will depend on the amount of sequence difference observed within samples. This means that a statistical method for describing molecular variation in space must take sequence difference into account, and must be applicable both within and between samples. Autocorrelation measures with these characteristics are called AIDAs, an acronym for *autocorrelation index for DNA analysis* (Bertorelle & Barbujani, 1995).

Preliminary processing of data

Autocorrelation analysis of molecular data requires preliminary construction of a binary data matrix (a flow-chart is in Table 8.1). Molecular population data currently belong to one of three categories: (a) RFLP data; (b) DNA sequences; and (c) number of repeats at a VNTR locus. For the sake of simplicity, we shall treat VNTR variation separately. For the other two

Table 8.1. *Steps necessary for calculating an AIDA correlogram*

0	Data coding	Alternative bases are coded as binary digits at all polymorphic sites.
1	Spatial distances	A matrix of spatial distances between individuals is computed.
2	Distance classes	Arbitrary distance class limits are chosen.
3	Average vector	The frequency of one allele is calculated at each polymorphic site.
4	Computation of *II*	An autocorrelation index is computed in each distance class.
5	Confidence limits	A confidence interval about the null expectation of *II* is evaluated.
6	Significance testing	Observed and expected *II* values are compared.

classes of data, we shall assume that only two alternative bases can occur at each polymorphic site; this assumption is consistent with Watterson's (1975) infinite-site model, but it does not necessarily imply it. A slightly more complicated procedure exists for dealing with the occurrence of three different bases at the same site (Bertorelle & Barbujani, 1995). In each case, the set of polymorphic sites in a DNA fragment will define a haplotype; only for mtDNA and Y chromosome polymorphisms will there be a one-to-one correspondence between individuals and haplotypes.

Following what Excoffier, Smouse & Quattro (1992) have termed a phenetic approach, a restriction haplotype is represented by an array of 0s and 1s, indicating, respectively, absence or presence of a restriction cut at the various sites. A DNA sequence can be coded in the same way, assigning arbitrarily the values 0 or 1 to either of the alternative bases observed at each polymorphic site. In this way, each haplotype will be coded as a binary vector $\mathbf{p}$ of this type:

$$\mathbf{p}_1 = [1\ 0\ 0\ 1\ 0\ 0\ 1\ 1\ 1\ 0]$$
$$\mathbf{p}_2 = [0\ 0\ 1\ 0\ 1\ 1\ 1\ 1\ 1\ 1]$$
$$\mathbf{p}_3 = [0\ 0\ 1\ 0\ 1\ 1\ 1\ 1\ 0\ 1]$$
$$\mathbf{p}_4 = [0\ 1\ 0\ 1\ 0\ 0\ 0\ 0\ 1\ 0]$$

In this example, the sample size is four, and there are $S = 10$ polymorphic sites. VNTRs are more complicated to deal with. Length variants can also be represented by strings of 0s and 1s, indicating respectively absence and presence of a repeat. The data can be described in this way if genetic diversity is essentially generated by a stepwise mutation process (Ohta & Kimura, 1983; see also Valdes, Slatkin & Freimer, 1993). In that case, differences between alleles at VNTR loci are the product of mutations modifying the allele length by one or two repeats. As a consequence, the more diverse two haplotypes, the more remote their closest common ancestor. Indeed, alleles of similar length also seem to be evolutionarily related (Weber & Wong, 1993; see also the study of cystic fibrosis mutations by Morral *et al.*, 1994), suggesting that the stepwise mutation model is reasonable. There might be cases, however, in which recombination or unequal crossing over may have generated alleles that substantially differ in size from the parental alleles. In these cases, an alternative, evolutionary criterion, based on construction of a dendrogram is preferable, as well as in the cases in which substantial levels of homoplasy are observed (Excoffier, Smouse & Quattro, 1992). In what follows, however, we shall refer only to the phenetic criterion described above.

AIDA analysis

As a first step (Table 8.1), spatial distances are computed between the localities where haplotypes have been sampled. Step 2 requires the identification of arbitrary distance class limits. Haplotypes sampled at the same locality will be at distance 0 from each other; by comparing them, one will obtain a measure of autocorrelation at distance 0, i.e. an estimate of intra-population diversity. Depending on the sample sizes and on the distribution of the samples in the area of interest, one can choose distance classes of equal width, or distance classes each containing approximately equal numbers of comparisons.

For each of the binary digits in the **p** vector, an average is then calculated across all n haplotypes in the sample (step 3). The elements of the average vector thus computed are the relative frequencies of each restriction site, or alternative base, or DNA repeat. II is then calculated separately for each distance class, as

$$II = \frac{n \sum_{i=1}^{n-1} \sum_{j>i}^{n} w_{ij} \sum_{k=1}^{S} (p_{ik} - \overline{p_k})(p_{jk} - \overline{p_k})}{W \sum_{i=1}^{n} \sum_{k=1}^{S} (p_{ik} - \overline{p_k})^2}$$

(step 4). In this formula, p_{ik} and p_{jk} are the values of the **p** vector in the i-th and j-th haplotype, at the kth site, and k is the corresponding element of the average vector. Summation is over the S sites and the n individuals. W is the number of pairwise comparisons in the class of interest; w_{ij} is 1 if the haplotypes i and j fall in the distance class of interest, and 0 if they do not. When two haplotypes share a restriction site or a nucleotide $(p_{ik} - \overline{p_k})$ and $(p_{jk} - \overline{p_k})$ will both be positive or negative, their product will be positive and it will contribute positively to II; when two haplotypes differ, the

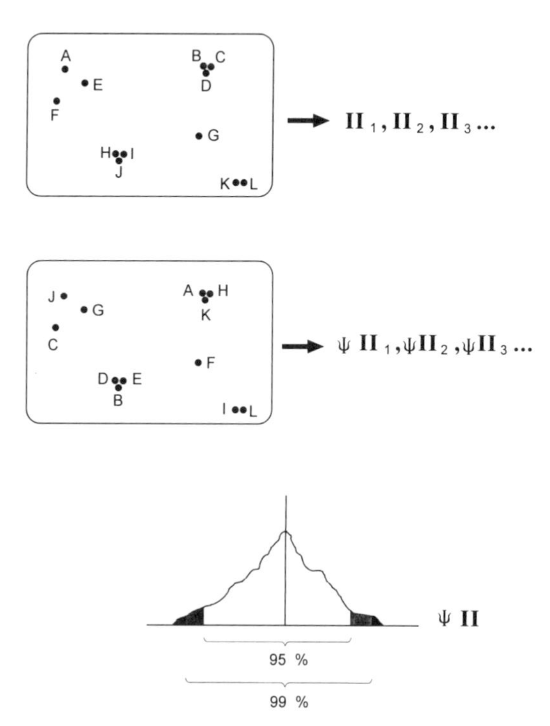

Fig. 8.2. Hypothetical distribution of haplotypes in seven samples (top), and one example of random assignment of the same haplotypes to the same localities (middle). Each haplotype is identified by a different letter, but they need not differ genetically from each other. If an observed *II* value falls outside the 95% or 99% confidence limits of the distribution of pseudo-*II* values (ψII), it is considered significant at the 5% or 1% level, respectively.

product of these quantities will be negative, resulting in a negative contribution to *II*.

Although analytical approaches to the estimation of the error of *II* are conceivable, they rely on unwarranted or unlikely assumptions, such as independence of the various sites in the DNA fragment. Randomization methods are a popular alternative in population biology (Manly, 1991). Consider the samples of Fig. 8.2. After calculating the observed *II* statistics, one can assign haplotypes to random locations in the area studied, keeping the sample sizes constant. A pseudo-II value can then be calculated, representing the value which one would obtain were haplotypes randomly distributed. This procedure can be repeated to obtain an empirical distribution of pseudo-II values, the one expected in the absence of any

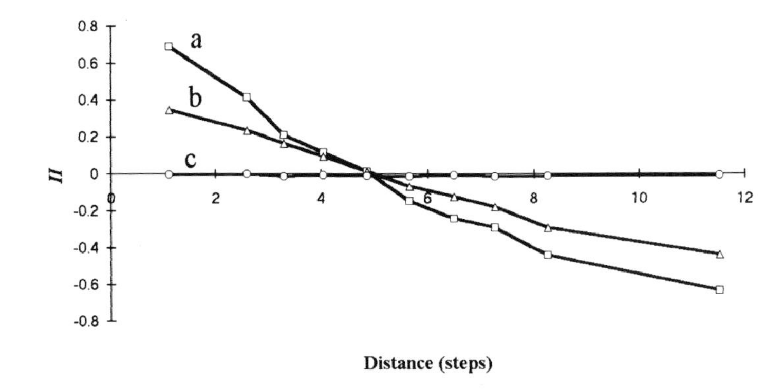

Fig. 8.3. Examples of *II* correlograms of simulated haplotype distributions: **a** pure gradient; **b** gradient on which individual mobility was simulated; **c** random distribution.

spatial pattern (step 5). Its confidence limits can be evaluated, and the statistical significance of the observed values (step 6) can be assessed by comparison.

Assessing the performance of AIDAs

Does *II* recognize a spatially patterned distribution of haplotypes from a random distribution? And, can one establish a one-to-one correspondence between observed patterns and evolutionary processes, so that pattern identification will lead to a precise and objective evolutionary inference? In the following paragraphs we shall explain why our answers to these questions are, respectively, yes and no.

Computer simulations show that *II* correctly identifies several spatial patterns of haplotype variation (Bertorelle & Barbujani, 1995). In those experiments, haploid individuals were placed on the nodes of a grid, simple spatial distributions of their haplotypes were generated, and we investigated whether the pattern inferred from AIDA corresponded to the simulated one. Gradients of haplotypes were always associated with a characteristic correlogram, whose values decline from positive significant to negative significant. This was the case even when the gradient was disturbed by other overlapping phenomena, such as the possibility of limited individual dispersal. In these cases, the *II* coefficients, whether positive or negative, were closer to 0, but the decreasing shape of the correlograms did not change; its slope was highest for no dispersal, and it decreased as the mobility of individuals increased. The highest mobility corresponds to random mating, and indeed the random patterns thus generated always resulted in sets of insignificant *II* values (Fig. 8.3).

Experiments are in progress to test whether the spatial patterns generated under isolation by distance yield the asymptotically decreasing correlograms predicted by theory (Barbujani, 1987) and observed for allele frequencies (Sokal & Jacquez, 1991).

Analyses of mtDNA data from natural populations seem to agree with the findings of the simulations. Highly mobile birds (Bertorelle & Barbujani, 1995), and some European human populations (unpublished data from our laboratory) do not show significant spatial structuring of haplotypes. On the other hand, desert rodents who underwent a directional expansion (Bertorelle & Barbujani, 1995) show a correlogram of the kind that Sokal *et al.* (1989) described as long-distance differentiation. In that kind of pattern, the extreme populations are highly differentiated, close populations are genetically similar, but the *II* coefficients fluctuate at intermediate distances.

In summary, both simulation experiments and preliminary analyses of some datasets suggest that AIDA faithfully describes spatial patterns. Unfortunately, this does not mean that, given one pattern, a specific evolutionary process can be inferred. On the contrary, we know that several distinct evolutionary pressures may lead to the same pattern. The simplest example is a cline, for which at least three classes of causes exist: (1) response to a selective gradient (see Endler, 1977); (2) demic diffusion, i.e. a population increase leading to dispersal and incomplete admixture (see Menozzi, Piazza & Cavalli-Sforza, 1978); (3) a range expansion accompanied by founder effects and followed by gene flow (see Easteal, 1988 and Barbujani, Sokal & Oden, 1995). Similarly, non-significant patterns may result from panmixia, or from a spatially uniform selective pressure preventing genetic differentiation. Other examples are possible. In conclusion, we agree with Slatkin & Arter (1991) who wrote that spatial autocorrelation is suitable for exploratory data analysis, but less so for inferring specific evolutionary phenomena. Still, we believe that autocorrelation indices may facilitate inferences, by ruling out evolutionary processes that are incompatible with the spatial pattern identified.

An application: mtDNA diversity in Italy

RFLP data on mtDNA have been published for 12 samples of Italian populations (Brega *et al.*, 1994, and references therein). Five restriction endonucleases (*Bam*HI, *Hae*II, *Msp*I, *Ava*II and *Hinc*II) have been employed in all studies, and so their results are comparable. Fig. 8.4 shows the location of the samples, and their sizes. Haplotype sequences were inferred from restriction morphs according to Excoffier (1990), and AIDA statistics were calculated for the twelve samples altogether (IS samples),

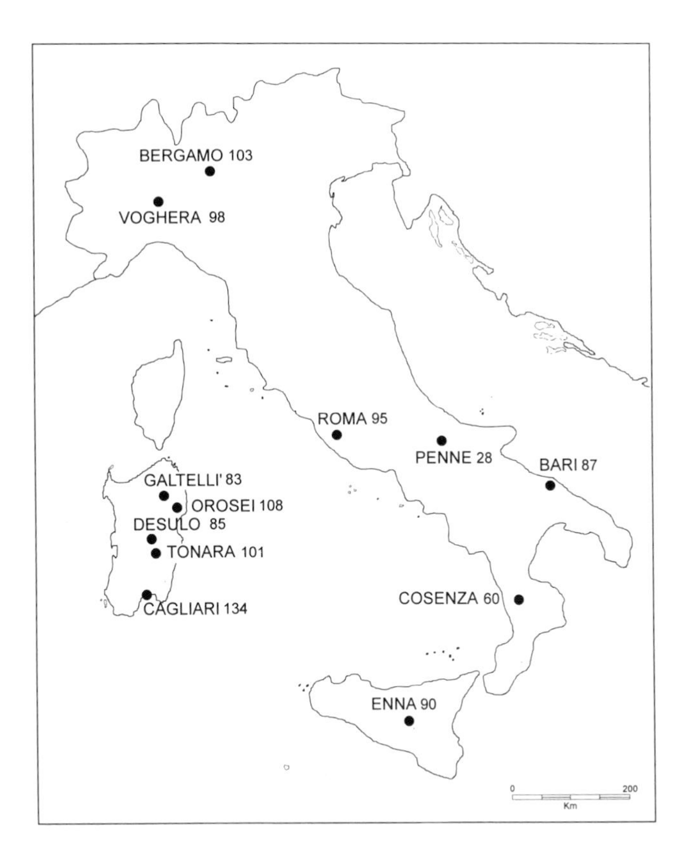

Fig. 8.4. Location of the 12 samples of mtDNA in Italy, and sample sizes. The Sicilian data come from several sampling locations, and have been attributed to a locality in the centre of the island.

and for the seven samples from peninsular Italy (IT samples), including Sicily, which does not show major differences from the rest of the peninsula at the allele-frequency level (Barbujani & Sokal, 1991). The overall sample size was 1072. Twelve distance classes were considered; their limits were chosen so as to distribute the data in a reasonably balanced way.

Geographical patterns

Twenty-six sites showed polymorphism. The correlograms calculated from the two sets of samples (Fig. 8.5) have something in common. Both show positive autocorrelation at distance 0, indicating, as expected, higher genetic homogeneity within than between samples. The level of significance of the first coefficient was higher in the IS samples, which means that the

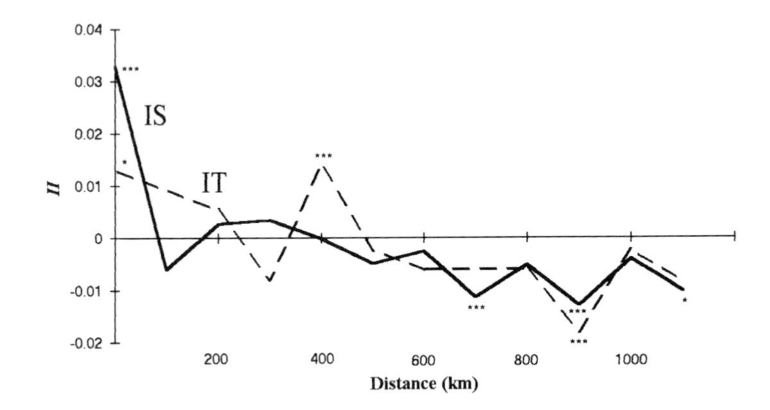

Fig. 8.5. AIDA correlograms of mtDNA variation in the Italian + Sardinian (IS) and in the Italian (IT) samples.

Sardinian individuals, on average, resembled each other more than the individuals in the peninsula. In fact, two Sardinian samples, Galtell® and Tonara, show only three distinct haplotypes, whereas the peninsular samples include, on the average, more than 11 haplotypes. The numbers of different haplotypes are not correlated with sample sizes, and so there is reason to believe that Sardinia really shows a lower mitochondrial diversity.

The IT samples show only three significant autocorrelation peaks. One is around 400 km, and it is due to a strong genetic similarity between Roma and Bari, the only two samples being separated by that distance; we have no explanation for their resemblance. Apart from that, there is negative autocorrelation in the two extreme distance classes, implying that there are differences between the mtDNA gene pools of the Northern samples on the one hand, and of the samples from Sicily and Calabria on the other. All in all, this correlogram describes a long-distance differentiation pattern, that is, a cline in which intermediate localities are not always genetically intermediate.

When the Sardinian data are added to the analysis (IS samples), the spatial structure becomes more significant. Positive autocorrelation disappears at 400 km, because now more comparisons fall in that class. The negative autocorrelation at 1000 and 1100 km is still due to North–South differentiation; an additional, highly significant negative peak occurs in the distance class (700 km) including most comparisons of Sardinian and Northern haplotypes. This is still a pattern compatible with long-distance differentiation, but its most remarkable feature is the negative intermediate-distance autocorrelation. Sokal (1979) defines patterns of this kind as depressions.

Evolutionary scenarios

At the allele-frequency level, Sardinia shows a large genetic difference from the mainland. In the peninsula, few allele frequencies are clinally distributed, because several barriers, geographic and linguistic, are associated with sharp gene-frequency change (Barbujani & Sokal, 1991). The pattern inferred from mtDNA RFLPs also points to genetic heterogeneity of Sardinians, but its main feature in the peninsula seems to be North–South differentiation.

Some 40 generations, or less, have elapsed since the development of Italian dialects (Rea, 1973). Therefore, linguistic barriers within Italy cannot have existed for more than a few centuries. Even if we imagine that some of them correspond to pre-existing linguistic barriers and are somewhat older, their association with areas of sharp gene-frequency change suggests that the latter are also recent. On the other hand, even 100 generations are a very short time for mtDNA haplotypes to differentiate. MtDNA studies, actually, tend to show only little genetic differentiation in Europe (Excoffier, 1990; Merriwether *et al.*, 1991), and most genetic differences between European populations seem to have arisen in the Pleistocene (Rogers & Jorde, 1995). Probably, the mtDNA patterns we described have also been established in a different, more remote time period than most allele-frequency patterns.

To investigate further in this direction, we applied a model put forward by Rogers & Harpending (1992). These authors proposed the use of the distribution of pairwise sequence differences (mismatch distribution) to draw inferences on past demographic expansions. Their method is suitable for this study, as demographic growth is among the plausible explanations for the patterns of genetic diversity we describe. We expect that such growth has occurred in a relatively remote past, since its effects seem visible at the mtDNA, but not at the allele frequency, level. Locating it in time would allow us to compare a time estimate based on genetic diversity with non-genetic, archaeological evidence.

More statistical tests

Demographic expansions result in departure of some population parameters from neutral, equilibrium expectations. Tajima (1989) derived the relationship between the number of polymorphic sites and the average mismatch between pairs of individuals expected at equilibrium. On the basis of this relationship, he introduced a statistic, D, for which he calculated confidence limits. Significant negative D values point to an excess of rare haplotypes, and this is what was observed in both the IS and the IT samples. We had $D = -1.76$ for Italy, -1.77 for Sardinia, and -1.91 for Italy and Sardinia, all values significant at the $P = 0.05$ level. A result of this type is compatible with several evolutionary phenomena, selection (Excoffier & Langaney, 1989) among them. However, a sudden population expansion, such as the one that probably occurred in humans (Takahata, 1993), is also expected to cause the appearance of several rare haplotypes, and thus to result in negative D values (Merriwhether *et al.*, 1991).

Tajima's tests do not prove that the particular pattern of mtDNA diversity we observe is due to a past expansion, but they are consistent with this view. Before proceeding to estimate the age of the expansion, we calculated a statistic called raggedness (Harpending *et al.*, 1993), which permits discrimination between populations that presumably expanded and populations that did not (Rogers, 1995). Raggedness is also based on the mismatch distribution. The values obtained are very different in Sardinia (0.355) and in peninsular Italy (0.075); only the latter is consistent with a major population growth. In other words, population growth cannot be considered among the likely explanations for the result of Tajima's test on Sardinia. On the contrary, raggedness of the mismatch distribution suggests the peninsular population has not been demographically stable. Its expansion has been tentatively dated as described below.

Under plausible evolutionary assumptions, demographic expansions are expected to be reflected in the mismatch distribution. If an expansion occurred S units of mutational time before the present, the average mismatch is expected to be S, minus a correction factor (Rogers, 1994). A mutational time unit, τ, is equal to $2ut$, where u is the mutation rate, and t is a number of generations (Harpending *et al.*, 1993). The estimate of the mutation rate is therefore crucial for calculating the age of the expansion, and it must refer to the fraction of DNA that restriction analysis surveyed. To quantify how much of mtDNA was screened by the five endonucleases we considered, we followed an approach put forward by Nei and Tajima (1981). For the mutation rate, we employed two values published by Rogers & Harpending (1992), 2.5 and 5×10^{-7} per nucleotide per generation. In this way, we obtained two values of u, 0.000184 and 0.000369. The corresponding estimates of τ are 821 and 410 generations. If a generation lasts 20 or 25 years, then a population expansion compatible with the mismatch distribution in the IT sample, whose average is 1.16, may have occurred any time between 8200 and 20 500 years ago. As a control, similar calculations, although not theoretically justified, were carried out for the five Sardinian samples as well. They placed an analogous population growth between 460 and 1100 years ago.

These dates cannot be considered precise. We are aware that the relevant parameters can be estimated only approximately, especially mutation rates. Also, the values we have calculated do not really define a confidence interval; rather, they indicate a range of possible dates. However, they at least provide a time frame for the hypothetical expansion that may be reflected in mtDNA diversity. This offers the possibility of checking whether processes of potential evolutionary relevance have affected the Italian population in that time interval. Two steps appear necessary at this stage. One is to verify by computer simulation whether an expansion may have had these effects on mtDNA diversity in Italy but not in Sardinia. The other is to compare the estimated dates with archaeological evidence on the demography of Italy.

A simulation experiment

The question we are asking here is whether a population expansion is likely to determine the characteristics that we observed, in two groups of samples of mtDNAs from Italy and from Sardinia. We chose to focus on a parameter that is easy to simulate, i.e. the number of polymorphic sites detected by RFLP analysis. The simulation algorithm (see Bertorelle & Slatkin, 1995) was based on the coalescent process, described by Harding (Chapter 6). The coalescent is a stochastic process in which the genealogy of a sample of genes (the entire mtDNA in our case) is reconstructed proceeding backwards, through a series of events of common ancestry, or coalescences (see Hudson, 1990). The probability that two individuals have a common ancestor at each generation can be evaluated under an ideal Wright–Fisher model, and the time during which there are j distinct lineages is exponentially distributed with mean $2N/j(j-1)$. The genealogy of the sample is then reconstructed up to the earliest common ancestor. Changes in population size can be incorporated, by stretching or squeezing the branches by a factor that depends on the factor of increase or decrease, respectively.

Once a genealogy has been simulated, mutations are added to the tree. A finite site model was used (Watterson, 1975), allowing for the possibility of reverse mutations and multiple changes per site. Since approximately one-tenth of the mtDNA (the control region) is known to be evolving rapidly, one-tenth of the simulated sites were allowed to mutate ten times as rapidly as the others.

Gene genealogies of the Italian and of the Sardinian samples were simulated 1000 times each, using the products Nu (where N represents the effective number of females in the population before the expansion) and ut, as estimated from the mismatch distributions of the appropriate group of

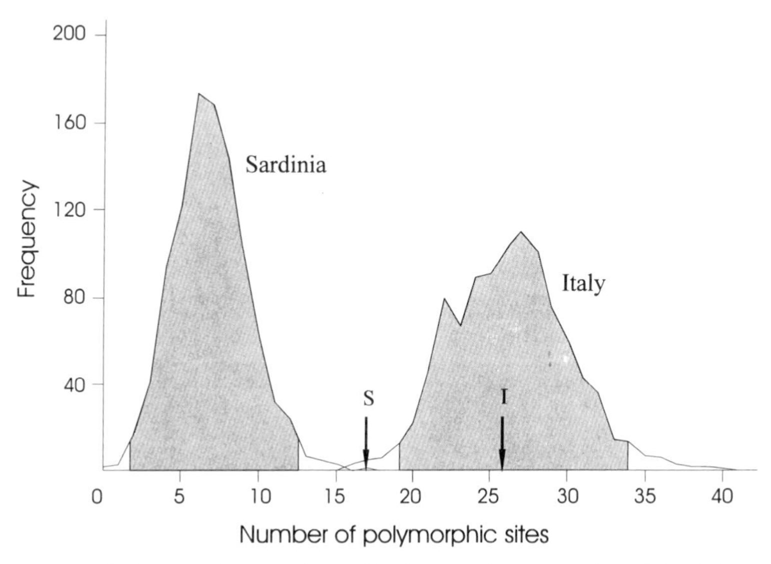

Fig. 8.6. Number of polymorphic sites (x-axis) and number of simulation replicates in which those numbers were observed, in 1000 simulations of the genealogies of the Italian and the Sardinian samples. The arrows show the observed numbers of polymorphic sites in the two samples, 26 for Italy and 17 for Sardinia. As is evident, there is a good agreement between the Italian data and the expectations based on the simulation of a 100-fold population growth. On the contrary, a much smaller number of polymorphic sites would be expected in Sardinia, had the population gone through a phase of rapid growth.

samples. Under Rogers's (1995) method, the current population size is assumed to approach infinity, and therefore it is not estimated exactly. We chose a 100-fold increase of population size; that factor is not expected to affect the simulation greatly, since any sudden demographic increase results in a star-like genealogy with most of the coalescences occurring just before the increase.

Fig. 8.6 shows that there is good agreement between the number of polymorphic sites observed in Italy, and the simulated consequences of a demographic increase. The value observed is very close to the mode of the simulated values, and is therefore fully compatible with the model underlying the simulation. That is not the case for Sardinia, where the expectations of a model of sudden increase are distributed to the left of the observed value. In fact, the number of polymorphic sites observed in Sardinia falls outside the 95% confidence interval of the simulated distribution. Therefore, there is no reason to presume that that population increased suddenly in numbers.

The age of the expansion that we estimated depends on the mutation rate, which cannot be measured exactly. However, the results of the simulation do not depend on it, because the number of polymorphic sites in the simulated genealogies is a function of the products Nu and ut, where the mutation rate is multiplied, respectively, by population size (N) and age of the expansion (t). This means that the simulation does not increase our confidence in our time estimates, but it suggests once more, on the basis of a different approach, that mtDNA diversity reflects an expansion only in Italy, and not in Sardinia.

Discussion

Evidence from autocorrelation analysis of mtDNA, from the distribution of mismatches, and from simulation of gene genealogies, agree in indicating that the populations of peninsular Italy and of Sardinia may have had different evolutionary histories. The distributions of haplotype diversity appear to depart from equilibrium in both samples. However, it is only for peninsular Italy that departure seems related to a population growth. Although necessarily approximate, estimates of its age place it in the late Palaeolithic or in the early Neolithic. The much more recent dates estimated for Sardinia do not mean that its population size was constant until the last millennium, but simply that mtDNA data provide no evidence for a sudden burst of demographic growth there.

The spread of cereal farming in the Neolithic was accompanied by a substantial increase in population sizes in Europe (Ammerman & Cavalli-Sforza) and probably Asia (Renfrew, 1992; Barbujani & Pilastro, 1993). The Italian population may have undergone a 100-fold increase (Rendine, Piazza & Cavalli-Sforza, 1986) around 7500 or 6500 years ago (Renfrew, 1987; Sokal, Oden & Wilson, 1991). This date is close to our minimum estimate of the age of an expansion inferred from mtDNA data. In addition, an expansion due to the dispersal of farmers should result in a gradient of haplotypes (Barbujani *et al.*, 1995), which is essentially what we have found in our spatial autocorrelation analysis. Therefore, a rapid Neolithic growth of the Italian population is compatible with the evidence from mtDNA data, and shows a nice, if rough, agreement with archaeological findings. Is that the only process accounting for the results of this study?

We think other processes should be considered as well. First, the earliest date we estimated is close to the latest estimate of the spread of Neolithic technologies in Italy; the temporal agreement is not perfect. Secondly, studies of mtDNA in other populations have only seldom, or never, suggested Neolithic expansions (see, e.g. Takahata, 1993 and Rogers & Jorde, 1995). There is a methodological reason for this: Rogers and Harpending's model is heavily influenced by the first expansion a population underwent. Successive events of the same kind are expected to leave only a minor mark on the mismatch distribution (Rogers & Harpending, 1992). Although this aspect of the model is not fully clear at the moment, it seems that a demographic growth could be traced to the Neolithic only if populations were substantially stable in size in previous periods, and notably in the Pleistocene. It is hard to tell if this is the case for Italy.

An alternative interpretation of our results may be based upon Mussi's (1990) reconstruction of the demography of Italy at the last glacial maximum. Archaeological findings suggest that human groups were small and dispersed, to the point that there is no evidence of settlements between 25 000 and 30 000 years ago. Only around 20 000 BP do population sizes seem to increase. Some sites appear to have been continuously occupied, and therefore that increase is attributed to an expansion of already existing groups, rather than to immigration. This is not consistent with our results, since it is hard to imagine how a quasi-clinal pattern may have been generated without substantial gene flow. However, the dating of this phenomenon overlaps with the oldest date estimated from the mtDNA mismatch distribution.

All in all, we see no strong reason to prefer either hypothesis. The view whereby the patterns of mtDNA diversity in Italy have been established in the Palaeolithic (20 500 years BP) has the advantage of incorporating the possibility of a later, Neolithic expansion (8200 years BP). The reverse is not true because, as we have said, the mismatch distribution seems rather insensitive to expansions other than the first major one. One might be tempted to average our estimates, and locate an episode of expansion around 14 000 years ago. That does not seem a good idea, though, because archaeologists tend to believe that population sizes in Europe were shrinking at that time (Gamble, 1983).

We think that these results, approximate though they may be, show how the analysis of geographic variation can help reconstructing evolutionary phenomena. Only after AIDA has been applied to other human groups will we know its potentials and limits better. As for the case considered in this chapter, AIDA contributes to indicating that population structure in Sardinia and in the rest of Italy may have been determined by different demographic events. In addition, AIDA demonstrates that the pattern of mtDNA variation does not exactly mirror allele frequency patterns, presumably because it reflects evolutionary factors acting in more remote times.

Acknowledgements

We thank Marta Mirazon Lahr for telling us about the Italian Palaeolithic, and Rosaria Scozzari and Giulia Capitani for collaborating with us in various phases of this research project. We are also grateful to Barbara Thomson and Clandio Friso for Figs. 8.1, 8.2 and 8.4. This study was partly supported by the Italian CNR grant No 94-2270.

References

Ammerman, A. J. & Cavalli-Sforza, L. L. (1984). *The Neolithic Transition and the Genetics of Populations in Europe*. Princeton, NJ: Princeton University Press.

Barbujani, G. (1987). Autocorrelation of gene frequencies under isolation by distance. *Genetics*, **117**, 777–83.

Barbujani, G. & Pilastro, A. (1993). Genetic evidence on origin and dispersal of human populations speaking languages of the Nostratic macrofamily. *Proceedings of the National Academy of Sciences, USA*, **90**, 4670–3.

Barbujani, G. & Sokal, R. R. (1991). Genetic population structure of Italy. II. Physical and cultural barriers to gene flow. *American Journal of Human Genetics*, **48**, 398–411.

Barbujani, G., Sokal, R. R. & Oden, N. L. (1995). Indo-European origins: a computer-simulation test of five hypotheses. *American Journal of Physical Anthropology*, **96**, 109–32.

Bertorelle, G. & Barbujani, G. (1995). Analysis of DNA diversity by spatial autocorrelation. *Genetics*, **139**, 811–19.

Bertorelle, G. & Slatkin, M. (1995). The number of segregating sites in expanding human populations, with implications for estimates of demographic parameters. *Molecular Biology and Evolution*, **12**, 887–92.

Brega, A., Mura, G., Caccio, S., Semino, O., Brdicka, R. and Santachiara-Benerecetti, A. S. (1994). MtDNA polymorphisms in a sample of Czechoslovaks and in two groups of Italians. *Gene Geography*, **8**, 45–54.

Easteal, S. (1988). Range expansion and its genetic consequences in populations of the giant toad, *Bufo marinus*. *Evolutionary Biology*, **23**, 49–84.

Endler, J. A. (1977). *Geographic Variation, Speciation, and Clines*. Princeton, NJ: Princeton University Press.

Excoffier, L. (1990). Evolution of human mitochondrial DNA: evidence for departure from a pure neutral model of populations at equilibrium. *Journal of Molecular Evolution*, **30**, 125–39.

Excoffier, L. & Langaney, A. (1989). Origin and differentiation of human mitochondrial DNA. *American Journal of Human Genetics*, **44**, 73–85.

Excoffier, L., Smouse, P. E. & Quattro, J. M. (1992). Analysis of molecular variance inferred from metric distances among DNA haplotypes: application to human mitochondrial DNA restriction data. *Genetics*, **131**, 479–91.

Gamble, C. (1983). Culture and society in the upper Palaeolithic of Europe. In *Hunter–gatherer Economy in Prehistory*, ed. G. Bailey, pp. 201–11. Cambridge: Cambridge University Press.

Harpending, H., Sherry, S. T., Rogers, A. L. & Stoneking, M. (1993). The genetic structure of ancient human populations. *Current Anthropology*, **34**, 483–96.

Hudson, R. R. (1990). Gene genealogies and the coalescent process. *Oxford Surveys in Evolutionary Biology*, **7**, 1–44.

Manly, B. J. F. (1991). *Randomization and Monte Carlo Methods in Biology*. London: Chapman and Hall.

Menozzi, P., Piazza, A. & Cavalli-Sforza, L. L. (1978). Synthetic maps of human gene frequencies in Europeans. *Science*, **201**, 786–92.

Merriwether, D. A., Clark A. G., Ballinger S. W., Schurr, T. G., Soodyall, H., Jenkins, T., Sherry, S. T. & Wallace, D. C. (1991). The structure of human mitochondrial DNA variation. *Journal of Molecular Evolution*, **33**, 543–55.

Morral, N., Bertranpetit J., Estivill X. *et al.* (1994). The origin of the major cystic fibrosis mutation (ΔF508) in European populations. *Nature Genetics*, **7**, 169–75.

Mussi, M. (1990). Continuity and change in Italy at the last glacial maximum. In *The World at 18 000 BP: High Latitudes*, ed. O. Soffer and C. Gamble, pp. 126–47. London: Unwin.

Nei, M. & Tajima, F. (1981). DNA polymorphism detectable by restriction endonucleases. *Genetics*, **97**, 145–63.

Ohta, T. & Kimura, M. (1983). The model of mutation appropriate to estimate the number of electrophoretically detectable alleles in a genetic population. *Genetical Research*, **22**, 201–4.

Rea, J. A. (1973). The Romance data of the pilot studies for glottochronology. In *Current Trends in Linguistics II: Diachronic and Typological Linguistics*, ed. T. S. Sebeok, pp. 355–68. The Hague: Mouton.

Rendine, S., Piazza, A. & Cavalli-Sforza, L. L. (1986). Simulation and separation by principal components of multiple demic expansions in Europe. *American Naturalist*, **128**, 681–706.

Renfrew, C. (1987). *Archaeology and Language: The Puzzle of Indo-European Origins*. London: Jonathan Cape.

Renfrew, C. (1992). Archaeology, genetics, and linguistic diversity. *Man*, **27**, 445–78.

Rogers, A. R. (1995). Genetic evidence for a Pleistocene population explosion. *Evolution*, **49**, 608–15.

Rogers, A. R. & Harpending, H. (1992). Population growth makes waves in the distribution of pairwise genetic differences. *Molecular Biology and Evolution*, **9**, 552–69.

Rogers, A. R. & Jorde, L. B. (1995). Genetic evidence on modern human origins. *Human Biology*, **67**, 1–36.

Slatkin, M. & Arter, H. E. (1991). Spatial autocorrelation methods in population genetics. *American Naturalist*, **138**, 499–517.

Sokal, R. R. (1979). Ecological parameters inferred from spatial correlograms. In *Contemporary Quantitative Ecology and Related Ecometrics*, ed. G. P. Patil and M. Rosenzweig, pp. 167–196. Fairland, MD, USA: International Co-operative Publishing House.

Sokal, R. R., Harding, R. M. & Oden, N. L. (1989). Spatial patterns of human gene frequencies in Europe. *American Journal of Physical Anthropology*, **80**, 267–94.

Sokal, R. R. & Jacquez, G. M. (1991). Testing inferences about microevolutionary processes by means of spatial autocorrelation analysis. *Evolution*, **45**, 152–68.

Sokal, R. R. & Oden, N. L. (1978). Spatial autocorrelation analysis in biology. 1. Methodology. *Biological Journal of the Linnean Society*, **10**, 199–228.

Sokal, R. R., Oden, N. L. & Wilson, C. (1991). New genetic evidence for the spread of agriculture in Europe by demic diffusion. *Nature*, **351**, 143–5.

Tajima, F. (1989). Statistical method to test the neutral mutation hypothesis by DNA polymorphism. *Genetics*, **123**, 585–95.

Takahata, N. (1993). Allelic genealogy and human evolution. *Molecular Biology and Evolution*, **10**, 2–22.
Valdes, A. M., Slatkin, M. & Freimer, N. B. (1993). Allele frequencies at microsatellite loci: the stepwise mutation model revisited. *Genetics*, **133**, 737–49.
Weber, J. L. & Wong, C. (1993). Mutation of human short tandem repeats. *Human Molecular Genetics*, **2**, 1123–8.
Watterson, G. A. (1975). On the number of segregating sites in genetical models without recombination. *Theoretical Population Biology*, **7**, 256–76.

9 *Mitochondrial DNA sequences in Europe: an insight into population history*

J. BERTRANPETIT, F. CALAFELL, D. COMAS,
A. PÉREZ-LEZAUN AND E. MATEU

Background: approaches to population history

The biological problem of unravelling the evolutionary tempo and mode of a given species and its evolutionary relationships to other species has been approached by several biological sciences, among which molecular biology has given a huge boost to the field. In the exploration of the past of human populations, several disciplines can contribute to our understanding, be it in the social sciences (archaeology and other historical sciences, historical linguistics) or in the biological sciences (palaeoanthropology, molecular biology, genetics). The time depth of the events under study and their geographical locations may be of key importance in order to recognize both which disciplines may be relevant and, within a given discipline, which are the appropriate tools to use.

Issues that may be addressed range from wide and global ones as, for example, the origin of the whole of humanity, to more specific ones, such as the origin of the settlers of a given area or ethnic group. Historical sciences have, in their long tradition of scholarship, used very different methods to investigate the social structure of the past according to the time period under scrutiny, from oral or written traditions to the sophisticated physical and chemical methods used in archaeology.

In the last few years molecular biology has been able to offer a wide range of approaches to analysing different molecules which may be used to search for different time depths; different regions of the genome evolve at different rates and thus differences between two genomic regions (in the same individual or in different individuals) may reflect events on different time scales. When dealing with genetic differences within humans, a single species with a short evolutionary history, it is not always clear which genomic regions are the most appropriate for solving specific questions, and the literature on human population genetics and evolution offers constantly new tools and possibilities for study. It is therefore, difficult *a priori* to recognize which genetic analysis may be the most appropriate to deal with a temporal framework which may be quite well defined in an archaeological context.

Here we propose to analyse the genetic variation in the sequence of nucleotides of the control region of the mitochondrial DNA (mtDNA) in various European and Western Asian populations in order to clarify aspects of interest in their population history, in the relationships among them and their origin. In approaching a problem posed by palaeoanthropology and archaeology, we will discuss some of available genetic results and will analyse the mtDNA data. Results will be discussed in the unique framework of the history of human populations.

Palaeoanthropology

Major morphological and cultural changes, which are likely to be related to the origin of modern populations, have been attested in Europe in the period between 40 000 and 30 000 years BP. It has long been recognized that there were marked morphological differences between Neanderthals and anatomically modern humans (AMH) in Europe. The significance of the differences is controversial, and while many scholars maintain that a replacement took place, probably related to the Upper Palaeolithic expansion (Stringer, 1989; Klein, 1989, 1994), others argue for a morphological continuity for certain traits, at least with the oldest AMH (Frayer, 1992) and thus the Neanderthal gene pool would have contributed to the generations living after the period of the typical Neanderthal morphology in the Upper Palaeolithic.

The replacement hypothesis should offer an explanation for the origins of the European AMH, and it is usually argued that this substitution is nothing but a local event within a more general process of a recent origin of all modern humans, postulated to be in Africa (Stringer, 1989; Klein 1989, 1994). For the European replacement, the various findings of very old remains with a putative modern morphology (Skhull, Qafzeh) in the Middle East have supported the hypothesis of a West Asian origin of AMH (Stringer, 1989; Mellars, 1993; Bar-Yosef, 1994) which would have spread in Europe, thus replacing the Neanderthals.

Archaeology

Such a replacement event had to be related to some form of better adaptation or pre-eminence of the new settlers. Archaeology has been able to detect a major event in the transition between the Middle Palaeolithic culture (Mousterian) and the Upper Palaeolithic, which is interpreted as a revolution by some archaeologists (Mellars, 1993). The Aurignacian, the first fully developed, widespread cultural tradition in Europe and the Middle East has been related to the morphological substitution referred above. None the less, there are many aspects which await explanation before the theory of a simultaneous replacement of people and culture in Europe can be supported. For example, there is no convincing evidence that the first Upper Palaeolithic culture originated in Asia (Lilianne Meignen, personal communication); and the spread of Aurignacian from central to southern Europe seems to have been very rapid (Straus, 1989, 1994) and it is unlikely that a demographic substitution could have happened as rapidly.

Genetics

Europe is, without doubt, the region of the world where most genetic studies have been undertaken (Cavalli-Sforza, Menozzi & Piazza, 1994). Results with classical genetic markers (blood groups and protein polymorphisms) may be divided into two categories. First, specific groups, such as the Basques, Saami (Lapps), Sardinians and Icelanders have been identified as outliers within the general genetic European landscape; in each case specific events in the population history have been alleged to explain the peculiarity. Secondly, the main pattern of the general variation in Europe and Middle East has been interpreted as having been produced by the Neolithic expansion from the Middle East into Europe (Ammermann & Cavalli-Sforza, 1984); simulation and autocorrelation analysis (Sokal, Oden & Wilson, 1991) are compatible with this hypothesis, which assumes that all over Europe the Neolithic expanded by demic diffusion (Ammermann & Cavalli-Sforza, 1984). Under this hypothesis the extant genetic variation as observed through gene frequency variation for neutral polymorphisms (classical genetic markers) would have been mainly shaped around 10 000 years ago and has persisted until now. The relationship of the Neolithic expansion with that of Indo-European languages is a matter of debate (Renfrew, 1987; Villar, 1991).

The peculiarity of some of the outliers, such as the Basques, may have an older origin (Bertranpetit & Cavalli-Sforza, 1991; Calafell & Bertranpetit, 1994), in Upper Palaeolithic or Mesolithic times; nevertheless, the assumptions necessary to permit times of divergence to be estimated from gene frequency analysis (for example, that there has been no migration between groups after splitting) cannot always be made and this renders this approach unsuitable for retrieving population history chronologies. Moreover, as recently shown (Sajantila *et al.*, 1995), gene frequency variation may be deeply influenced by episodes of strong genetic drift due to low population size, a constant in most human history (Hassan, 1981).

The time depth of demographic events might be better ascertained through direct analysis of DNA. In a European-wide survey of cystic fibrosis (Morral *et al.*, 1994) it was proposed that part of the geographic pattern of variation of several microsatellites in the cystic fibrosis gene was established with the Palaeolithic settlement of Europe.

mtDNA

Human mitochondrial DNA (mtDNA) is a circular molecule, comprising 16 569 nucleotide pairs. It codes for 13 polypeptides that are essential to the energy-producing machinery of the cell, as well as the ribosomal and transfer RNAs needed to synthesize those proteins. Unlike nuclear DNA, coding regions in mtDNA are tightly packed, and few tracts are not translated. The longest non-coding region in mtDNA is the Control Region, where the origin of replication (e.g. the point from which mtDNA replication starts) is found. The complete sequence of human mtDNA has been known for nearly 15 years (Anderson *et al.*, 1981).

Two different techniques are used to detect nucleotide diversity in mtDNA. Nucleotide changes may create or destroy restriction endonuclease target sequences; digestion with a large number of restriction enzymes reveals such mutations throughout the mtDNA molecule. Alternatively, the most polymorphic regions, such as two segments in the control region, can be sequenced.

The value of mtDNA for population genetics

Mitochondrial DNA harbours a set of features that renders it particularly useful for human population genetic studies. Every mitochondrion contains several mtDNA copies, and every cell holds hundreds of mitochondria; therefore, mtDNA is found in a high copy number, which facilitates amplification, even from small amounts of sample or from degraded sources. Thus, mtDNA is the marker of choice for forensic cases or for genetic studies of ancient human remains.

Every human zygote inherits its mtDNA exclusively from its mother. Although spermatozoa contain large numbers of mitochondria in their middle segments, these do not seem to be passed on to the zygote. Maternal inheritance has important consequences for population genetics: effective population size, a parameter which determines the extent of genetic drift, is lower for mtDNA than for biparentally inherited nuclear DNA. Migration patterns are often different for men and women: invasions by male hordes may leave a significant nuclear genetic impact, but no mitochondrial traces.

In most cases, all the mtDNA molecules in an individual are identical (a phenomenon called *homoplasmy*). When a somatic mutation arises, the number of mutant mtDNAs that are passed on to daughter cells drifts by chance, leading eventually to fixation after many replication rounds. However, control region heteroplasmic individuals have been described by Gill *et al.* (1994) and by Comas, Pääbo & Bertranpetit (1995).

Mitochondrial DNA exhibits a high mutation rate, especially in the non-coding regions. Therefore, mtDNA is extremely polymorphic. In non-isolated populations, few unrelated individuals bear the same sequence. Ancient mutations can be used to trace back the links between populations: for instance, a deletion of nine base pairs has shown the Asian origins of Polynesians (Stoneking *et al.*, 1990).

Recombination has not been seen in mitochondrial DNA: it is inherited as a single block, or haplotype. This feature allows us to reconstruct the genealogy of the sequences observed, and also permits us to build simple models of the distribution of the pairwise nucleotide differences. It also means that all human mtDNAs have a unique ancestor, and that, under certain conditions, it may be possible to determine when the first split in the genealogical tree of human mtDNAs occurred. Knowledge of the mutation rate in mtDNA is one of those conditions. One of the techniques used to calibrate the mitochondrial clock is the comparison between human and chimpanzee sequences; assuming that humans and chimpanzees diverged 5 to 7 million years ago, a crude estimate can be obtained for the mutation rate. In the control region, several mutation rate estimates are available, ranging from about 5×10^{-6} to 10^{-6} mutations per nucleotide per generation (Vigilant *et al.*, 1991; Ward *et al.*, 1991), at least two orders of magnitude higher than nuclear DNA mutation rates.

After sequencing a large number of individuals, Vigilant *et al.* (1991) showed that African populations were the most internally heterogeneous, and suggested that human mtDNAs diverged around 150 000 years ago. These results agree with a recent human origin in Africa, as postulated by palaeoanthropologists such as Stringer (1989). The methods used by Vigilant *et al.* for the dating of mtDNA coalescence have been strongly criticized, but the high African diversity remains undisputed. Mitochondrial DNA sequences have also been used to gain regional perspectives on human population history, such as the peopling of the Pacific (Lum *et al.*, 1994) and of the Americas (Ward *et al.*, 1991, 1993; Santos, Ward & Barrantes, 1994) (see Chapters 13 and 15).

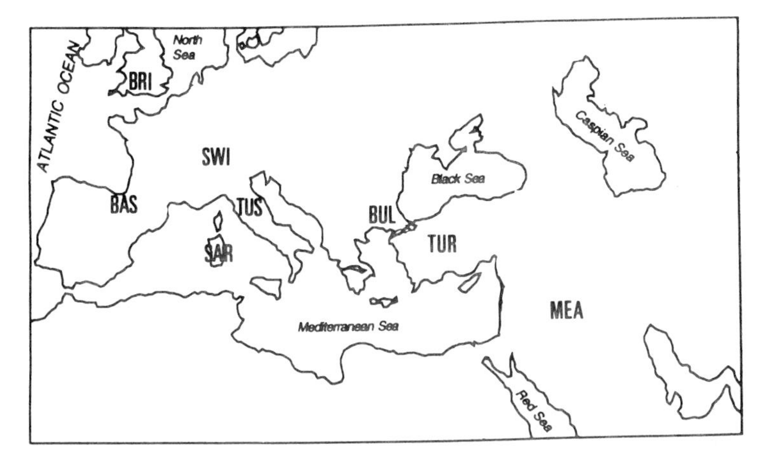

Fig. 9.1. Map of location of the samples included in the analysis. BAS, Basques; BRI, British; BUL, Bulgarians., MEA, Middle Easterns; SAR, Sardinians; SWI, Swiss; TUR, Turks; TUS, Tuscans.

The present study: control region mtDNA in Europe and Middle East

We have undertaken an analysis of mtDNA sequences in Europe and the Middle East. The sequence used for comparison comprises 360 nucleotides (positions 16 024 to 16 383 in the notation of Anderson *et al.*, 1981) in segment I of the control region. Populations included are Basques, British, Swiss, Tuscans, Sardinians, Bulgarians, two independent samples of Turks, and Middle Easterners (Fig. 9.1); they cover most of the southern and western flanks of the continent. The Basque sample (Bertranpetit *et al.*, 1995) was collected in schools from a few villages in Gipuzkoa; autochthony was sought. It comprises 45 individuals bearing 27 different haplotypes. The British sample (Piercy *et al.*, 1993), in contrast, was a random sample of 100 English and Welsh Caucasoid individuals gathered for forensic purposes; they were found to present 72 different sequences. The Swiss sample (Pult *et al.*, 1994) consisted of 74 individuals with 43 different haplotypes. Francalacci *et al.* (in press) collected samples from 49 autochthonous south Tuscans, who showed 40 different sequences. The Sardinians came from all over the island (Di Rienzo & Wilson, 1991): 69 individuals showed 46 haplotypes. Bulgarians and one of the Turkish samples (Calafell *et al.*, in press) were healthy members of families who had sought genetic counselling; the Bulgarians were 30 individuals with 24 haplotypes, and the Turks were 29, with 27 haplotypes. The other Turkish sample (Comas *et al.*, submitted) comprised 45 college students from most of the country; they showed 40 different sequences. Finally, the Middle Easterners (Di Rienzo & Wilson, 1991) included 42 Saudi Bedouins, Palestinians and Yemenite Jews with 38 different sequences. We will use the term Caucasoid to refer to all these populations, irrespective of their European or Asian location.

In total, the data set consists of 360 nucleotide sequences from 483 individuals; there are 279 different sequences and 139 variable sites. Among the many possible analyses of this information we examine here the genetic variability found and its geographic stratification, and use a phylogenetic approach to the analysis of variation. All the genetic results are discussed in relation to the population history framework.

The low genetic variation of Caucasoids

European populations show a low level of diversity as compared to other continents, especially Africa. This has been observed though various parameters in several studies (Bertranpetit *et al.*, 1995; Francalacci *et al.*, in press). A low level of variation may be due either to a recent origin or to a high migration rate. It is practically impossible to compare the migration pattern among Europeans with other groups for prehistoric times. However, it seems unlikely that, in a hunter–gatherer economy, there could be important behavioural differences in geographic places with similar ecosystems. It is therefore highly likely that a relatively recent origin of Europeans (or Caucasoids) caused their low genetic variability.

Despite this relative homogeneity, a geographic pattern of genetic variation can be observed among the populations studied. This can be established through different parameters, one of the most informative of which is the mean and distribution of the pairwise differences among the sequences for each location, which also helps to uncover past population expansions.

The pairwise difference distribution

Rogers and Harpending (1992) and Harpending *et al.* (1993) showed that population expansions generate smooth, bell-shaped pairwise difference distributions, whereas stationary situations produce irregular, multimodal distributions. Moreover, as time after the expansion goes by, the modal value increases; when distributions from recent and ancient populations are displayed together the more ancient distributions appear to the right of the younger ones.

Each of the nine samples analysed shows a smooth, bell-shaped pairwise difference distribution (Fig. 9.2). Moreover, modal values follow a clear

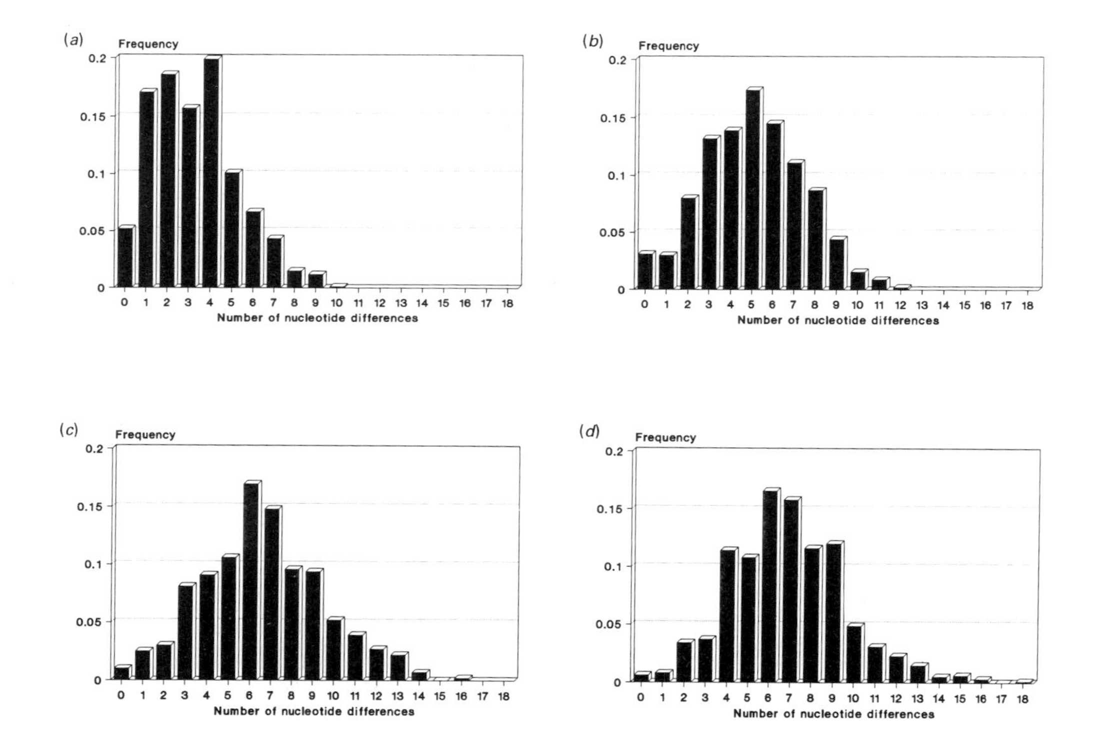

Fig. 9.2. Nucleotide pairwise difference distributions in four European and West Asian populations. A, Basques; B, Tuscans; C, Turks; D, Middle Easterners. Note that the distributions are shifted to the right in the eastern (presumably older) populations.

cline from the high values in the Middle East to a minimum in the Basques. The mean pairwise difference for each population has a highly significant negative correlation with geographical distance from the Middle East ($r = -0.884$, $p < 0.001$). The interpretation is straightforward: there has been a very marked population expansion all over the area analysed, first in the Middle East, and gradually later in populations to the West. This pattern seems compatible with the hypothesis of an expansion from the Middle East.

The timing of these population expansions can be estimated using the parameter tau (τ), related to the position of the peak of the distribution and a function of the time since the expansion in mutation units (Rogers & Harpending, 1992). Interestingly, the τ value for the populations studied decreases, as expected, from east to west, although very slowly: the value for the Basques (2.14) is several times smaller than the Middle East maximum (7.02); other populations present intermediate values. This result means that the expansion was not a rapid event, but took a long time, of the order of several tens of thousands of years.

Building a gene tree

A precise way to understand the genetic complexity, in terms of mutational events, of the 279 different sequences found is through the computation and graphic representation of a distance matrix among them all. Although more complicated distances between each pair of sequences could be computed (Nei, 1987), a 2-parameter model (Kimura, 1980) that takes into account the different probabilities of transitions and transversions seems adequate for our purpose, as Kimura's model gives distance estimates almost proportional to the number of differences between two sequences. Calculations were performed using the program DNADIST from the PHYLIP 3.5c package (Felsenstein, 1989) and the ratio of transition to transversions was set to 10:1, according to available estimations. The result is a very large 279×279 symmetric matrix, which cannot be analysed directly.

In order to visualize the distance matrix, a neighbour-joining tree (Saitou & Nei, 1987) was built. The neighbour-joining algorithm does not assume a constant rate of evolution, and provides unrooted trees (i.e. it does not indicate by itself which sequences are ancestral and which are derived). Contrary to the preferences of other authors (see, for example, Torroni *et al.*, 1993) we do not fold the neighbour-joining tree to conform to a cladogram-like figure; incorrect interpretations may be produced if the graphical display of a neighbour-joining tree is distorted to render it less complex.

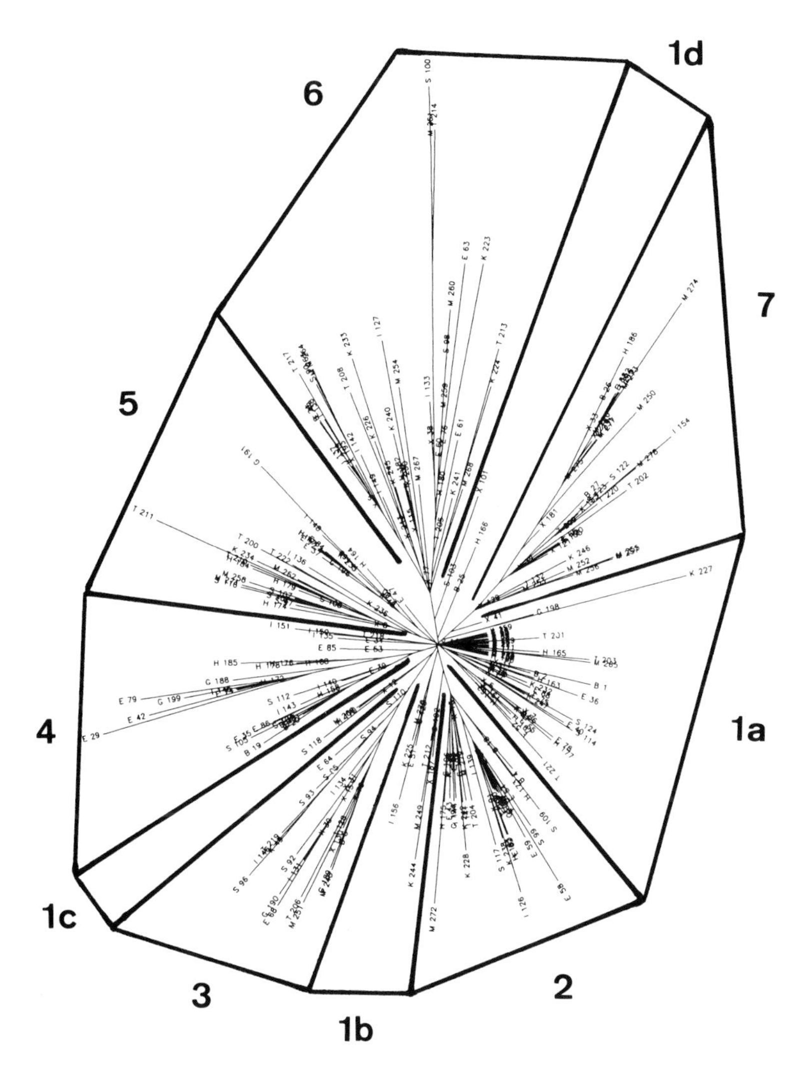

Fig. 9.3. Neighbour-joining tree of 279 Caucasoid sequences. Distances are based on the Kimura 2-parameter model. The main haplogroups detected are shown and are numbered arbitrarily, according to the text.

Haplogroups

The result is shown in Fig. 9.3, a rather intricate representation, where, none the less, it is possible to recognize a certain structure in the different branches into which the tree is organised; each branch represents an haplogroup and consists of several sequences which arise from a presumably ancestral one, which we will call the *stem* of the haplogroup. In turn,

Table 9.1. *Description of the stem sequence for the main seven haplogroups found in European populations*

Haplogroup	Definition	% Caucasoids	Distribution
1	Reference	(a)	Clinal, high in the West
2	C at 16 311	20.1	No pattern; very high in Africa
3	T at 16 294	11.8	No pattern; high in Africa
4	T at 16 270	9.3	Mostly Caucasoid. Clinal, high in the West
5	C at 16 189	20.1	No pattern
6	T at 16 223	17.6	Clinal, high in the East; higher elsewhere
7	C at 16 126	20.8	Exclusively Caucasoid. Clinal, high in the East

(a) as haplogroup 1 is defined by the reference segment, with no specific substitution, no frequency is given.

every stem presents one or a few characteristic mutations which link in to the centre of the tree. Several haplogroups may be easily recognized; the naming of each of them is totally arbitrary; however, the mutations that define the stem sequence are specific to a haplogroup and define it (Table 9.1). We proceed to describe briefly the most important haplogroups.

Haplogroup 1 arises from the centre of the tree, where the reference sequence (e.g. the first mtDNA published, by Anderson *et al.* in 1981) is found. This is the most common sequence in Caucasoids and was reported in all the populations in our sample. No clear structure can be ascertained within this haplogroup as the various sequences are derived from the reference by one or a few, often unique, mutations. Western European sequences (Basques, Sardinians and British) are over-represented in this haplogroup, specially when compared to the Middle Eastern sequences. Subgroups la to ld are only a product of a graphical representation of the tree and do not share common mutations.

A sequence bearing a C (instead of a T) at position 16 311 is the stem of haplogroup 2. This mutation is quite common, and was found in 56 out of 279 different Caucasoid sequences. (20.1%). Its worldwide distribution ranges from complete absence in Northern and Central Amerindians (Ward *et al.*, 1991, 1993; Santos *et al.*, 1994) to high frequencies in Africa, reaching fixation in Pygmies and !Kung (Vigilant *et al.*, 1991). No clear geographic pattern is found within Caucasoids.

Haplogroup 3 is defined by a T at 16 294. This mutation was found in 33 different sequences (11.8%) and shares the distribution pattern of 16 311 C: absence in Amerindians, high frequencies in Africa, lack of a pattern within Europeans.

All sequences in haplogroup 4 share a T at 16 270, which is found in 26 (9.3%) sequences. Save for a few isolated cases (in the Yoruba and in one

Amerindian), which may be recent occurrences, it is a Caucasoid mutation. Its frequency rises clinally from absence in Turkey and the Middle East to a maximum in the Basques. Wakeley (1993) did not consider 16 270 to be a *fast* mutation site. It is likely that this mutation was present in the most recent common ancestral group of modern Europeans.

Haplogroup 5 is defined by a C at 16 189. This site seems to have a high mutation rate, according to Wakeley (1993). 20.1% of European sequences have a C in this position; no geographic structure can be discerned.

Haplogroup 6 is defined by a T at 16 223, which is found in 17.6% of European sequences. The frequency of this mutation declines from high values in the Middle East to a minimum (4.4%) in the Basques. Most other world populations show high 16 223 T frequencies, up to 90% in some African and Amerindian groups.

All sequences in haplogroup 7 share a C at 16 126. This is the site that most often (58/279) differs from the reference sequence in European sequences. Mutation frequency in Caucasoids is clearly clinal, from close to 50% in the Middle East to 6.7% among Basque sequences. Few other occurrences of this substitution have been reported in non-European populations: i) in the Yoruba from Nigeria, but not in other Africans (Vigilant *et al.*, 1991), which may be explained by a recent occurrence, and ii) in the high-caste Havik from SW India (Mountain *et al.*, 1995), but not in the low-caste Mukri. Indian higher castes seem to be descendants of Neolithic, or later Indo-European expansions from the Middle East, while lower castes may have more ancient origins. Thus this substitution is a specific Caucasoid marker: the variation within the haplogroup it defines must have arisen after the ancestors of Caucasoids split from other human groups.

All haplogroup stems differ from the reference sequence in a single transition. In many cases, new transitions from the stem generate, within a haplogroup, second order-haplogroups. The substitutions that define the various haplogroups are the most frequent substitutions in the European populations; high frequencies and the production of new haplotypes are consistent with an ancient origin of these mutations.

Age of the haplogroups

The time since the origin of a given gene tree can be estimated if mutation rates are available. There are several difficulties in this: i) The tree obtained is unrooted and the ancestral sequence may not be known. ii) At the time of colonization, a certain amount of variability could have existed among the initial settlers. In this case, the calculations would yield an upper boundary and the expansion may be more recent. iii) There are many difficulties in the estimation of the mutation rate, and several estimates of the mean mutation rate for hypervariable region I have been published (Vigilant *et al.*, 1991; Ward *et al.*, 1991; Tamura & Nei, 1993), with the highest estimate several times the lowest.

For the evolutionary process of generating new haplotypes through successive mutations, it is possible to calculate the time needed to produce a certain amount of diversity. The number of mutations relative to the original sequence follows a Poisson distribution with a mean number of mutations $\lambda = \mu lt$, where μ is the mutation rate per nucleotide and generation, l is the sequence length (360 nucleotides in this case) and t is the number of generations elapsed since the origin.

From the gene tree obtained, two approaches may be used to evaluate the time depth. On one hand, as the reference sequence occupies a central position it is plausible to assume that it was the ancestor of all other sequences, and thus it is possible to calculate the time depth needed to generate all the variation observed. On the other hand, it is possible to analyse part of the tree, in this case haplogroup 7 (defined by a C at 16 126); this is the only substitution that is almost exclusively European and, moreover, its frequency is clinal, with high values in the Middle East and low for Basques; variation within this group should coalesce after the group from which Caucasoids originated had split from other human populations. The results for both cases are very similar: $\lambda = 3.30$ for the whole tree and $\lambda = 3.40$ for the sequences within haplogroup 7. Both results give a fairly accurate estimate of the time (in mutation units) for the origin of Europeans because the values obtained for the whole tree and for haplogroup 7 are likely to be, respectively, an upper and a lower estimate. The variation in haplogroup 7 is likely to have been produced mainly in the Middle East and have spread to the west into Europe as other analyses of the same data set suggest (Bertranpetit *et al.*, in preparation).

The estimation of age hinges on the mutation rate estimates, which range from 4.14×10^{-6} (Ward *et al.*, 1991) to 1.44×10^{-6} mutations per nucleotide per generation (Vigilant *et al.*, 1991). According to these values the variation found would have accumulated in about 2300 to 6550 generations, which, for a generation time of 20 years would give an age between 46 000 and 130 000 years. Needless to say, it would be valuable to have a more accurate estimate of the mean mutation rate for the mtDNA region analysed. Despite the wide range of values, this result fully supports the replacement hypothesis, as the ancestors of Neanderthals would have control region sequences with a much older coalescence time.

Are there any Neanderthal sequences left?

The values obtained are based on mean number of substitutions in the gene tree assuming a Poisson distribution. In fact, in both cases (the total tree or haplogroup 7) the distribution of the frequency of sequences according to the number of substitutions from the stem sequence (the reference in the first case and with a C at 16 126 in the second) follows a Poisson distribution according to a Kolmogorov–Smirnov goodness of fit test. This result, besides supporting the previous results, may be used to discuss the possible existence of outliers in the distribution.

One of the most controversial issues in the replacement–continuity debate is the possible persistence of Neanderthal features (and genes) after the Upper Palaeolithic transition. The data set analysed here focuses on present-day populations and does not allow us to infer the nature of this transition. The data do, however, allow us to ask whether a Neanderthal sequence has been found in the present set. In fact, it is not known what a Neanderthal control region sequence would look like, but the distribution of differences found agrees on a single and relatively recent event. Neanderthals seem to have been derived *in situ* in Europe and their ancestors could be the first hominids to settle in Europe, at a very early date (Carbonell *et al.*, 1995); in this case the divergence time in relation to groups in other places would be very large and thus high numbers of substitutions would be expected. The mutation distribution, nonetheless, does not show a long tail or even a single case with a high number of substitutions, and the tail frequencies agree totally with a single Poisson distribution. It thus seems unlikely that sequences coalescing at a much earlier time than an upper estimate of 130 000 years are present in our large Caucasoid sample.

Conclusions

The above results should be all interpreted within a multidisciplinary framework, and can be used to confirm or reject hypotheses; they may also lead to new questions for future research. The main conclusions can be summarized in four points:

i) The present results support the hypothesis of a recent origin of modern Europeans and Caucasoids, and are clearly in favour of the replacement model. The time of coalescence calculated, although with a broad interval, is not compatible with an origin at the time of the first hominid settlement of Europe. It seems plausible therefore, given the antiquity of the first colonization, that some kind of population replacement occurred.

ii) It does not seem likely that, within the observed genetic variation observed, there are remains of the first hominid settlers, or, in other words, no *Neanderthal sequences* persist in the gene pool of present populations. From i) and ii) together, we can state that, in some point in time, a complete replacement took place.

iii) The origin of the population that settled Europe seems to be in the Middle East, and is, therefore, in agreement with palaeoanthropological evidence. There are traces in the genes of a population expansion that seem to have originated (at least for the populations analysed here) in the Middle East. The population expansion could easily be related with a population replacement, although an expansion by itself does not entail the disappearance of the previous population; demographic disparity and random processes may have led to their eventual demise. It is impossible to prove when and how the genetic traces of Neanderthals vanished from the European gene pool.

iv) The pace of spread of the new populations is much slower than has been postulated, according to archaeological knowledge, for the cultural changes. The hypothesis of a replacement of the Neanderthals by AMH in relation to the Upper Palaeolithic cultural transition does not seem possible according to the present evidence. The pace of morphological change is not known due to the lack of fossil remains. However, archaeological data point to a rapid spread of the Aurignacian culture, which is not reflected in the mtDNA evidence. We thus postulate a slow process of population replacement not tied directly to the archaeological evidence of cultural changes.

The main genetic characteristics of Europeans seem to have been shaped in Palaeolithic times. Later expansions and migrations, although of some importance (and well established, like that related to the Neolithic expansion), seem to have had only a minor impact on the genetic make-up of present day Caucasoids, at least when the analysis is based in mtDNA sequences. The relative contribution of these later expansions and migrations, none the less, remains an open question.

References

Ammerman, A. J. & Cavalli-Sforza L. L. (1984). *The Neolithic Transition and the Genetics of Populations in Europe*. Princeton, NJ: Princeton University Press.

Anderson, S., Bankier, A. T., Barrell, B. G., De Bruijn, M. H., Coulson, A. R., Sanger, F., Schreier, P. H., Smith, A. J. H., Staden, R. & Young, G. (1981). Sequence and organization of the human mitochondrial genome. *Nature*, **290**, 457–65.

Bar-Yosef, O. (1994). The contribution of Southwest Asia to the study of the origin of modern humans. In *Origins of Anatomically Modern Humans*, ed. M. H. Nitecki and D. V. Nitecki, pp.23–66. New York and London: Plenum Press.

Bertranpetit, J. & Cavalli-Sforza, L. L. (1991). A genetic reconstruction of the history of the population of the Iberian Peninsula. *Annals of Human Genetics*, **55**, 51–67.

Bertranpetit, J., Sala, J., Calafell, F., Underhill, P., Moral, P. & Comas, D. (1995). Human mitochondrial DNA variation and the origin of the Basques. *Annals of Human Genetics*, **59**, 63–81.

Calafell, F. & Bertranpetit, J. (1994). Principal component analysis of gene frequencies and the origin of Basques. *American Journal of Physical Anthropology*, **93**, 201–15.

Calafell, F., Underhill, P., Tolun, A., Angelicheva, D. & Kalaydjieva, L. (1996). From Asia to Europe: mitochondrial DNA sequence variability in Bulgarians and Turks. *Annals of Human Genetics*, **60**, 35–49.

Carbonell, E. Bermudez de Castro, J. M., Arsuaga J. L., Díez, J. C., Rosas, A., Cuenca-Bescós, G., Sala, R., Mosquera, M., & Rodríguez, X. P. (1995). Lower Pleistocene hominids and artefacts from Atapuerca-TD6 (Spain). *Science*, **269**, 826–30.

Cavalli-Sforza, L. L., Menozzi, P. & Piazza, A. (1994). *History and Geography of Human Genes*. Princeton, NJ: Princeton University Press.

Comas, D., Pääbo, S. & Bertranpetit, J. (1995). Heteroplasmy in the control region of human mitochondrial DNA. *Genome Research*, **5**, 89–90.

Comas, D., Calafell, F., Mateu, E., & Bertranpetit, J. Geographic variation in human mitochondrial DNA control region sequence: the population history of Turkey and its relationship to the European populations. Submitted.

Di Rienzo, A., & Wilson, A. C. (1991). Branching pattern in the evolutionary tree for human mitochondrial DNA. *Proceedings of the National Academy of Sciences, USA*, **88**, 1597–601.

Felsenstein, J. (1989). PHYLIP – Phylogeny Inference Package (Version 3.2). *Cladistics*, **5**, 164–6.

Francalacci, P., Bertranpetit, J., Calafell, F. & Underhill, P. (1996). Sequence diversity of the control region of mitochondrial DNA in Tuscany and its implications for the peopling of Europe. *American Journal of Physical Anthropology*, in press.

Frayer, D. (1992). Evolution at the European edge: Neanderthal and Upper Paleolithic relationships. *European Prehistory*, **2**, 9–69.

Gill, P., Ivanov, P. L., Kimpton, C., Piercy, R., Benson, N., Tully, G., Evett, I., Hagelberg, E. & Sullivan, K. (1994). Identification of the remains of the Romanov family by DNA analysis. *Nature Genetics*, **6**, 130–5.

Harpending, H. C., Sherry, S. T., Rogers, A. R. & Stoneking, M. (1993). The genetic structure of ancient human populations. *Current Anthropology*, **34**, 483–96.

Hassan, F. A. (1981). *Demographic Archaeology*. New York: Academic Press.

Kimura, M. (1980). A simple method for estimating evolutionary rate of base substitutions through comparative studies of nucleotide sequences. *Journal of Molecular Evolution*, **16**, 111–20.

Klein, R. G. (1989). *The Human Career*. Chicago: Chicago University Press.

Klein, R. G. (1994). The problem of modern human origins. In *Origins of Anatomically Modern Humans*, ed. M. H. Nitecki and D. V. Nitecki, pp. 3–17.

New York and London: Plenum Press.
Lum, J. K., Rickards, O., Ching, C. & Cann, R. L. (1994). Polynesian mitochondrial DNAs reveal three deep maternal lineage clusters. *Human Biology*, **66**, 567–90.
Mellars, P. (1993). Archaeology and modern human origins in Europe. *Proceedings of the British Academy*, **82**, 1–35.
Morral, N., Bertranpetit, J., Estivill, X. *et al.* (1994). Tracing the origin of the major cystic fibrosis mutation (DeltaF508) in European populations. *Nature Genetics*, **7**, 169–75.
Mountain, J. L., Hebert, J. M., Bhattacharyya, S. S., Underhill, P., Ottolenghi, C., Gadgil, M. & Cavalli-Sforza, L. L. (1995). Demographic history of India and mitochondrial DNA sequence diversity. *American Journal of Human Genetics*, **56**, 979–92.
Nei, M. (1987). *Molecular Evolutionary Genetics*. New York: Columbia University Press.
Piercy, R., Sullivan, K. M., Benson N. & Gill, P. (1993). The application of mitochondrial DNA typing to the study of white Caucasian genetic identification. *International Journal of Legal Medicine*, **106**, 85–90.
Pult, I., Sajantila, A., Simanainen, J., Georgiev, O., Schaffner, W. & Pääbo, S. (1994). Mitochondrial DNA sequences from Switzerland reveal striking homogeneity of European populations. *Biological Chemistry Hoppe-Seyler*, **375**, 837–40.
Renfrew, C. (1987). *Archaeology and Language. The Puzzle of Indoeuropean origins.* London: Jonathan Cape.
Rogers, A. R. & Harpending, H. (1992). Population growth makes waves in the distribution of pairwise genetic differences. *Molecular Biology and Evolution*, **9**, 552–69.
Saitou, N. & Nei, M. (1987). The neighbor-joining method: a new method for reconstructing phylogenetic trees. *Molecular Biology and Evolution*, **4**, 406–25.
Sajantila, A., Lahermo, P., Anttinen, T., Lukka, M., Sistonen, P., Savontaus, M. J., Aula, P., Beckman, L., Tranebjaerg, L., Gedde-Dahl, T., Issel-Tarver, L., Di Rienzo, A. & Pääbo, S. (1995). Genes and languages in Europe: an analysis of mitochondrial lineages. *Genome Research*, **5**, 42–52.
Santos, M., Ward R. H. & Barrantes R. (1994). mtDNA variation in the Chibeha Amerindian Huetar from Costa Rica. *Human Biology*, **66**, 963–77.
Sokal, R. R., Oden, N. L. & Wilson, C. (1991). Genetic evidence for the spread of agriculture in Europe by demic diffusion. *Nature*, **351**, 143–5.
Stoneking, M., Jorde, L. B., Bhatia, K. & Wilson, A.C. (1990). Geographic variation in human mitochondrial DNA from Papua New Guinea. *Genetics*, **124**, 717–33.
Straus, L. G. (1989). Age of modern Europeans. *Nature*, **342**, 476–7.
Straus, L. G. (1994). Upper Palaeolithic origins and radiocarbon calibration: more new evidence from Spain. *Evolutionary Anthropology*, **2**, 195–8.
Stringer, C. B. (1989). The origin of early modern humans: A comparison of the European and non-European evidence. In *The Human Revolution: Behavioural and Biological Perspectives on the Origins of Modern Humans*, ed. P. Mellars and C. Stringer, pp.232–44. Princeton, NJ: Princeton University Press.
Tamura, A., & Nei, M. (1993). Estimation of the number of nucleotide substitutions in the control region of mitochondrial DNA in humans and chimpanzees.

Molecular Biology and Evolution, **10**, 512–23.

Torroni, A., Schurr, T. G., Cabell, M. F., Brown, M. D., Neel, J. V., Larsen, M., Smith, D. G., Vullo, C. M. & Wallace, D. C. (1993). Asian affinities and continental radiation of the four founding Native American mtDNAs. *American Journal of Human Genetics*, **53**, 563–90.

Vigilant, L., Stoneking, M., Harpending, H., Hawkes, K. & Wilson, A. C. (1991). African populations and the evolution of mitochondrial DNA. *Science*, **253**, 1503–7.

Villar, F. (1991). *Los indoeuropeos y los origenes de Europa. Lenguaje e Historia.* Madrid: Editorial Gredos.

Wakeley, J. (1993). Substitution rate variation among sites in the hypervariable region I of human mitochondrial DNA. *Journal of Molecular Evolution*, **37**, 613–23.

Ward, R. H., Frazier, B. L., Dew-Jager, K. & Pääbo, S. (1991). Extensive mitochondrial diversity within a single Amerindian tribe. *Proceedings of the National Academy of Sciences, USA*, **88**, 8720–4.

Ward, R. H., Redd, A., Valencia, D., Frazier, V. & Pääbo, S. (1993). Genetic and linguistic differentiation in the Americas. *Proceedings of the National Academy of Sciences, USA*, **90**, 10663–7.

10 *Palaeolithic and neolithic contributions to the European mitochondrial gene pool*

B. SYKES, H. CÔRTE-REAL AND M. RICHARDS

Introduction

Mitochondrial DNA (mtDNA) variation has been used extensively for examining large-scale questions of human evolution, population structure and demographic history (Cann, Stoneking & Wilson 1987, Vigilant *et al.*, 1991). Recently, however, the ability of mtDNA to resolve the close genetic relationships within relatively homogeneous populations such as the Europeans has been questioned (Bertranpetit *et al.*, 1994; Pult *et al.*, 1994). Using a new phylogenetic method we believe, on the contrary, that it is possible to reveal considerable structure within Europe and, by comparing geographical affinities and divergences between groups of mitochondrial lineages, examine different hypotheses about European colonization.

Mitochondria are maternally inherited and non-recombining, and the effectively haploid genome accumulates mutations faster than nuclear DNA. The most variable region of the mitochondrial genome is the 1122 bp non-coding control region between bp 16024 and 00576 (numbering after Anderson *et al.*, 1981) within which the variation is concentrated in two regions (I and II) (Stoneking *et al.*, 1991). As a population genetics method it differs from allele-frequency based surveys of nuclear encoded variants in several respects: i) the variation is very extensive and is not scrambled by recombination; ii) the effective population size is roughly one quarter that for nuclear variants which enhances the effect of drift; iii) being maternally inherited, only female lineages are relevant; iv) deducing the phylogenetic relationships within and between haplotype clusters is relatively straightforward and allows divergence and expansion times to be estimated.

For the analysis, we have used region I control region sequences from 536 individuals from northern and western Europe and from Turkey, taking the birthplace of the maternal grandmother for geographical

Table 10.1. *Sample sizes, number of different haplotypes and diversity estimates for Europe and the Middle East*

Location	Sample size	Number of haplotypes	Mean pairwise difference ± SE	Simple diversity (*h*) ± SE
Middle East	42	36	7.13 ± 0.09	0.993 ± 0.002
Turkey	22	18	4.29 ± 0.14	0.957 ± 0.023
Finland	29	21	3.48 ± 0.09	0.956 ± 0.018
Switzerland	74	42	3.22 ± 0.03	0.960 ± 0.009
Bavaria	49	35	3.86 ± 0.05	0.981 ± 0.005
North Germany	107	69	4.04 ± 0.03	0.973 ± 0.006
Denmark	33	19	3.32 ± 0.09	0.930 ± 0.021
Iceland	14	12	3.87 ± 0.20	0.962 ± 0.025
Wales	92	45	3.25 ± 0.03	0.926 ± 0.014
Cornwall	69	42	3.66 ± 0.05	0.955 ± 0.012
Sardinia	69	43	4.30 ± 0.06	0.935 ± 0.017
Basques	61	34	2.69 ± 0.04	0.926 ± 0.019
Spain	30	26	4.16 ± 0.09	0.984 ± 0.009
Portugal	30	20	3.96 ± 0.10	0.926 ± 0.028
All (less Middle East)	679	284	3.75 ± 0.01	0.956 ± 0.004

assignment. In addition, 185 published control region sequences from Europe and the Middle East (di Rienzo & Wilson, 1991; Pult *et al.*, 1994) were incorporated into the analysis. A further 100 published sequences from UK Caucasians with unknown maternal origins (Piercy *et al.*, 1993) were used for some comparisons.

Mitochondrial diversity

721 individuals carried a total of 313 different haplotypes of which 233 occurred only once in the sample. The simple diversity and mean pairwise sequence difference estimates in Table 10.1 show the high diversity within all European populations (range 0.93–0.98; average 0.96) but with relatively low values for the mean of the pairwise sequence differences (range 2.66–4.01; average 3.76). As found by others (Bertranpetit *et al.*, 1994), mean pairwise comparisons and genetic distance estimates were also comparatively low between most populations with relatively few interpopulation diversities being statistically significant by a permutation test (Table 10.2) (Hudson, Boos & Kaplan, 1992). Unsurprisingly, there were differences between the Middle East sample and all others in Europe. The only consistently significantly different European population by this test were the Basques who also had the lowest intra-population pairwise diversity (2.66). However, it should be noted that two populations descended from a common ancestor will not necessarily be discriminated

Table 10.2. *Mean pairwise comparisons within populations (on the diagonal) and between populations (above the diagonal). Significance of the inter-populational diversifies tested by the permutation test (** $0.05<p<0.01$; ** $0.01<p<0.001$; *** $p<0.001$). *Genetic distances (Ds) between populations are below the diagonal*

	M.East	Turkey	Finland	Switz.	Bavaria	Germ.	Denm.	Iceland	Wales	Conw.	Sardinia	Basque	Spain	Portugal
M.East	6.76	5.58*	5.33***	5.21***	5.38**	5.50***	5.24***	5.50*	5.23***	5.31***	5.61***	5.09***	5.59*	5.49**
Turkey	0.154	4.09	3.70	3.62	3.92	3.99	3.77*	3.92	3.69	3.88	4.10	3.45**	4.08	4.02
Finland	0.269	0.007	3.32	3.23	3.55	3.59	3.41*	3.58	3.33*	3.52*	3.73	3.06*	3.76*	3.67*
Switz.	0.257	0.007	0.004	3.15	3.45	3.51	3.27*	3.49	3.19	3.40	3.65*	2.94*	3.66	3.55*
Bavaria	0.122	0.005	0.011	0.004	3.75	3.80	3.54	3.77	3.50	3.66	3.92	3.27**	3.93	3.79
Germ.	0.186	0.011	0.001	0.004	0.011	3.87	3.64*	3.85	3.58**	3.75	3.98	3.33**	4.01	3.89
Denm.	0.212	0.087	0.111	0.053	0.024	0.059	3.29	3.54	3.26	3.46	3.76*	3.02*	3.76	3.62
Iceland	0.193	0.051	0.002	0.008	0.032	0.004	0.026	3.85	3.48	3.66	3.98	3.22	3.96	3.82
Wales	0.247	0.039	0.065	0.015	0.023	0.043	0.012	0.052	3.21	3.42	3.69**	2.95	3.70	3.55
Cornw.	0.135	0.037	0.059	0.029	0.015	0.020	0.018	0.055	0.017	3.59	3.85	3.18**	3.86	3.71
Sardinia	0.188	0.012	0.030	0.032	0.004	0.007	0.074	0.011	0.050	0.015	4.08	3.42**	4.11	3.96
Basque	0.380	0.080	0.066	0.034	0.063	0.061	0.048	0.028	0.021	0.054	0.052	2.66	3.45**	3.30*
Spain	0.152	0.015	0.050	0.035	0.004	0.017	0.063	0.015	0.043	0.005	0.014	0.067	4.11	3.98
Portugal	0.201	0.063	0.099	0.070	0.004	0.045	0.066	0.016	0.031	0.007	0.012	0.061	0.019	3.82

by intermatch pairwise comparisons if the branch lengths are the same even if the variants accumulated are different. This helps to explain the lack of discrimination between different European populations using intermatch comparisons and the permutation test reported by Pult *et al.*, 1994.

Phylogenetic relationships

We have used a median network approach to explore the phylogenetic relationships between haplotypes (Bandelt *et al.*, 1995). Networks retain ambiguities caused by character conflicts in the data which would otherwise be resolved arbitrarily into very large numbers of maximum parsimony (MP) trees by inter-specific phylogenetic packages such as PAUP (Swofford, 1993). Fig. 10.1 shows the reduced median network for Europe using our data and that of others (Sardinia: di Rienzo & Wilson, 1991; Switzerland: Pult *et al.*, 1994). Since the entire data set is too large to be displayed in a single diagram we present here only those haplotypes represented more than once in Europe. We then explored unfilled nodes in the network and included single haplotypes which fell onto these nodes. The result is a network with only a few missing intermediates and provides a skeleton with which to examine the general topological features of European mtDNA diversity. Middle Eastern haplotypes (Di Rienzo & Wilson, 1991) which fit into the network have also been included for comparison.

The skeleton illustrates that European populations have certain features in common, which are not shared with the Middle East or other parts of the world (for a network analysis of a number of other data sets see Bandelt *et al.*, 1995). The most immediately obvious is the star-like phylogeny, in which the root haplotype (i.e. that with the greatest number of radiating edges) is the Cambridge Reference Segment (CRS: Anderson *et al.*, 1981). This haplotype is the consensus sequence and also has the highest frequency in all European populations, at around 20%, with the exception of Iceland.

Identification of different lineage groups

Although most haplotypes in this skeleton network are within two mutation events of the CRS, others appear along branches which protrude significantly beyond this limit. In order to examine whether these branches had any meaning, for example, whether they represented diversity in the population prior to the expansion from the CRS or matched other sequences outside Europe, we analysed them as four separate groups with the remaining central sequences denoted as Group I. Two of these groups were indistinguishable geographically from the distribution of group I,

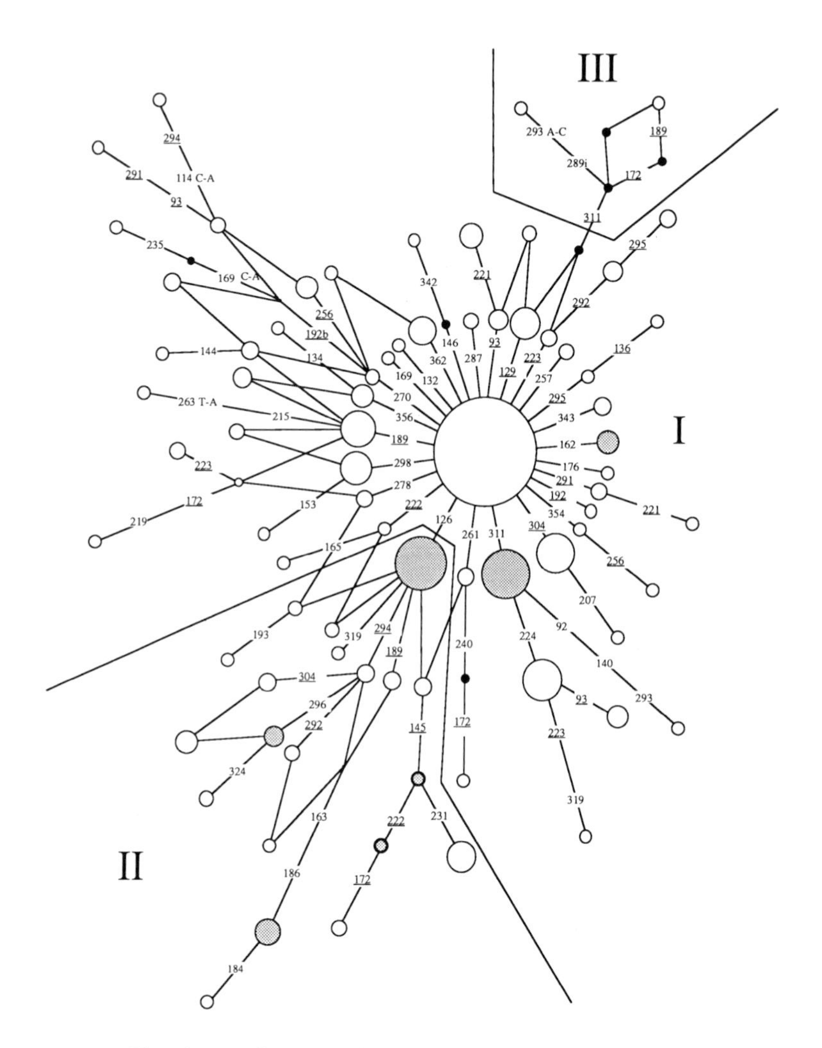

Fig. 10.1. Median network constructed from the 75 haplotypes occurring at a frequency of two or more in Europe (including Turkey). Unfilled circles are haplotypes and their areas are proportional to the number of individuals sharing that haplotype. Filled circles are nodes occupied by haplotypes present only once in the data (including the data of Piercy *et al.*, 1993). Shaded circles are European haplotypes that also occur in the Middle East data of Di Rienzo & Wilson (1991). Shaded circles outlined in bold are intermediate Middle Eastern haplotypes not found in Europe. The branches are evolutionary events which separate the haplotypes and the events are shown as numbers. These numbers correspond to base positions, less 16 000, in the Cambridge Reference Sequence at which events have occurred. Transversions are defined but transitions do not require definition. The single insertion of an adenine 3′ to position 16 289 is denoted 289i. Parallel events at the same position are underlined. Reticulations in the network are an indication of parallel events which cannot be resolved. Notation of the base is only given once in any reticulation but is the same in parallel connections within it. Lines separate the lineage groups as described in the text.

Table 10.3. *Mean pairwise differences (mpd) and estimated divergence time (yrs BP) for lineage groups in Europe and, where they occur, in the Middle East*

Group	Europe[a]		Middle East	
	mpd	div.time[b]	mpd	div. time
I	3.05	32 000	4.80	50 000
II	3.59	38 000	5.47	57 500
IIA	1.07	11 000	3.85	40 400
IIB-W	1.20	12 500		
IIB-E	0.57	6000		
IIC	3.46	36 500	4.00	42 000
III	2.73	29 000	3.03	32 000

[a]Includes Turkey.
[b]Using a divergence rate of 1 per 10 500 years and given to the nearest 500 years.

being restricted to European populations, and were incorporated within it. However, two groups were marked by considerable overlap with Middle Eastern populations, and were denoted as Groups II and III.

Intra-group diversities

We next examined the diversity which has accumulated within these groups (Table 10.3). This can give an estimate of the age of the group in conjunction with an appropriate mutation rate calibration. For this purpose we used a minimum divergence rate of 1 per 10 500 yr for bp 16 092–16 365 in region I of the mtDNA control region, This was calculated from the transversional divergence between humans and chimpanzees (Morin *et al.*, 1994) and using a transversion: transition ratio of 30:1 (Ward *et al.*, 1991). This shows that the ages of both Group I and Group II lineages are much older in the Middle East than in Europe, confirming the westward direction of European settlement, at least for these two groups. Group III had similar ages in both Europe and the Middle East but these estimates are based on low sample sizes. On first inspection, the divergence times of the main groups (I and II) appear to have similar ages within Europe dating back to the early Upper Paleolithic at 30 000 to 35 000 years. However, these would be overestimates if there were pre-existing diversity within the colonizing population.

To evaluate this, we examined the European and Middle East populations for shared haplotypes and nodes since in most cases a haplotype evolves only once and each case then represents an instance of migration. Among Group I lineages there was a conspicuous absence of the root (CRS) haplotype in the Middle East. In contrast, there are a number of separate Group II nodes shared between Europe and the Middle East suggesting that multiple Group II lineages colonised Europe. By comparing nodes shared between the two regions, we distinguished three separate clusters (II A,B,C) rooting to nodes at CRS + 126, CRS + 126, 145, 261 and CRS + 126, 294, respectively. When analysed in this way, IIC has a divergence time in Europe roughly similar to that of Group I while IIA and IIB are much younger in Europe than in the Middle East suggesting they might be later arrivals (Table 10.3).

Current theories on European population structure

Genetic interpretations of European prehistory have been dominated by the pioneering work of Cavalli-Sforza and others (Ammerman & Cavalli-Sforza, 1984; Cavalli-Sforza, Menozzi & Piazza, 1993, 1994; Sokal, Oden & Wilson, 1991). Using extensive accumulated data on the allele frequencies of classical, nuclear-encoded protein polymorphisms, they have compiled synthetic gene maps, which demonstrate geographical clines in allele frequencies. The overall topological similarity between the map produced by the first principal component, which accounts for 28% of the total variance, and an isochronous construct dating the spread of farming from the Middle East led to the formulation of the 'wave of advance' model for the prehistoric settlement of Europe (Ammerman & Cavalli-Sforza, 1984). This is the most severe version of the demic diffusion model in which there is an expansion of people from the Neolithic source population into Europe driven by population growth resulting from agricultural surpluses nd absorbing the less numerous Mesolithic hunter–gatherer populations as it proceeds (Zvelebil, 1986). This model minimizes the role of the indigenous communities of Europe, and an implied consequence is that the majority of modern Europeans are descended from Middle Eastern farmers rather than from the indigenous Palaeolithic population.

Its opposite, indigenous development, proposes that, on the contrary, there was minimal diffusion of peoples from the Middle East but that some of the local hunter–gatherer groups in Europe moved into the Neolithic either independently or as a result of the diffusion of ideas and the trade of crops (Dennell, 1983). The third model, pioneer colonization, assumes some role for migrations out of western Asia into Europe, but sees this in terms of selective colonisation by fairly small groups (Zvelebil, 1986). Here we examine which, if any, of these models best explains the observed distribution of mtDNA lineages in Europe.

The approximate coincidence of divergence times for Groups I (32 000 yr BP) and IIC (32 500 yr BP) within the Upper Palaeolithic makes it much more likely that these lineage groups were brought into Europe, from the Middle East, either by the first anatomically modern humans or perhaps with the spread of the Aurignacian industry. Taken together, they account for 85% of the modern European sample. The divergence time for the, much rarer, Group III is about the same in Europe and the Middle East at about 30 000 yr BP. The European Group III consensus is 129 223 311. Since this is intermediate between an African consensus (which for Pygmies and !Kung is 129 187 189 223 294 311: Vigilant *et al.*, 1991) and the CRS, it is possible that Group III lineages represent descendants of intermediate haplotypes that arose during the evolution of the CRS from a common African ancestor. Interestingly, Papuans also have a very similar consensus [129, 148, 223, 311 (Sykes *et al.*, 1995)] and it is conceivable that the same expansion, perhaps 50 000 yr BP, carried ancestral Group III lineages from Africa both to New Guinea and to Europe.

Colonization during the spread of agriculture between 10 000 and 6000 yr BP seems to be the most likely explanation for the unusual distribution of the IIA and IIB haplogroups (ancient in the Middle East, but young in Europe). Interestingly, different regions of Europe have different frequencies of these haplogroups – about 12% in north and north-central Europe, 16% in north-western Europe, 6% in Sardinia and only about 5% in the Iberian peninsula. The frequency is at its lowest among the Basques at only 2% (this is supported by the data of Bertranpetit *et al.*, 1994 which we have not re-analysed here). This compares with 20% in modern Turkey and 33% in the Middle East where both haplogroups originate. The lack of evidence for a clear northwest–southeast gradient may imply that, as archaeologists have suggested, this movement was much more heterogeneous than the 'wave of advance' model suggests.

The IIB haplogroup is of particular interest with respect to the colonization routes. It is also much more common in the Middle East (16%) than in European populations (0–5%). It forms two clusters in Europe, a western cluster (IIB-W) found in Wales and Cornwall, which retains the variant at position 222 and includes another at 172, and an eastern cluster (IIB-E) found primarily in Switzerland, Germany, Denmark and Finland which lacks the variant at position 222 but has a transition at position 231 (Fig. 10.2). These two clusters have a common ancestor about 23 000 years ago, but separately they have ages of 12 500 and 6000 years. The split between the western and eastern clusters seems to predate the departure from the Middle East. We suggest that this distribution may correspond to the two proposed Neolithic colonization routes through Europe, one through central to northern Europe (IIB-E) and another around the west Mediterranean littoral and across France to Britain (IIB-W), which are linked to the development of the archaeologi-

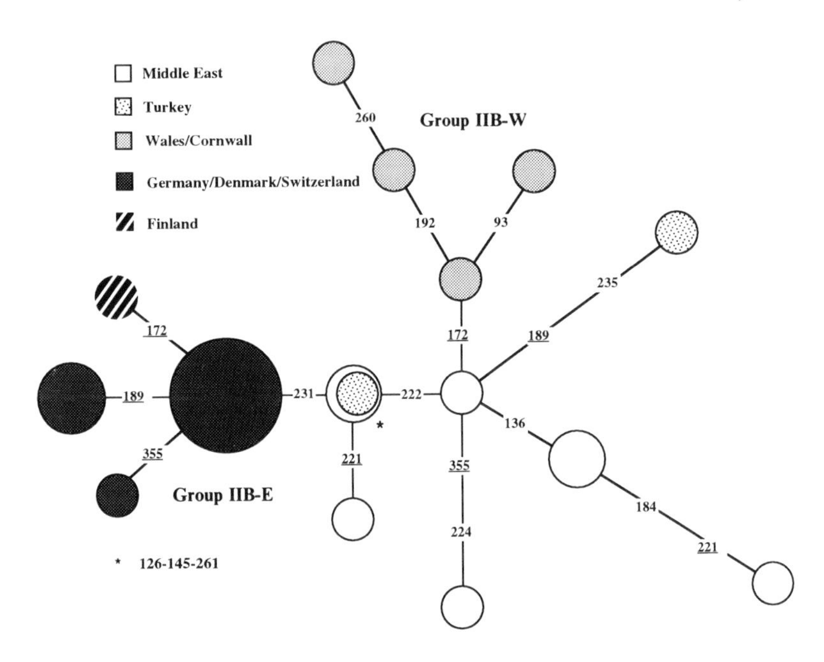

Fig. 10.2. Median network of Group IIB haplotypes in Europe and including the data of Piercy *et al.* (1993). Notation as for Fig. 10.1.

cally defined Linear Ware cultures of central Europe and Impressed Ware cultures in the Atlantic west (Whittle, 1985). IIB haplotypes are absent altogether from the Iberian sample, supporting the view that there was minimal immigration into this region during the Neolithic.

Our interpretation of the mtDNA data is that the majority of modern Europeans are descended from the early settlement of Europe, from the Middle East, during the early Upper Palaeolithic. We see evidence of later colonization from the Middle East which might coincide with the spread of agriculture but its distribution is patchy and the overall demographic influence on modern Europeans is relatively small. Of the three models outlined earlier, our interpretation favours the pioneer colonization model whereby there was selective penetration by fairly small groups of Middle Eastern agriculturalists of a Europe numerically dominated by the descendants of the original Palaeolithic settlements and that the ensuing conversion of this population from a hunter–gatherer–fishing to an agricultural economy was achieved by technology transfer rather than large-scale population replacement.

Overall this work shows that mtDNA control region haplotype analysis, in conjunction with appropriate phylogenetic treatment, is not only capable of revealing population structure and origins on the widest inter-continental scale but also has the resolving power, with adequate sampling, to disentangle even relatively homogeneous populations such as that of Europe.

Acknowledgements

We thank Peter Forster, Andy Demaine, Ann Millward, Debbie Fewster, Surindar Papiha, Christophe Theopold, Risto Pentinnen, Mr and Mrs David Peterson, Antje Arfsten and Aysegul Taylan Ozkan for help in sample collection, Kate Smalley, Susannah Tetzner for technical help, Hans-Jurgen Bandelt, Peter Forster, Vincent Macaulay, Kingsley Micklem and Rosalind Harding for help with computation and phylogenetics analysis; Anna Di Rienzo and Irmgaard Pult for access to unpublished information. This work was supported by grants from the Wellcome Trust, Natural Environment Research Council and the European Union Human Capital and Mobility Network (Biological History of European Populations Contract No:ERBCHBGCT920032).

References

Ammerman, A. J. & Cavalli-Sforza, L. L. (1984). *The Neolithic Transition and the Genetics of Populations in Europe*. New Jersey: Princeton University Press.

Anderson, S., Bankier, A. T., Barrell, B. G., de Bruijn, M. H. L., Coulson, A. R., Drouin, J., Eperon, I. C. *et al.* (1981). Sequence and organisation of the human mitochondrial genome. *Nature*, **290**, 457–65.

Bandelt, H-J., Forster.P., Sykes, B. C. & Richards M. B. (1995). Mitochondrial portraits of human populations using median networks *Genetics*, **141**, 743–53.

Bertranpetit, J., Sala, J., Calafell, F., Underhill P. A., Moral, P. & Comas, D. (1994). Human mitochondrial DNA variation and the origin of Basques. *Annals of Human Genetics*, **59**, 63–81.

Butzer, K. W. & Freeman, L. G. (1988). *Foragers and Farmers: Population Interaction and Agricultural Expansion in Prehistoric Europe*. Chicago: University of Chicago Press.

Cann, R. L., Stoneking, M. & Wilson, A. C. (1987). Mitochondrial DNA and human evolution. *Nature*, **325**, 31–6.

Cavalli-Sforza, L. L., Menozzi, P. & Piazza, A. (1993). Demic expansions and human evolution. *Science*, **259**, 639–46.

Cavalli-Sforza, L. L., Menozzi, P.& Piazza, A. (1994). *The History and Geography of Human Genes*. New Jersey: Princeton University Press.

Dennell, R. (1983). *European Economic Prehistory: A New Approach*. London: Academic Press.

Di Rienzo, A. & Wilson, A. C. (1991). Branching pattern in the evolutionary tree for human mitochondrial DNA. *Proceedings of the National Academy of Sciences, USA*, **88**, 1597–601.

Hudson, R. R., Boos, D. D. & Kaplan, N. L. (1992). A statistical test for detecting geographic subdivision. *Molecular Biology and Evolution*, **9**, 138–51.

Morin, P. A., Moore, J. J., Chakraborty, R., Jin, L., Goodell, J. & Woodruff, D. S. (1994). Kin selection, social structure, gene flow and the evolution of chimpanzees. *Science*, **265**, 1193–201.

Piercy, R., Sullivan, K. M., Benson, N. & Gill, P. (1993). The application of mitochondrial DNA typing to the study of white Caucasian genetic identification. *International Journal of Legal Medicine*, **106**, 85–9.

Pult, I., Sajantila, A., Simanainen, J., Georgiev, O., Schaffner, W. & Pääbo, S. (1994). Mitochondrial DNA sequences from Switzerland reveal striking homogeneity of European populations. *Biological Chemistry Hoppe-Seyler*, **375**, 837–40.

Sokal, R. R., Oden, N. L. & Wilson, C. (1991). Genetic evidence for the spread of agriculture by demic diffusion. *Nature*, **351**, 143–5.

Stoneking, M., Hedgecock, D., Huguchi, R. G., Vigilant, L. & Erlich, H. A. (1991). Geographic variation in human mitochondrial DNA from Papua New Guinea. *American Journal of Human Genetics*, **48**, 370–82.

Swofford, D. L. (1993). *PAUP: Phylogenetic Analysis Using Parsimony, Version* 3.1.1. Champaign, Illinois: Natural History Survey.

Sykes, B. C., Leiboff, A., Low-Beer, J., Tetzner, S. & Richards, M. R. (1995). The origins of the Polynesians:an interpretation from mitochondrial lineage analysis. *American Journal of Human Genetics* **56**, 1463–75.

Vigilant, L., Stoneking, M., Harpending, H., Hawkes, K. & Wilson, A. C. (1991). African populations and the evolution of human mitochondrial DNA. *Science*, **253**, 503–7.

Ward, R. H., Frazier, B. L., Dew-Jager, K. & Pääbo, S. (1991). Extensive mitochondrial diversity within a single Amerindian tribe. *Proceedings of the National Academy of Sciences, USA*, **88**, 8720–4.

Whittle, A. (1985). *Neolithic Europe: A Survey*. Cambridge: Cambridge University Press.

Zvelebil, M. (1986). Postglacial foraging in the forests of Europe. *Scientific American*, **254**, 104.

11 *The molecular diversity of the Niokholo Mandenkalu from Eastern Senegal: an insight into West Africa genetic history*

L. EXCOFFIER, E. S. POLONI,
S. SANTACHIARA-BENERECETTI, O. SEMINO
AND A. LANGANEY

Introduction

The Senegalese Niokholo Mandenka population is located in the Niokholo Hills, 30 km North-West of Kedougou (Fig. 11.1). This population includes some 3000 individuals distributed in a dozen villages constituting an endogamous group with rare exchanges with two surrounding ethnic groups (Fulani and Bedik), and more commonly with other Mandenka populations living along the Gambia river and in the Saraya region (de Montal, 1985). The Niokholo Mandenkalu (Mandenkalu is the plural form of Mandenka) belong to the large Mande linguistic group, including several millions of speakers distributed between Mali and the Atlantic Ocean. The Mande is a subfamily of the large Niger–Congo family spreading from West Africa to Southern Africa. The Mandenkalu are thought to have settled in Eastern Senegal between the fourteenth and sixteenth century during the decline of the Mali Empire, coming from the East. A field trip in January and February 1990 allowed us to collect 203 blood samples in six villages a few kilometres distant from each other. Two hundred lymphoblastoid cell lines were successfully established, providing a continuous source of nuclear and mitochondrial DNA. The DNA polymorphism of several loci, such as α- and β-globin, HLA class I and class II, Y chromosome, mitochondrial DNA (mtDNA), and 80 random nuclear loci was then assessed in collaboration with several Swiss and international laboratories. In this chapter we will summarize the results of the analyses of mtDNA RFLPs and control region DNA sequences, of Y chromosome *Taq* I/p49a,f polymorphisms, and of 80 random nuclear RFLP markers in the Man-

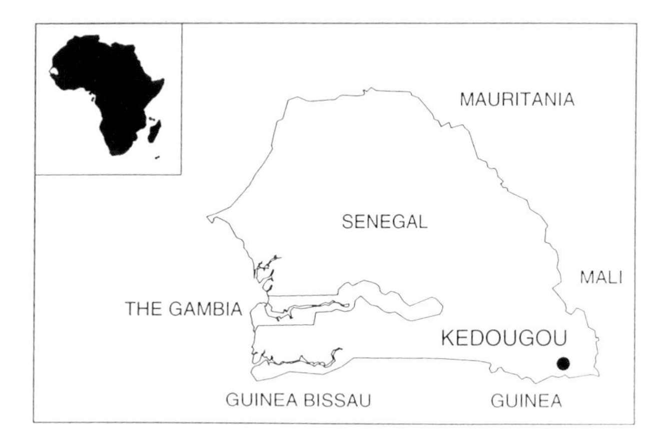

Fig. 11.1. Map of Senegal and location of the Niokholo Mandenka populations.

denka population. We will compare the Niokholo Mandenka's genetic diversity to that observed among other sub-Saharan African and worldwide populations. These results will be discussed in the context of the genetic history of the Niokholo and West African peoples.

Mitochondrial DNA

RFLPs

mtDNA diversity of the Mandenka has been studied for six RFLPs (*Hpa* I, *Hae II*, *Bam* HI, *Msp* I, *Ava* II, and *Hinc* II) in 119 maternally unrelated individuals (Graven *et al.*, 1995). The Mandenka sample presents a very high frequency of morph *Hpa* I-3, mainly found in Africa, and a virtual absence of polymorphism for *Bam* HI and *Msp* I, which is a common situation in Western Africa. Most of the detected polymorphism is due to the enzyme *Ava* II, the most common morph being *Ava* II-3. Two new *Ava* II morphs have been identified (*Ava* II-36 and *Ava* II-37) (Graven *et al.* 1995). At the haplotype level, a total of 12 different haplotypes have been recognized (Table 11.1), among which two low frequency haplotypes (207-2 and 208-2) had not previously been reported. Haplotype 2-2 is the most frequent haplotype (49.2%), followed by haplotype 7-2 (24.2%) and 1-2 (10%).

The molecular diversity and the frequencies of mtDNA RFLP haplotypes defined by 5 enzymes (*Hpa* I, *Bam* HI, *Msp* I, *Ava* II and *Hinc* II) were compared among a set of 54 worldwide populations by computing modified pairwise F_{ST} values (Reynolds, Weir & Cockerham, 1983) using the program AMOVA (Excoffier, Smouse & Quattro, 1992). In Fig. 11.2,

Table 11.1. *Mitochondrial RFLP haplotypes found in the Niokholo Mandenka population*

Haplotype	RFLP Haplotypes[a]	n	Mean nucleotide Diversity[b]	Mean coalescent time (years)[c]
1-2	2.1.1.1.1.2	12	0.0106	65 587 ± 24 827
2-2	3.1.1.1.3.2	59	0.0115	70 333 ± 25 710
7-2	3.1.1.1.1.2	29	0.0089	56 556 ± 23 055
21-2	2.1.1.1.2.2	2	0	9399 ± 9399
39-2	2.1.4.1.1.2	4	0.0045	33 241 ± 17 625
47-2	2.1.1.1.3.2	2	0	9399 ± 9399
52-2	2.1.1.1.17.2	5	0.0025	22 642 ± 14 517
66-2	2.3.1.1.9*.2	1	0	9399 ± 9399
131-2	3.1.2.1.1.2	2	0	9399 ± 9399
207-2	3.1.4.1.3.2	1	0	9399 ± 9399
208-2	3.1.1.1.36.2	2	0	9399 ± 9399

[a]The numbers correspond to enzyme morphs in the following order : *Hpa* I, *Bam* HI, *Hae* II, *Msp* I, *Ava* II, and *Hinc* II.
[b]Computed on associated control region sequences.
[c]The mean coalescence time was computed using a substitution rate of 17.3% per million years (Vigilant *et al.*, 1991) for the control region, assuming humans and chimpanzees diverged 4 million years ago.

we show a multi-dimensional scaling analysis based on these population pairwise genetic distances. For their mitochondrial genome, the Niokholo Mandenka are most closely related to the Senegalese Tukulor and to other Senegalese populations. Three well-differentiated clusters can be seen in Fig. 11.2: a cluster of African populations speaking Niger–Congo languages, another cluster of African populations speaking Khoisan languages, and a third 'Eurasian' cluster grouping all other European, North-African, Asian and Amerindian populations. The cluster differences are mainly due to the predominance of different haplotypes in the population samples belonging to different clusters. Haplotype 2-2 is very common among the Niger–Congo speaking populations, haplotypes 3-2 and 4-2 are very frequent among the Khoisan-speaking populations, whereas haplotype 1-2 is the most common haplotype everywhere else. Note that a Falasha sample from Ethiopia shows intermediate features between the Niger–Congo group and the Eurasian group, as it possesses both haplotypes 1-2 and 2-2 at high frequencies. In Fig. 11.3, we have linked population samples by great-circle lines if they were not statistically different at the 5% level. As can be seen, the Caucasoid population samples are closely related to each other and present a very low, but significant, amount of genetic differentiation ($F_{ST} = 0.059$, $P < 0.001$) compared to

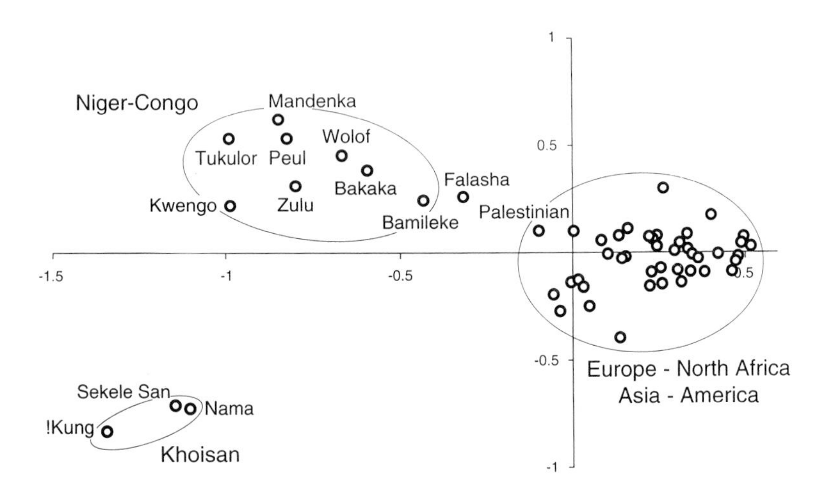

Fig. 11.2. Multi-dimensional scaling analysis of 54 populations samples analysed for mtDNA RFLP haplotypes.

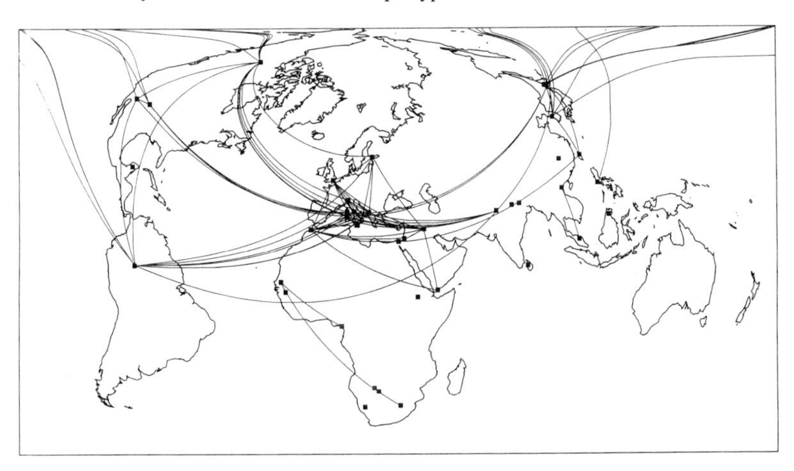

Fig. 11.3. Location and genetic affinities of 54 populations samples analysed for mtDNA RFLP haplotypes. Two samples are connected by great-circle lines if they are genetically not significantly different at the 5% level.

sub-Saharan Africans ($F_{ST} = 0.312$, $P < 0.001$). Surprisingly, Amerindian population do not appear significantly different from several European and Asian populations. This may be due to their low internal genetic diversity ($F_{ST} = 0.049$, $P < 0.004$) and the very high frequencies of haplotype 1-2 (generally $> 90\%$) in these populations. Note also that a Japanese sample is not significantly different from a Hindu sample of New Dehli and from several European samples. These unexpected long-distance genetic relationships are in agreement with the observation in Eurasian populations of an mtDNA diversity significantly lower than that expected under neutrality and population equilibrium (Excoffier, 1990).

Control region DNA sequence polymorphism

Two hypervariable segments (HVS) of the control region have been sequenced for the same 119 individuals already analysed for mtDNA RFLPs (Graven *et al.*, 1995). An average of 727 nucleotides have been sequenced for each individual. Ninety one polymorphic sites have been found in the sample (58 for HVS I and 33 for HVS II). The distribution of these polymorphic sites is found to be significantly non-random (Graven *et al.*, 1993), suggesting variable levels of functional constraint on this non-coding region or the involvement of the secondary structure of the control region in the mutation pattern. Combining sequence information from the two HVS, a total of 60 different sequences were found (Graven *et al.*, 1995). The average nucleotide diversity ($\hat{\theta}$) combined over the two regions is equal to 0.0196, comparable to that found in the East Pygmies ($\hat{\theta} = 0.0177$), but significantly larger than that found in previously studied samples (Graven *et al.* 1995). In the Mandenka population, HVS II ($\hat{\theta} = 0.0176$) was nearly as diversified as is HVS I ($\hat{\theta} = 0.0218$), whereas in previously studied populations HVS I is generally much more polymorphic than HVS II (Graven *et al.*, 1995; Vigilant *et al.*, 1991).

Comparison between mtDNA RFLPs and D-Loop sequences

The pattern of polymorphism for mtDNA RFLPs (at low resolution for the whole mtDNA molecule) and control region sequences (high resolution for a small fraction of the mtDNA genome) has been compared in 119 Mandenka individuals. The neighbour-joining tree (Saitou & Nei, 1987) in Fig. 11.5 shows that control region sequences associated with the restriction haplotypes 2-2 and 7-2 are evolutionarily closely related and form, with a single exception, monophyletic groups. The previously inferred phylogenetic relationships among common African RFLP haplotypes (haplotype 7-2 being intermediate between haplotypes 1-2 and 2-2, Johnson *et al.*, 1983) should nevertheless be revised in view of the sequence information, as haplotype 7-2 appears to be derived from haplotype 2-2. This interpretation is in agreement with the estimation of the mean coalescent times of sequences associated with given RFLP haplotypes shown in Table 11.1. In the Mandenka population, haplotype 2-2 appears to be slightly older ($70\,333 \pm 25\,710$ years) than haplotype 1-2 ($65\,587 \pm 24\,827$ years), and considerably older than haplotype 7-2 ($56\,556 \pm 23\,055$ years). An analysis of molecular variance (AMOVA)

(Excoffier *et al.*, 1992) was performed on the 60 control region sequences, using the RFLP haplotypes as the defining structure to test. This analysis reveals that 59.6% of the total molecular variance of the control region sequences is due to differences among sequences associated with different restriction haplotypes, whereas the rest is due to differences among sequences associated with the same RFLP haplotype. The association between control-region sequences and RFLP haplotypes is shown to be highly significant ($P < 0.002$) by the permutation test described in Excoffier *et al.* (1992). This association suggests that the use of mtDNA low resolution RFLPs is perfectly valid for studies on human evolution. The use of RFLP haplotype information is also helpful in structuring and understanding the sequence differentiation pattern. It has allowed us to organize the sequence molecular phylogeny in a meaningful way. This approach should prove useful when comparing control region sequences drawn from different populations.

Y chromosome

The *Taq* I polymorphism detected by the three Y chromosome probes p12f2, p49a and p49f has been examined for 64 independent Niokholo Mandenka males (Poloni *et al.*, submitted). The Mandenka population seems monomorphic for probe p12f2 and presents low levels of polymorphism for probes p49a and p49f. For these latter two probes, bands C0 and D0 are associated in all tested individuals. The association of bands A1C0D0 is very common (84%), a high frequency already found in other Senegalese populations (75%) (Torroni *et al.*, 1990). Haplotype 4 (A1C0D0F1I1) is the most frequent haplotype (69%) among the 12 identified Mandenka haplotypes (Table 11.2), in keeping with the situation prevailing in other Senegalese, West African, and Southern African populations. The G band is absent in approximately 8% of the individuals, a situation which is rarely found in other populations with the exception of some African populations.

Y chromosome p49a,f/*Taq* I frequencies have been compared among 46 populations. Pairwise genetic distances (Reynolds *et al.*, 1983) were computed and taken as input for a multidimensional scaling analysis. This multivariate analysis (Fig. 11.4) reveals close affinities between the Mandenka population and other Niger–Congo speaking populations. The Mandenkalu are found to be very similar to the Bamileke from Cameroon, the Wolof from Senegal and to several other Niger–Congo populations from West and Southern Africa. A majority (83/153) of pairwise genetic distances among the Niger–Congo populations were not significantly different from zero. It results in a lower level of internal diversity

Table 11.2. *Y chromosome p49a,f/Taq I haplotype frequencies in the Niokholo Mandenka sample*

	Fragments					
Haplotype[a]	A	C	D	F	I	*n*
3	1	0	0	1	0	1
4	1	0	0	1	1	44
5	2	0	0	1	1	2
11	3	0	0	1	1	1
25	0	0	0	0	1	2
27	1	0	0	0	1	5
28[b]	1	0	0	1	0	4
30	2	0	0	1	0	1
55	4	0	0	1	1	1
116[c]	6	0	0	1	1	1
117[c,d]	2	0	0	1	0	1
118[b,c]	3	0	0	1	0	1

[a]Nomenclature follows the numbering order adopted in our laboratory (see Poloni *et al.* submitted).
[b]Lacks band G.
[c]First described in this sample.
[d]Has an additional band of about 2.5 kb in the region of the I fragments.

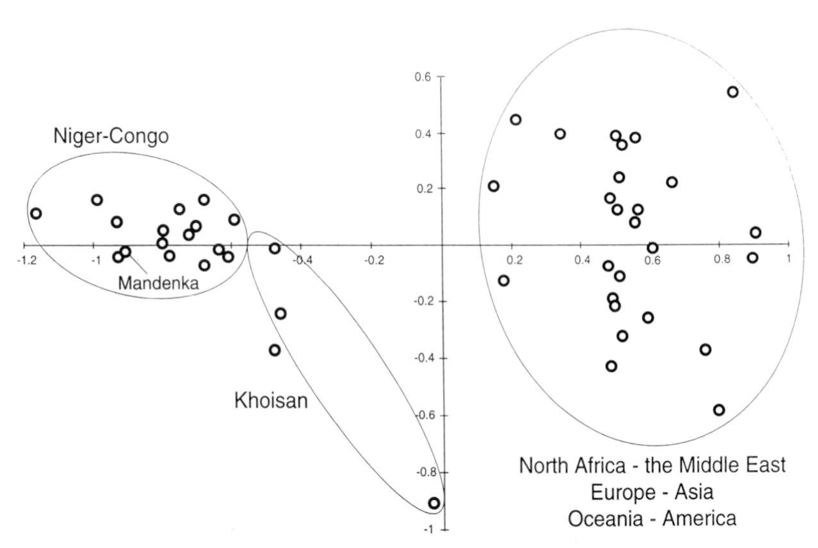

Fig. 11.4. Multi-dimensional scaling analysis of 46 populations samples analysed for *Taq* I/p49a,f polymorphism.

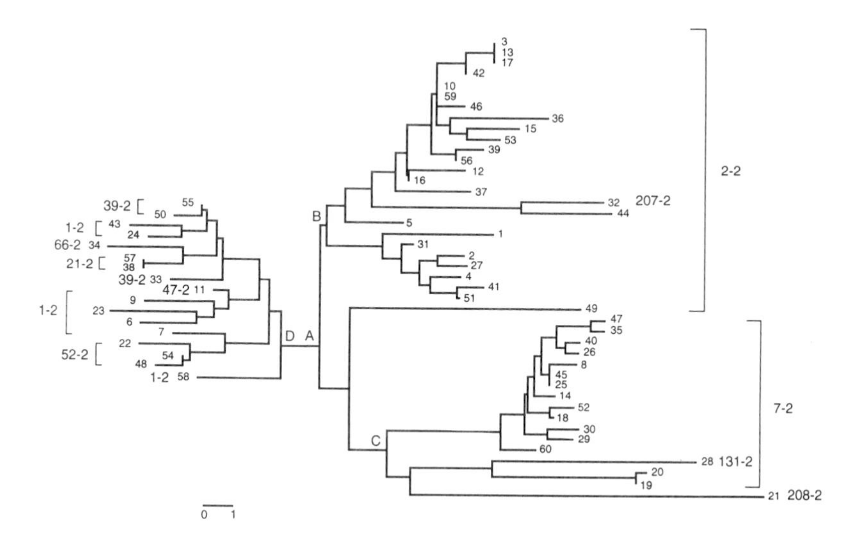

Fig. 11.5. Neighbour-joining tree of 60 control region sequences found among 119 Niokholo Mandenka individuals. All sequences have been analysed for the same 564 nucleotides. The RFLP haplotypes associated with given sequences are shown outside brackets.

($F_{ST} = 0.106$; $P < 0.001$) for this geographically widespread group compared to that observed in the less widespread Khoisan group ($F_{ST} = 0.144$; $P < 0.001$). The Khoisan speaking populations are found to be statistically different from each other and from the Niger–Congo populations. A hierarchical AMOVA (Excoffier *et al.*, 1992) shows that the Khoisan group as a whole is significantly different from the Niger–Congo group ($F_{CT} = 0.137$; $P < 0.001$). No distinction can be made between the Oriental and the Caucasoid populations embedded in the Eurasian group shown in Fig. 11.5. Globally, the Oriental populations are indeed not found to be significantly different from the Caucasoid populations ($F_{ST} = 0.013$; NS), even though there is significant genetic structure at the population level in the Eurasian cluster ($F_{CT} = 0.132$; $P < 0.001$). Note that these Eurasian populations are well differentiated from all sub-Saharan African populations, with the exception of an Egyptian sample which shows no statistical difference from an Ethiopian Amharic sample (Poloni *et al.* submitted). Interestingly, this Ethiopian sample has a haplotypic configuration more similar to that seen in Caucasoid populations than in Western and Southern Africa, and thus clusters with the Eurasian group.

Random nuclear RFLPs

A total of 80 random RFLP markers were tested on an average of 70 unrelated Mandenka individuals (Poloni *et al.*, 1995). The average

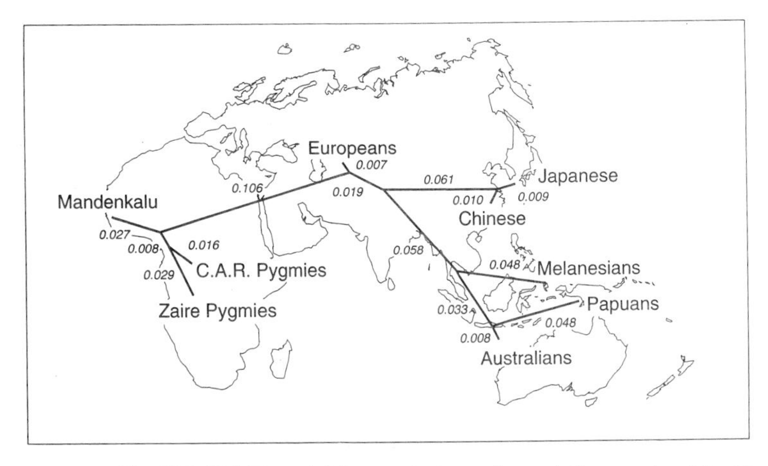

Fig. 11.6. Neighbour-joining tree between 9 populations analysed for 78 random nuclear markers (after Poloni *et al.*, 1995).

heterozygosity level for these markers is 0.306, a value comparable to those reported in previously analysed population samples (Bowcock *et al.*, 1991*a*, *b*). Comparison of the Mandenka nuclear diversity with that of eight other populations revealed that 77 out of 78 tested markers led to a significant amount of population structure as measured by *F*-statistics. The average F_{ST} value on these 78 markers is 0.148. Although only three African samples have been analysed for this set of nuclear markers, the Mandenkalu are found to be genetically close to, but significantly different from, two Pygmy samples, forming a cluster on the neighbour-joining tree shown in Fig. 11.6. These three African populations are well differentiated from a European sample and even more from other Oriental and Oceanian populations. Interestingly, an excellent agreement is found between geographical (great-circle) and genetic distances at a worldwide scale ($r = 0.62$, $P < 0.001$), a relationship easily seen on Fig. 11.6. This result confirms that geography has been a very important factor in differentiating human populations.

Discussion

The small Mandenka population presents high levels of genetic diversity for mtDNA, but also for other nuclear markers such as α-globin (Martinson *et al.* 1995), HLA class I (Dard *et al.* 1992*b*) and HLA class II (Tiercy *et al.*, 1992). This large amount of polymorphism suggests that the Mandenka ancestors have not passed through a bottleneck when migrating to the Niokholo. The effective female population size (about 5700 women, Graven *et al.*, 1995) of the Niokholo Mandenkalu is apparently larger than their total census size (3000 individuals), implying that they have maintained some gene flow with other surrounding populations. The ancient estimated age of several mtDNA RFLPs (> 50 000 years, Table 11.1) suggests that most of the Mandenkalu genetic diversity was established before they migrated to the Niokholo region 400–500 hundred years ago. It thus appears likely that the Niokholo Mandenka's gene pool is a representative sample of broader West African genetic diversity and that our conclusions concerning the time depth of the mitochondrial differentiation process also apply to West Africa if not to the whole Niger–Congo group.

Both mitochondrial and Y chromosome markers show that the Niokholo Mandenkalu are indeed genetically similar to other populations from West Africa and from the Niger–Congo linguistic group (Fig. 11.2 and 11.4), in keeping with earlier observations based on conventional genetic markers (Blanc *et al.*, 1990; Excoffier *et al.*, 1987, 1991). Although its homeland is unknown, linguistic evidence suggests that the proto Niger-Congo language began its differentiation some 8000 years ago (Ehret, 1984; Greenberg, 1964), and that people speaking Niger–Congo languages were already present in West Africa six to seven thousand years ago (Ehret, 1984). Such a recent common origin and a subsequent expansion of West African and Niger–Congo languages could explain both the very close genetic affinities seen among these people for Y chromosome polymorphism and the similarities observed in their mitochondrial gene pool. The genetic differentiation process is nevertheless much older than the linguistic differentiation process in West Africa (Table 11.1). It implies that most of the present genetic polymorphism has been created before the expansion of the Niger–Congo speakers and has been carried over by migrants probably under a form of demic diffusion.

As previously recognized for non-molecular markers (Blanc *et al.*, 1990; Dard *et al.*, 1992*a*; Excoffier *et al.*, 1987, 1991), the Mandenkalu and other Niger-Congo speakers appear clearly divergent not only from Khoisan speakers but also from East Africans, even though genetic data are still limited in this oriental region of Africa. We thus have to recognize in sub-Saharan Africa the existence of several distinct genetic subgroups closely corresponding to those defined by linguistics (Excoffier *et al.*, 1991), and the concept of a single African genetic entity has to be abandoned. This is not to say that African populations do not share a common ancestry, but that African samples belonging to different linguistic families cannot be pooled as a single group when being genetically compared to other population groups. This seems especially true for mtDNA RFLPs, where there is no simple opposition between an African group and the rest of the world (Cann, Stoneking & Wilson, 1987; Vigilant *et al.*, 1991), but where Niger–Congo and Khoisan clusters appear as different from each other as they are from the rest of the world (Fig. 11.2). Due to the close association observed between mtDNA RFLP and sequence polymorphism in the Mandenka population, we postulate that the same three clusters should generally be observed for the control region DNA sequence when data become available in additional populations.

Although it is always difficult to compare two different sets of polymorphisms assayed by different techniques on different locations of the genome, it appears valuable to relate strictly maternally (mtDNA) to strictly paternally (Y chromosome) transmitted polymorphisms. Although the analysis of these two polymorphisms leads to essentially identical population clusters (Fig. 11.2 and 11.4), they lead to very different conclusions about the amount of genetic variance present in these clusters. We would, of course, expect the mitochondrial genome to be much more variable than the Y chromosome, due to the high mtDNA substitution rate (Brown *et al.*, 1982). However, Eurasian populations appear more heterogeneous than Niger–Congo populations for the Y chromosome (Fig. 11.4), but show little differentiation for mtDNA RFLPs (Fig. 11.2) or DNA sequences (Bertranpetit *et al.*, 1995; Di Rienzo & Wilson, 1991). This contrast between mtDNA and Y chromosome evolution is difficult to explain by a mere mutation rate difference. Several genetic and demographic factors have been advanced to account for these differences. A recent demographic expansion has been proposed to explain lower mtDNA diversity in Eurasia (Harpending *et al.*, 1993; Rogers & Harpending, 1992; Slatkin & Hudson, 1991), and the same explanation could be advanced for the Y chromosome in Niger–Congo populations. However, the effect of demographic expansions should be reflected in both the mitochondrial and nuclear genomes, and it seems difficult to understand the different and antagonistic behaviour of mtDNA and Y chromosome in the Niger–Congo population. Note that on the basis of mtDNA evidence, the Mandenkalu do not seem to have gone through a large and recent demographic expansion phase (Graven *et al.*, 1995), but the study of additional populations and additional loci will be necessary before this conclusion can be extended to the whole Niger–Congo group. Other differentiating factors which affect males or females differently or which act specifically on the mitochondrial or the Y chromosome genomes could in principle also be involved. Among such factors, some unknown and probably different selective processes could account for mtDNA low diversity in Eurasia (Excoffier, 1990; Slatkin & Hudson, 1991), and low Y chromosome diversity among Niger–Congo speakers. Alternatively, social behaviour such as polygamy resulting in increased variance of male progeny size and thus in reduced male effective population size (Kimura and Crow, 1963) could account for the reduced genetic diversity of the Y chromosome in

Niger–Congo populations. Note, however, that this latter effect would be transient: larger genetic variance should ultimately result because of increased genetic drift in smaller populations. Finally, unequal migration rates between males and females could also lead to different patterns in strictly maternally or paternally transmitted polymorphisms. In this respect, note that most of the incoming gene flow is due to women in the patrilocal Mandenka population, as spouses are quite regularly brought in from neighbouring populations (Lalouel & Langaney, 1976). All these factors may have contributed to the divergent mitochondrial and Y chromosome diversity patterns. One cannot even dismiss the possibility that these differences could have resulted from a purely stochastic process. Nevertheless, additional molecular studies of sub-Saharan African populations are thus clearly required. Mitochondrial DNA sequence (Graven *et al.*, 1995; Vigilant *et al.*, 1989) and new Y chromosome polymorphisms (Mathias, Bayés & Tyler-Smith, 1994; Spurdle, Hammer & Jenkins, 1994 and see Chapter 2) will soon provide the very detailed genetic information needed to assess the nature of the differences between the evolution of female and male lineages in Africa.

Nuclear DNA studies provide yet another and certainly more balanced picture of human diversity. They reveal divergence of tested African populations, but they seem to have much more resolving power in separating European populations from Oriental populations. They also allow distinct clusters within Asia to be recognized (Fig. 11.6), a distinction which seems impossible on the basis of either mtDNA or Y chromosomes RFLPs. The high correlation observed between geographic and genetic distances ($r = 0.62$) computed on many random nuclear autosomal loci (Fig. 11.6) underlines the important role of geographical barriers in controlling levels of gene flow. In this view, part of the African divergence can certainly be attributed to the relative isolation of sub-Saharan Africa, but the size of this effect remains to be quantified.

Acknowledgement

We thank the Mandenka people from the Niokholo, and more particularly the people from Bantata, Banyon, Barraboy, Batranké, Tikankali and Tenkoto without whom this study would not have been possible. We thank Sékhou Camara for his continuous assistance and Dr. Alain Epelboin for his medical expertise during the fieldwork. We are grateful to the Senegalese authorities for their research authorizations and their co-operation. This work was supported by Swiss NSF grants No 32-28784.90 and 31-39847.93.

References

Bertranpetit, J., Sala, J., Calafell, F., Underhill, P. A., Moral, P. & Comas, D. (1995). Human mitochondrial DNA variation and the origin of Basques. *Annals of Human Genetics*, **59**, 63–81.

Blanc, M., Sanchez-Mazas, A., Hubert van Blyenburgh, N., Sevin, A., Pison, G. & Langaney, A. (1990). Inter-ethnic genetic differentiation: Gm polymorphism in eastern Senegal. *American Journal of Human Genetics*, **46**, 383–92.

Bowcock, A. M., Hebert, J. M., Mountain, J. L., Kidd, J. R., Rogers, J., Kidd, K. K. & Cavalli-Sforza, L. L. (1991*a*). Study of an additional 58 DNA markers in five human populations from four continents. *Gene Geography*, **5**, 151–73.

Bowcock, A. M., Kidd, J. R., Mountain, J. L., Hebert, J. M., Carotenuto, L., Kidd, K. K. & Cavalli-Sforza, L. L. (1991*b*). Drift, admixture, and selection in human evolution: a study with DNA polymorphism. *Proceedings of the National Academy of Sciences, USA*, **88**, 839–43.

Brown, W. M., Prager, E. M., Wang, A. & Wilson, A. C. (1982). Mitochondrial DNA sequences of primates: tempo and mode of evolution. *Journal of Molecular Evolution*, **18**, 225–39.

Cann, R. L., Stoneking, M. & Wilson, A. C. (1987). Mitochondrial DNA and human evolution. *Nature*, **325**, 31–6.

Dard, P., Sanchez-Mazas, A., Tiercy, J-. M., Magzoub, M. M. A., Verduyn, W., Goumaz, C., Jeannet, M., Pellegrini, B., Excoffier, L. & Langaney, A. (1992*a*). Joint Report : HLA-A, -B, and DR differentiations among North and West African populations. In *HLA* 1991; *Proceedings of the Eleventh International Workshop and Conference*, ed. K. Tsuji, M. Aizawa and T. Sasazuki, pp. 632–36. Oxford: Oxford University Press.

Dard, P., Schreiber, Y., Excoffier, L., Sanchez-Mazas, A., Shi-Isaac, X. W., Epelbouin, A., Langaney, A. & Jeannet, M. (1992*b*). Polymorphisme des loci de classe I : HLA-A, -B, et -C, dans la population Mandenka du Sénégal Oriental. *Comptes Rendus l'Académie des Sciences Serie III - Sciences de la Vie*, **314**, 573–8.

de Montal (1985). Les Malinké. In *Les Habitants du Département de Kédougou Sénégal*, Anonymous, pp. 56–64. Paris: Centre de Recherches Anthropologiques du Musée de l'Homme.

Di Rienzo, A. & Wilson, A. C. (1991). Branching pattern in the evolutionary tree for human mitochondrial DNA. *Proceedings of the National Academy of Sciences, USA*, **88**, 1597–1601.

Ehret, C. (1984). Historical/linguistic evidence for early African food productions. In *Hunters to Farmers. The Causes and Consequences of Food Production in Africa*, ed. J. D. Clark and S. A. Brandt, pp. 26–35. Berkeley: University of California Press.

Excoffier, L. (1990). Evolution of human mitochondrial DNA: evidence for departures from a pure neutral model of populations at equilibrium. *Journal of Molecular Evolution*, **30**, 125–39.

Excoffier, L., Harding, R. M., Sokal, R. R., Pellegrini, B. & Sanchez-Mazas, A. (1991). Spatial differentiation of RH and GM haplotype frequencies in sub-Saharan Africa and its relation to linguistic affinities. *Human Biology*, **63**, 273–97.

Excoffier, L., Smouse, P. & Quattro, J. (1992). Analysis of molecular variance inferred from metric distances among DNA haplotypes: application to human mitochondrial DNA restriction data. *Genetics*, **131**, 479–91.

Excoffier, L., Pellegrini, B., Sanchez-Mazas, A., Simon, C. & Langaney, A. (1987). Genetics and history of Sub-Saharan Africa. *Yearbook of Physical Anthropology*, **30**, 151–94.

Graven *et al.* (1993). High level of diversity in the mtDNA control region of the Senegalese Mandenkalu. Seventeenth International Congress of Genetics. Birmingham, UK, p. 158.

Graven, L., Passarino, G., Semino, O., Boursot, P., Santachiara-Benerecetti, A. S., Langaney, A. & Excoffier, L. (1995). Evolutionary correlation between control region sequence and RFLP diversity pattern in the mitochondrial genome of a Senegalese sample. *Molecular Biology and Evolution*, **12**, 334–45.

Greenberg, J. H. (1964). Historical inferences from linguistic research in sub-Saharan Africa. In *Boston University Papers in African History I*, ed. J. Butler, pp. 1–15. Boston: Boston University.

Harpending, H., Sherry, S. T., Rogers, A. & Stoneking, M. (1993). The genetic structure of ancient human populations. *Current Anthropology*, **34**, 483–96.

Johnson, M. J., Wallace, D. C., Ferris, S. D., Rattazzi, M. C. & Cavalli-Sforza, L. L. (1983). Radiation of human mitochondrial DNA types analyzed by restriction endonuclease cleavage patterns. *Journal of Molecular Evolution*, **19**, 255–71.

Kimura, M. & Crow, J. F. (1963). The measurement of effective population number. *Evolution*, **17**, 279–88.

Lalouel, J. M. & Langaney, A. (1976). Bedik and Niokholonko of Senegal: inter-village relationship inferred from migration data. *American Journal of Human Genetics*, **45**, 453–66.

Martinson, J. J., Swinburn, C., Boyce, A. J., Harding, R. M., Langaney, A., Excoffier, L. & Clegg, J. B. (1995). High diversity of alpha-globin haplotypes in a Senegalese population, including many previously-unreported variants. *American Journal of Human Genetics*, **57**, 1186–98.

Mathias, N., Bayés, M. & Tyler-Smith, C. (1994). Highly informative compound haplotypes for the human Y chromosome. *Human Molecular Genetics*, **3**, 115–23.

Poloni, E. S., Excoffier, L., Mountain, J. L., Langaney, A. & Cavalli-Sforza, L. L. (1995). Nuclear DNA polymorphism in a Mandenka population from Senegal: comparison with eight other human populations. *Annals of Human Genetics*, **59**, 43–61.

Poloni, E. S., Passarino, G., Santachiara-Benerecetti, A. S., Semino, O., Langaney, A. & Excoffier, L. (submitted). Human population structure as revealed by a Y chromosome specific polymorphism.

Reynolds, J., Weir, B. & Cockerham, C. (1983). Estimation of the co-ancestry coefficient: basis for a short-term genetic distance. *Genetics*, **105**, 767–79.

Rogers, A. R. & Harpending, H. (1992). Population growth makes waves in the distribution of pairwise genetic differences. *Molecular Biology and Evolution*, **9**, 552–69.

Saitou, N. & Nei, M. (1987). The neighbor-joining method: a new method for reconstructing phylogenetic trees. *Molecular Biology and Evolution*, **4**, 406–25.

Slatkin, M. & Hudson, R. R. (1991). Pairwise comparisons of mitochondrial DNA sequences in stable and exponentially growing populations. *Genetics*, **129**, 555–62.

Spurdle, A. B., Hammer, M. F. & Jenkins, T. (1994). The Y *Alu* polymorphism in

Southern African populations and its relationship to other Y-specific polymorphisms. *American Journal of Human Genetics*, **54**, 319–30.

Tiercy, J. -M., Sanchez-Mazas, A., Excoffier, L., Xi-Owen, S. -I., Jeannet, M., Mach, B. & Langaney, A. (1992). HLA-DR polymorphism in a Senegalese Mandenka population: DNA oligotyping and population genetics of DRB1 specificities. *American Journal of Human Genetics*, **51**, 592–608.

Torroni, A., Semino, O., Scozzari, R., Sirugo, G., Spedini, G., Abbas, N., Fellous, M. & Santachiara-Benerecetti, A. S. (1990). Y chromosome DNA polymorphisms in human populations: differences between Caucasoids and Africans detected by 49a and 49f probes. *Annals of Human Genetics*, **54**, 287–96.

Vigilant, L., Pennington, R., Harpending, H., Kocher, H. & Wilson, A. C. (1989). Mitochondrial DNA sequences in single hairs from a southern African population. *Proceedings of the National Academy of Sciences, USA*, **86**, 9350–4.

Vigilant, L., Stoneking, M., Harpending, H., Hawkes, K. & Wilson, A. C. (1991). African populations and the evolution of mitochondrial DNA. *Science*, **253**, 1503–7.

12 *The peopling of Madagascar*

H. SOODYALL, T. JENKINS, R. HEWITT, A. KRAUSE AND M. STONEKING

Introduction

Madagascar, located in the western Indian Ocean between latitude 12 and 26 degrees south and longitude 43 and 47 degrees east, is one of the last major islands to have been colonized by humans. It is separated by approximately 400 km from the Mozambique coastline of Africa to the west and by almost 6400 km from Indonesia to the east. The landscape and climate vary due to changes in altitude and latitude, and at least five major ecological regions referred to as Western, Southwestern, Central Highland, Eastern Central and Northern may be distinguished (Fig. 12.1). A chain of mountains which runs from north to south makes up the Central Highland region. These mountains descend sharply to the Indian Ocean creating a narrow eastern part and a wider western part. Madagascar has a tropical climate and experiences a rainy and a dry season. The eastern mountain slopes bear the remains of the dense rain forest, while the western plain is drier and supports forests of deciduous trees and savannah grassland. The southwest region is the driest while the northern region is prone to monsoon conditions (Brown, 1978).

In the absence of written records, scholars have made use of linguistic, archaeological and historical data to try to elucidate the origins of the proto-Malagasy. The favoured hypothesis based on linguistic (Dahl, 1951, 1977; Vérin, Kottak & Gorlin, 1970) and archaeological (Verin, 1986; Dewar & Wright 1993) evidence suggests that the original settlers arrived from Indonesia around the fourth century AD (Dahl, 1951) and that Africans came later (Vérin, 1986). However, Ferrand (1908) suggested that the first colonists were Africans and that it was the Indonesians who came later.

Although the biological diversity of the present-day Malagasy people is thought to be derived from a number of immigrant populations, the contributions are not easily distinguished, partly because of the linguistic unity evident in the people. Their language, Malagasy, belongs to the

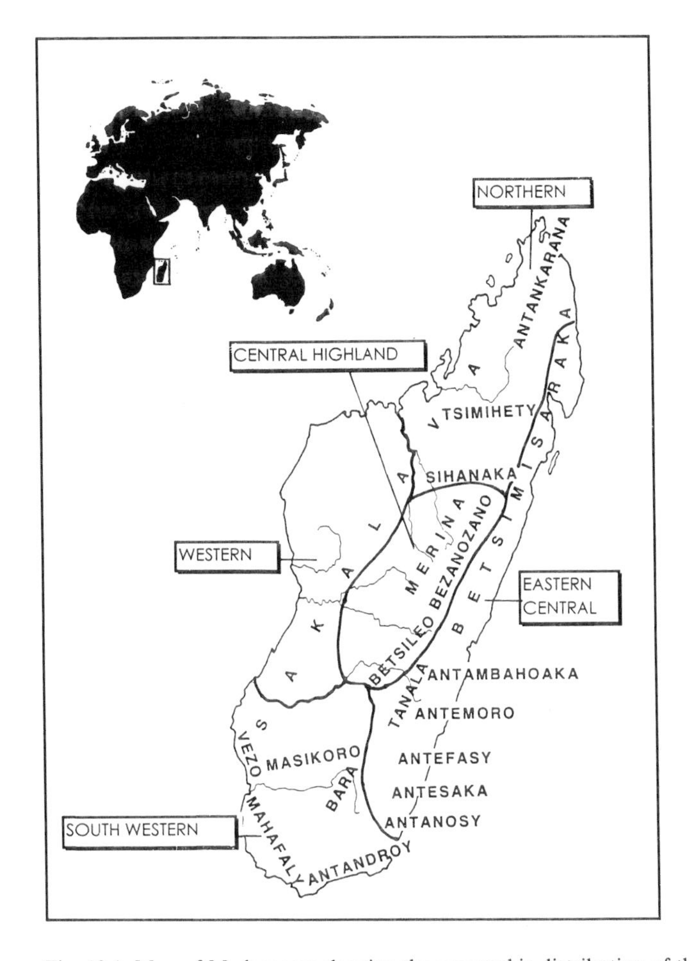

Fig. 12.1. Map of Madagascar showing the geographic distribution of the 18 *fuka* in the Northern, Central Highland, Western, Southwestern and Eastern Central regions.

Austronesian family of languages (Dyen, 1965). Minor dialectical differences exist in various regions and many features are thought to reflect Arabic and African influences (Ferrand, 1908; Brown, 1978; Vérin, 1986). Malagasy is most closely related to Maanyan, spoken in the Barito River region of Borneo. This region has therefore been postulated to be the geographic origin of the earliest colonizers (Dahl, 1951, 1977). The present population of Madagascar, estimated to be approximately 11 million, consists of 18 loosely defined ethnic groups or *fuko*, and their present distribution in Madagascar is shown in Fig. 12.1 (Brown, 1978).

Several hypotheses concerning the origins of the proto-Malagasy and the time of colonisation of Madagascar have been proposed (Vérin, 1986). However, the paucity of archaeological studies and a lack of linguistic and cultural diversity in the Malagasy has made firm conclusions difficult. Thus serogenetic studies have been resorted to in an attempt to determine the nature of the biological diversity of present-day Malagasy and to shed light on the origins of the earliest colonists. Apart from limited blood group, serum protein and red cell enzyme polymorphism studies (Singer *et al.*, 1957; Beuttner-Janusch & Beuttner-Janusch, 1964; Fourgnet *et al.*, 1964; Pigache, 1976), the Malagasy have not previously been studied genetically.

In this study the genetic affinities of the Malagasy have been investigated using variation in mitochondrial DNA (mtDNA) and the β^S globin (sickle cell) gene to shed light on the biological processes involved in the peopling of Madagascar. Since mtDNA is maternally inherited and does not recombine (Giles *et al.*, 1980; Zuckerman *et al.*, 1984), it can be used to trace the source of the female founders as well as to estimate the time of colonization of Madagascar. The β^S gene is not found in Indonesia, but does occur at high frequency in many parts of Africa (Nagel & Fleming, 1992). As there are distinct haplotypes associated with the β^S gene in different regions of Africa (Pagnier *et al.*, 1984), β^S haplotypes may be an appropriate marker to trace the source of the genetic contribution of Africans to the gene pool of the Malagasy.

Material and methods

Subjects

Blood was obtained during four field trips to Madagascar during 1992–93 from over 2000 individuals including 1425 unrelated Malagasy individuals, who claimed to have the same patrilineal and matrilineal *fuko*, and, in addition, in order to elucidate the β^S haplotypes, from 22 families, comprising 72 individuals.

Studies on the distribution of the β^S gene according to geographical region, stated *fuko*, places of birth and residence have been initiated. The 1425 unrelated individuals were classified into those originating from highland and those originating from lowland regions, and pooled into the following categories representing geographic regions: Central Highland, Northern, Eastern Central, Western, and Southwestern, as shown in Fig. 12.1. The 22 families all had at least one individual with sickle cell anaemia or sickle cell trait and were useful for the determination of β^S haplotypes.

For the mtDNA survey, 280 unrelated Malagasy were selected from the sample of 1425. *Fuko* were pooled into the geographic regions outlined above (excluding Western).

Methods

The β^S status of the unrelated individuals was determined by cellulose agarose electrophoresis and confirmed with citrate agar electrophoresis, which would exclude HbD which has the same mobility as HbS. DNA was extracted from the buffy coats of approximately 20 ml of whole blood using the method described by Miller *et al.*, (1988).

mtDNA studies

The intergenic 9 base pair (bp) deletion located between the cytochrome oxidase II and transfer RNA for lysine (COII/tRNA^{Lys}) genes was identified using the method outlined by Wrischnick *et al.* (1987). Twenty-four sequence specific oligonucleotide (SSO) probes were used to study variation within nine regions of the mtDNA control region using the dot-blot methodology described by Stoneking *et al.* (1991), Melton *et al.* (1995) and Soodyall *et al.* (in preparation). An SSO-type was derived for each individual by combining the variants detected by SSO-typing for regions IA, IB, IE, IC, ID, IIA, IIB, IIC and IID as discussed in Melton *et al.* (1995), Soodyall *et al.* (in preparation). Blanks, scored as zero (0) in the present nomenclature, were obtained when none of the SSO probes from the defined region hybridized with the DNA blot. This suggested that mutations in addition to those tested for were contained within the SSO region thereby preventing hybridization of the probes designed to detect specific sequences.

β^S haplotypes

In the 22 families, a 9 site β^S-associated haplotype was determined. The β^S status of each individual was confirmed by PCR amplification of the 5′ region of the β-globin gene followed by *Dde* I digestion (Old & Ludlam, 1991). Six haplotype sites were determined after PCR amplification and restriction enzyme digestion, namely *Hinc* II/ε, *Hind* III/$^G\gamma$, *Hinc* II/5'$\psi\beta$, *Hinc* II/3'$\psi\beta$, *Ava* II/β, *Hinf* I/β). Southern blotting was used for the remaining 3 sites, namely *Xmn* I/γ, *Hind* III/$^A\gamma$, *Hpa* I/β. The methods are described in Hewitt *et al.* (in preparation).

Results and discussion

Distribution of the β^S gene in Madagascar

The β^S frequency varies from 0.0156 in the highland regions to 0.0744 in the lowlands, with a marked disturbance of Hardy–Weinberg equilibrium in the lowland regions, due to the selective effects of malaria. Within the lowland groups, the frequencies were highest on the east coast. These data have been presented in detail in Hewitt *et al.* (in preparation).

β^S *associated haplotypes*

Four different haplotypes have been commonly associated with the β^S gene in four geographically distinct regions of Africa, the 'Senegal' haplotype in Atlantic West Africa, the 'Benin' haplotype in Central West Africa, the 'Bantu' haplotype in Bantu-speaking Africa (Pagnier *et al.*, 1984) and the 'Cameroon' haplotype in the Eton group of Cameroon (Lapoumeroulie *et al.*, 1992). A fifth haplotype is associated with the gene in east Saudi Arabia and India (Kulozik *et al.*, 1986). In the present study, 25 of 28 β^S chromosomes had the 'Bantu' haplotype, 1 the 'Senegal' and 2 rare haplotypes. This provides strong evidence for a major African, more specifically Bantu, contribution to the Malagasy ancestry. The detailed analysis is presented in Hewitt *et al.* (in preparation).

Frequency of the mitochondrial 9 bp deletion in the Malagasy

The intergenic COII/tRNA^{Lys} 9 bp deletion has been found at varying frequencies in Southeast Asians (Horai, Gojobori & Matsunaga, 1987; Stoneking & Wilson 1989; Ballinger *et al.*, 1992; Harihara *et al.*, 1992; Horai 1991*a*,*b*; Passarino *et al.*, 1993; Melton *et al.*, 1995), Pacific Island populations (Hertzberg *et al.*, 1989; Hagelberg & Clegg 1993; Hagelberg *et al.*, 1994; Lum *et al.*, 1994; Redd *et al.*, 1995) as well as in Native Americans (Schurr *et al.*, 1990; Ward *et al.*, 1991, 1993; Shields *et al.*, 1992, 1993; Torroni *et al.*, 1992, 1994; Horai *et al.*, 1993; Lorenz & Smith, 1994). These studies have shown that, when the 9 bp deletion is used in conjunction with mtDNA control region sequence variation, the 9 bp deletion is a useful marker to trace migrations out of Asia which resulted in the eventual colonisation of the Pacific and the New World. The 9 bp deletion is often referred to as an 'Asian-specific' marker, but this is incorrect since it is also found in some African populations (Vigilant 1990; Chen, Torroni & Wallace, 1994; Merriwether *et al.*, 1994; Soodyall *et al.*, 1994, 1996).

In Madagascar, the 9 bp deletion was found in 75 out of 280 (26.8%) individuals and its frequencies in the different *fuko* from the four regions represented in this study are shown in Fig. 12.2. The deletion was found throughout Madagascar, being more frequent among inhabitants from the Central Highland, Northern, and Eastern Central regions compared with *fuko* from the Southwestern region (Fig. 12.2).

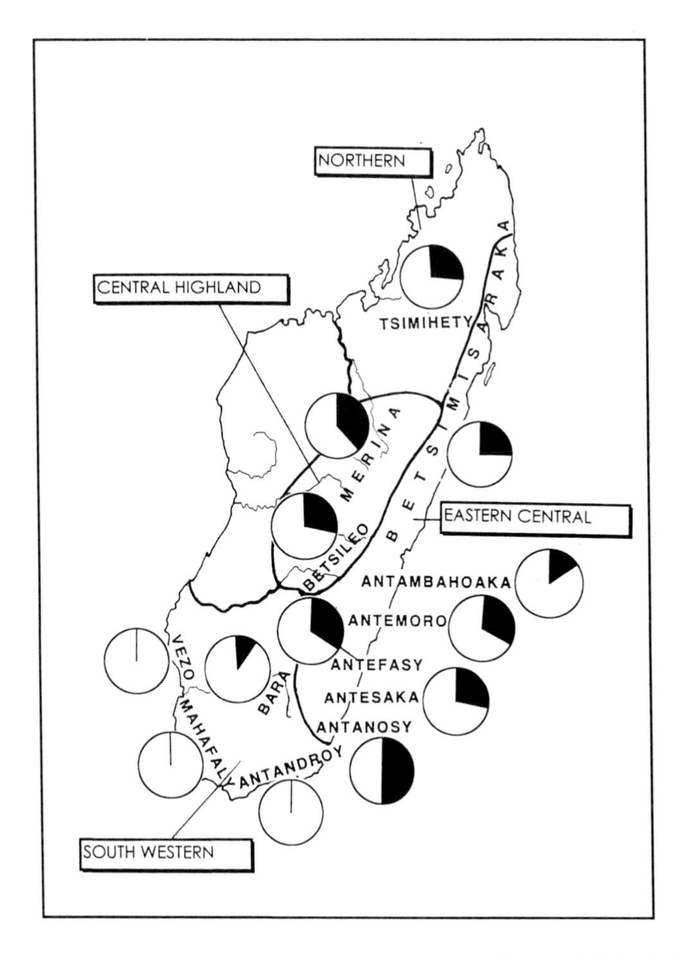

Fig. 12.2. Frequencies of the intergenic COII/tRNA^{Lys} 9 bp deletion in *fuka* from four regions of Madagascar.

SSO-type mtDNA in the Malagasy

Several lines of evidence gleaned from the historic, linguistic and archaeological data have suggested that the present-day Malagasy are derived from varying degrees of Indonesian and African admixture. Genetic studies based on blood groups and the sickle-cell trait have estimated that approximately two-thirds of the gene pool in the Malagasy is derived from Africa while only one-third is derived from an Asian source (Singer *et al.*, 1957). The source of the African and Asian components has not been defined.

Since the 9 bp deletion has originated independently in Asians and

Table 12.1. *Frequencies of SSO-types found in Malagasy with the 9 bp deletion given as a percentage of the total sample per region*

Type number	Sample size SSO-TYPE	Central Highland 89	Northern 22	Eastern Central 128	South Western 41	Total 280
1	121122111	1.1				0.4
2	121122231			0.8		0.4
3	121311002	1.1		1.6		1.1
4	121311302	3.4	9.1	8.6	2.4	6.1
5	130112211	1.1				0.4
6	132112211	1.1		4.5		0.7
7	133112011	2.2				0.4
8	133112201		4.5			1.1
9	133112210			1.6		0.7
10	133112211	20.2	9.1	15.6	4.9	15.0
11	133122211			0.8		0.4
12	133311312	1.1				0.4

Africans (Soodyall *et al.*, 1994; Redd *et al.*, 1995), the associated mtDNA control region SSO-types, which are exclusive to Asians (Melton *et al.*, 1995) and Africans (unpublished), can be used to estimate the proportion of mtDNA genes contributed by Asians and Africans to the gene pool of the Malagasy. From our survey of SSO-type variation, we found that 12 of the 59 unique SSO-types were found only in individuals with the 9-bp deletion (Soodyall *et al.*, in prep.). The SSO-types and their frequencies in the different groups with the 9 bp deletion are given in Table 12.1.

SSO-type 10 (133112211), which was found in 56% of Malagasy individuals with the 9 bp deletion, was also the commonest type found in individuals from Samoa and coastal Papua New Guinea with the 9 bp deletion (Redd *et al.*, 1995). This SSO-type was rarely (1.6%) found in Southeast Asians and Indonesians (Melton *et al.*, 1995). SSO-type 4 (121311302), found at a frequency of 22.6% in Malagasy with the deletion, was also found in Africans with the 9 bp deletion at a frequency of approximately 30% (unpublished). SSO-types 4 and 10, together, make up 80% of the SSO-types found in Malagasy who have the 9 bp deletion. The remaining SSO-types derived from the Asian and African source are closely related to SSO-types 10 and 4, respectively, and have been confirmed by sequencing both hypervariable regions of the control region (unpublished).

Using the SSO-types found in the Malagasy (Table 12.1), we can trace types 1–4 to an African source and types 5–12 to an Asian source. The Asian

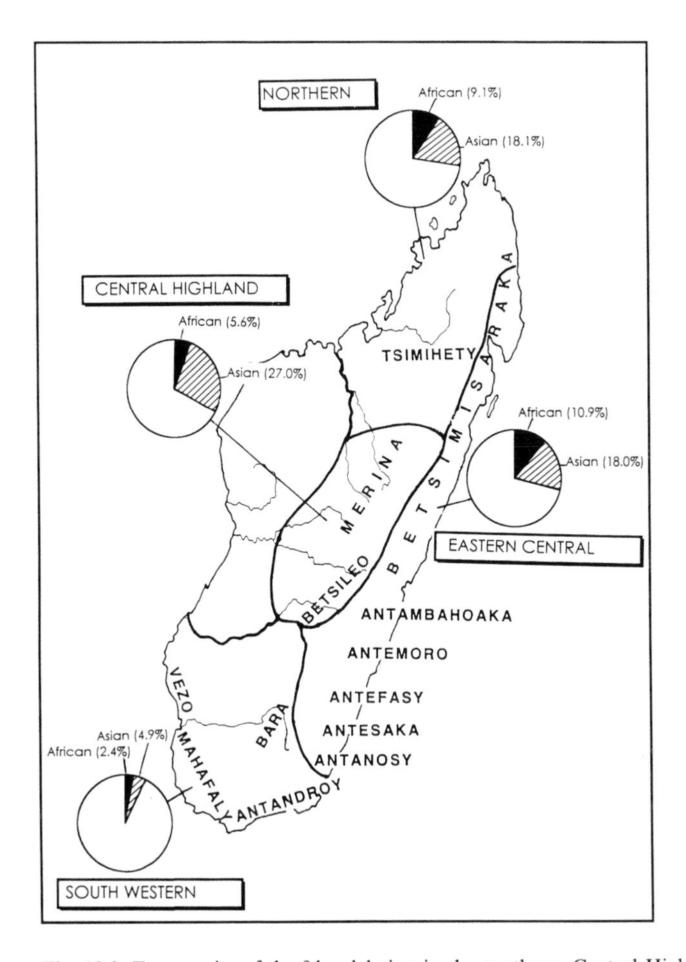

Fig. 12.3. Frequencies of the 9 bp deletion in the northern, Central Highland, Southwestern and Eastern Central regions of Madagascar showing the contributions of the African (bold) and Asian (shaded) source of the deletion in the Malagasy.

contribution is highest among the Central Highland group (83%); here the Asian to African contribution is approximately 4:1 compared with approximately 2:1 in the Northern, Eastern Central and Southwestern regions (Fig. 12.3). Previous studies based on anthropometrics and culture have suggested a stronger Indonesian influence in the Merina and Betsileo who occupy the Central Highland region (Brown 1978; Singer *et al.*, 1957). In contrast, the Southwestern region which is occupied by the Bara, Antandroy, Mahafaly and Vezo in this study (Fig. 12.1), have more African-like traits (Brown, 1978). The observation that the deletion is found throughout the island (Fig. 12.1), and is associated with both Asian and

African SSO-types (Fig. 12.3), suggests that both the African and Asian contributions have co-existed in Madagascar for centuries.

Thus far, the 9 bp deletion has been detected in Pygmies from Zaire and the Central African Republic, in the Lese who are Bantu-speaking Negroids living in close association with the Pygmies from Zaire, in Malawians and central and southeastern Bantu-speakers from southern Africa (Soodyall *et al.*, 1996). The deletion has not been found in 113 west African Negroids and 171 Khoisan individuals from southern Africa (Merriwether *et al.*, 1994), but it was detected in 6/189 Nigerians and 3/115 individuals from the Ivory Coast. The deletion was found in only one Luo from the east African sample of 80 individuals (unpublished). Control region sequence analyses suggest that the Malagasy are most closely related to Bantu-speaking Negroids from southeastern Africa with the African form of the 9 bp deletion (unpublished). Since the β^S gene is extremely rare in Bantu people who settled south of the Limpopo river, possibly because they migrated into the southern African subcontinent before the β^S gene arose (Jenkins, 1982), the ancestors of the Malagasy must have originated from southeastern Bantu between the Limpopo and Zambezi rivers, possibly Malawi or Mozambique.

The 'Polynesian' mtDNA motif

Some informative mtDNA variants found in the Malagasy have also been used to try to trace the 'Asian' contribution. Three substitutions located at positions 16 217 (T→C), 16 247 (A→G) and 16 261 (C→T) (relative to the published reference sequence, Anderson *et al.*, 1981) have been found in conjunction with the 9 bp deletion in some populations derived from Asia (Hagelberg & Clegg 1993; Hagelberg *et al.*, 1994; Lum *et al.*, 1994; Redd *et al.*, 1995; Melton *et al.*, 1995). These three substitutions are referred to in individuals with the 9 bp deletion as the 'Polynesian' motif (Soodyall *et al.*, in preparation) since it is found at high frequencies in both contemporary and ancient Polynesians (Hagelberg & Clegg 1993; Hagelberg *et al.*, 1994; Lum *et al.*, 1994; Redd *et al.*, 1995). The 'Polynesian' motif was found in 96% (51/53) of Malagasy individuals who have SSO-types derived from the Asian source (Soodyall *et al.*, 1995). The frequencies of the Polynesian motif in the Malagasy and other populations who are derived from Southeast Asia are shown in Fig. 12.4. Using the Polynesian motif as an index of the Asian source of the 9 bp deletion, the Malagasy show a closer genetic affinity to Polynesians than they do to Indonesians (Soodyall *et al.*, 1995).

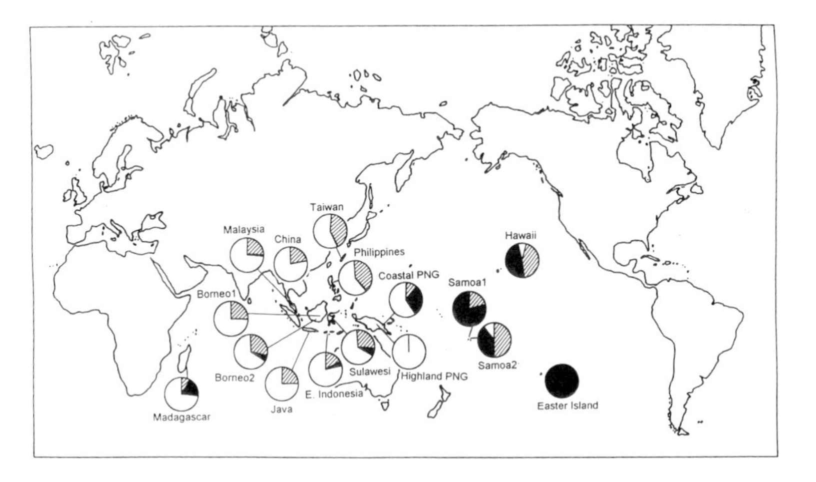

Fig. 12.4. Frequencies of the 9 bp deletion (shaded) and the Polynesian motif (bold) in Southeast Asia, Indonesia, Papua New Guinea, Polynesia and Madagascar.

Origins of the Malagasy

Using the 9 bp deletion and the associated mtDNA SSO-types, we were able to estimate the proportion of African and Asian mtDNA genes in the present-day Malagasy. Moreover, the Polynesian motif is not found in Indonesians from the Barito River area, the geographic origin of the Malagasy hypothesised on the basis of linguistic evidence. However, the Polynesian motif does occur elsewhere in Borneo as well as in east Indonesia at low frequencies (Fig. 12.4). One possible explanation for the presence of the Polynesian motif in Madagascar is that it arose independently. Sequence analysis suggests that the order of mutations leading to the Polynesian motif was the 9 bp deletion first, followed by substitutions at 16 217, 16 261 and last at 16 247 (Redd *et al.*, 1995). MtDNAs with the 9 bp deletion and the mutations at 16 217 and 16 261 are widespread across Southeast Asia (Melton *et al.*, 1995); perhaps the original colonists of Madagascar carried these mtDNAs, and the 16247 mutation subsequently arose independently in Madagascar (Soodyall *et al.*, 1995). However, the probability that a newly arisen, independent mutation could increase in frequency to nearly 20% by genetic drift in just 1500 years (about 75 generations) is very low (A.G. Clark, personal communication).

A much more likely explanation for these results is that the original colonists of Madagascar either came directly from Polynesia, or from an Indonesian population who eventually colonized Pacific islands. These results also shed light on the probable route of colonization to Madagascar.

The absence of the Polynesian motif and mtDNA types found in peoples of Asian descent from Indian (Melton *et al.*, 1995) and African (Soodyall *et al.*, in prep) populations lends more support to the hypothesis that Madagascar was colonised directly from the south Pacific region (Vérin, 1986) and not by an indirect route via India and Africa, as has been previously postulated (Deschamps, 1963; Brown, 1978). As the β^S gene does not occur in Indonesia and Southeast Asia, little information on the Asian contributions can be obtained from the study of this gene. The absence of the Arab–Indian β^S haplotype, however, lends some support to the hypothesis that there was no major Arab–Indian contribution to the Malagasy gene pool. Studies of nuclear DNA polymorphisms in the Malagasy, including a number of globin markers that are common in Polynesians are in progress and should elucidate the source and route of colonization of Madagascar. In addition, sequence analysis of mtDNA control region variation should shed light on the time of colonization by the first settlers of Madagascar.

Acknowledgements

We are indebted to all the Malagasy individuals who consented to participate in this study. We should like to thank Drs Thomy de Ravel, Pascale Willem, Gwyn Campbell, A. Rakouth, Stan Quambek and Jean Roux for their assistance in the collection of the blood specimens and our colleagues in the Department of Human Genetics Department, South African Institute for Medical Research, for help in processing them. This study was supported by a National Science Foundation grant to M. Stoneking; a Fogarty International Fellowship award to H. Soodyall and from grants to T. Jenkins from the South African Institute for Medical Research, The University of the Witwatersrand and The National Geographic Society.

References

Anderson, S., Bankier, A. T., Barrel, B. G., de Bruijn, M. H. L., Coulson, A. R., Drouin, J., Eperon, I. C., Nierlich, D. P., Roe, B. A., Sanger, F., Schreier, H., Smith, A. J. H., Staden, R. & Young, I. G. (1981). Sequence and organization of the mitochondrial genome. *Nature*, **290**, 457–65.

Ballinger, S. W., Schurr, T. G., Torroni, A., Gan, Y. Y., Hodge, J. A., Hassan, K., Chen, K-H. & Wallace, D. C. (1992). Southeast Asian mitochondrial DNA analysis reveals genetic continuity of ancient Mongoloid migrations. *Genetics*, **130**, 139–52.

Beuttner-Janusch, J. & Beuttner-Janusch, V. (1964). Hemoglobins, haptoglobins and transferrins in the peoples of Madagascar. *American Journal of Physical Anthropology*, **22**, 163–9.

Brown, M. (1978). *Madagascar Rediscovered*. London: Damien Tunnacliffe.

Chen, Y-S., Torroni, A. & Wallace, D. C. (1994). Mitochondrial DNA variation in African populations. *American Journal of Human Genetics (suppl.)*, **55**, 852.
Dahl, O. C. (1951). *Malgache et Maanyan: Une comparison linguistique*. Oslo: Egede Institutett.
Dahl, O. C. (1977). La subdivision de la famille Barito et la place du Malgache. *Acta Orientalia*, **38**, 77–134.
Deschamps, H. (1960). *Histoire de Madagascar*. Paris: Berger Levrault.
Dewar, R. E. & Wright, H. T. (1993). The Cultural History of Madagascar. *Journal of World Prehistory*, **7**, 417–66.
Dyen, I. (1965). A lexicostatistical classification of the Austronesian languages. *International Journal of American Linguistics* **17**, Memoir 19 (Vol. 31, no. 1).
Ferrand, G. (1908). L'Origine africaine des Malgaches. *Journal Asiatique*, **10**, 353–500.
Fourgnet, R., Sarthou, J-L., Roux, J. & Acri, K. (1974). Hemoglobine 'S' et origines du peuplement de Madagascar. Nouvelle hypothèse sur son introduction en Afrique. *Archiv Institut Pasteur Madagascar*, **411**, 185–220.
Giles, R. E., Blanc, H., Cann, H. M. & Wallace, D. C. (1980). Maternal inheritance of human mitochondrial DNA. *Proceedings of the National Academy of Sciences, USA*, **77**, 6715–9.
Hagelberg, E. & Clegg, J. B. (1993). Genetic polymorphism in prehistoric Pacific islanders determined by analysis of ancient bone DNA. *Proceedings of the Royal Society of London B*, **252**, 163–70.
Hagelberg, E., Quevedo, S., Turbon, D. & Clegg, J. B. (1994). DNA from Easter Islanders. *Nature*, **369**, 25–6.
Harihara, S., Momoki, H., Suutou, Y., Shimizu, K. & Omoto, K. (1992). Frequency of a 9-bp deletion in the mitochondrial DNA among Asian populations. *Human Biology*, **64**, 161–6.
Hertzberg, M., Mickleson, K. N. P., Serjeantson, S. W., Prior, J. F. & Trent, R. J. (1989). An Asian-specific 9-bp deletion of mitochondrial DNA is frequently found in Polynesians. *American Journal of Human Genetics*, **44**, 504–10.
Hewitt, R., Goldman, A., Krause, A. & Jenkins, T. (in preparation). β-globin haplotype analysis suggests that a major source of Malagasy ancestry is derived from Bantu speaking Negroids.
Horai, S. (1991*a*). A genetic trail of human mitochondrial DNA. In *New Era of Bioenergetics*, ed. Y. Mukohata, pp. 273–99. Tokyo: Academic Press.
Horai, S. (1991*b*). Molecular phylogeny and evolution of human mitochondrial DNA. In *New Aspects of the Genetics of Molecular Evolution*, ed. M. Kimura and N. Takahata, pp. 135–52. Tokyo: Japan Scientific Societies Press; Berlin: Springer.
Horai, S., Gojobori, T. & Matsunaga, E. (1987). Evolutionary implications of mitochondrial DNA polymorphisms in human populations. In *Human Genetics: Proceedings of the 7th International Congress*, ed. F. Vogel and K. Sperling, pp. 177–81. Heidelberg: Springer.
Horai, S., Kondo, R., Nakagawa-Hattori, Y., Hayashi, S., Sonoda, S. & Tajima, K. (1993). Peopling of the Americas, founded by four major lineages of mitochondrial DNA. *Molecular Biology and Evolution*, **10**, 23–47.
Jenkins, T. (1982). Human evolution in Southern Africa. In: *Human Genetics, Part A: The Unfolding Genome*, ed. B. Bonné-Tamir, pp. 227–53. New York: Alan R. Liss Inc.

Kulozik, A. E., Wainscoat, J. S., Serjeant, G. R., Kar, B. C., Al-Awamy, B., Essan, G. J. F., Falusi, A. G., Haque, S. K., Hilali, A. M., Kate, S., Ranasinghe, W. A. E. P. & Weatherall, D. J. (1986). Geographical survey of β^S-globin gene haplotypes: evidence for an independent Asian origin of the sickle-cell mutation. *American Journal of Human Genetics*, **39**, 239–44.

Lapouméroulie, C., Dunda, O., Ducrocq, R., Trabuchet, G., Mony-Lob, M., Bodo, J. M., Carnevale, P., Labie, D., Elion, J. & Krishnamoorthy, R. (1992). A novel sickle cell mutation of yet another origin in Africa: the Cameroon type. *Human Genetics*, **89**, 333–7.

Lorenz, J. G. & Smith, D. G. (1994). Distribution of the 9-bp mitochondrial DNA region V deletion among North American Indians. *Human Biology*, **66**, 777–88.

Lum, J. K., Rickards, O., Ching, C. & Cann, R. L. (1994). Polynesian mitochondrial DNAs reveal three deep maternal lineage clusters. *Human Biology*, **66**, 567–90.

Melton, T., Peterson, R., Redd, A., Saha, N., Sofro, A. S. M., Martinson, J. J. & Stoneking, M. (1995). Polynesian genetic affinities with Southeast Asian populations as identified by mitochondrial DNA analysis. *American Journal of Human Genetics*, **57**, 403–14.

Merriwether, D. A., Huston, S. L., Bunker, C. A. & Ferrell, R. E. (1994). Origins and dispersal of the mitochondrial DNA region V 9-bp deletion and insertion in Nigeria and the Ivory Coast. *American Journal of Human Genetics*, **55** (Supplement), Abstract 912.

Nagel, R. L. & Fleming, A. F. (1992). Genetic epidemiology of the β^S gene. *Baillière's Clinical Haematology*, **5**, 331–65.

Old, J. M. & Ludlam, C. A. (1991). Antenatal diagnosis. *Baillière's Clinical Haematology*, **4**, 391–428.

Pagnier, J., Mears, J. G., Dunda-Belkhodja, O., Schaefer-Rego, K. E., Beldjord, C., Nagel, R. L. & Labie, D. (1984). Evidence for the multicentric origin of the sickle cell hemoglobin gene in Africa. *Proceedings of the National Academy of Sciences, USA*, **81**, 1771–3.

Passarino, G., Semino, O., Modiano, G. & Santachiara-Benerecetti, A. S. (1993). COII/tRNA^{Lys} intergenic 9-bp deletion and other mtDNA markers clearly reveal that the Tharus (southern Nepal) have Oriental affinities. *American Journal of Human Genetics*, **53**, 609–18.

Pigache, J. P. (1969). Contribution à l'etude de l'hémoanthropologie des malagaches. Thesis: quoted by Mourant, A.E., Kopéc, A. C. and Domaniewska-Sobczak, K. (1976). *The Distribution of the Human Blood Groups*. London: Oxford University Press.

Redd, A. J., Takezaki, N., Sherry, S. T., McGarvey, T., Sofro, A. S. M. & Stoneking, M. (1995). Evolutionary history of the COII/tRNA^{Lys} intergenic 9-bp deletion in human mitochondrial DNAs from the Pacific. *Molecular Biology and Evolution*, **12**, 604–15.

Schurr, T. G., Ballinger, S. W., Gan, Y-Y., Hodge, J. A., Merriwether, D. A., Lawrence, D. N., Knowler, W. C., Weiss, K. M. & Wallace, D. C. (1990). Amerindian mitochondrial DNAs have rare Asian mutations at high frequencies, suggesting they derived from four primary maternal lineages. *American Journal of Human Genetics*, **46**, 613–23.

Shields, G. F., Hecker, K., Voevoda, M. I. & Reed, J. K. (1992). Absence of the

Asian-specific region V mitochondrial marker in Native Beringians. *American Journal of Human Genetics*, **50**, 758–65.
Shields, G. F., Schmiechen, A. M., Frazier, B. L., Redd, A., Voevoda, M. I., Reed, J. K. & Ward, R. H. (1993). MtDNA sequences suggest a recent evolutionary divergence for Beringian and northern North American populations. *American Journal of Human Genetics*, **53**, 549–62.
Singer, R., Budtz-Olzen, O. E., Brain, P. & Saugrain, J. (1957). Physical features, sickling and serology of the Malagasy of Madagascar. *American Journal of Physical Anthropology*, **15**, 91–124.
Soodyall, H., Jenkins, T. & Stoneking, M. (1995). A high frequency of a 'Polynesian' mitochondrial DNA trait in the Malagasy. *Nature Genetics*, **10**, 377–8.
Soodyall, H., Redd, A. J., Vigilant, L. & Stoneking, M. (1994). Mitochondrial DNA control region sequence variation suggests an independent origin of an 'Asian-specific' 9-bp deletion in Africans. *American Journal of Human Genetics (supplement)*, **55**, 956.
Soodyall, H., Vigilant, L., Hin, A. V., Stoneking, M. & 'Jenkins, T. (1996). MtDNA control-region segment variation suggests multiple independent origins of an 'Asian-specific' 9-bp deletion in sub-Saharan Africans. *American Journal of Human Genetics*, **58**, 595–608.
Stoneking, M., Hedgecock, D., Higuchi, R. G., Vigilant, L. & Erlich, H. A. (1991). Population variation of human mtDNA control region sequences detected by enzymatic amplification and sequence-specific oligonucleotide probes. *American Journal of Human Genetics*, **48**, 370–82.
Stoneking, M. & Wilson, A. C. (1989). Mitochondrial DNA. In *Colonization of the Pacific: A Genetic Trail*, ed. A. V. S. Hill and S. W. Serjeantson, pp. 215–245. Oxford: Clarendon Press.
Torroni, A., Schurr, T. G., Yang, C-C., Szathmary, E. J. E., Williams, R. C., Schanield, M. S., Troup, G. A., Knowler, W. C., Lawrence, D. N., Weiss, K. M. & Wallace, D. C. (1992). Native American mitochondrial DNA analysis indicates that the Amerind and the NaDene populations were founded by two independent migrations. *Genetics*, **130**, 153–62.
Torroni, A., Chen, Y-S., Semino, O., Santachiara-Benerecetti, A. S., Scott, C. R., Lott, M. T., Winter, M. & Wallace, D. C. (1994). MtDNA and Y-chromosome polymorphisms in four Native American populations from southern Mexico. *American Journal of Human Genetics*, **54**, 303–18.
Vérin, P. (1986). *The History of Civilization in North Madagascar*. Rotterdam and Boston: Balkema.
Vérin, P., Kottak, C. & Gorlin, P. (1970). The glottochronology of Malagasy speech communities. *Oceanic Linguistics*, **8**, 26–31.
Vigilant, L. (1990). Control region sequences from African Populations and the evolution of human mitochondrial DNA. Unpublished PhD. Thesis. University of California at Berkeley.
Ward, R. H., Frazier, B. L., Dew-Jager, K. & Pääbo, S. (1991). Extensive mitochondrial diversity within a single Amerindian tribe. *Proceedings of the National Academy of Sciences, USA*, **88**, 8720–4.
Ward, R. H., Redd, A., Valencia, D., Frazier, B. & Pääbo, S. (1993). Genetic and linguistic differentiation in the Americas. *Proceedings of the National Academy of Sciences, USA*, **90**, 10663–7.

Wrischnik, L. A., Higuichi, R. G., Stoneking, M., Erlich, H. A., Arnheim, N. & Wilson, A. C. (1987). Length mutations in human mitochondrial DNA: direct sequencing of enzymatically amplified DNA. *Nucleic Acids Research*, **15**, 529–42.

Zuckerman, S. H., Solus, J. F., Gillespie, F. P. & Eisenstadt, J. M. (1984). Retention of both parental mitochondrial DNA species in mouse-Chinese hamster somatic cell hybrids. *Somatic Cell and Molecular Genetics*, **10**, 85–92.

13 *Molecular perspectives on the colonisation of the Pacific*

J. J. MARTINSON

Introduction

Human biologists interested in population diversity have long been receptive to the potentials offered by newly developed techniques of biological analysis. As a consequence, the populations of the world have been characterized using a variety of systems including morphological, anthropometric, dermatoglyphic, antigenic and electrophoretic characters.

In recent years, analysis of nucleotide sequence variation has added another layer of detail to the body of information available. The particular merit of molecular genetic data, if one is interested in the characterization of the fundamentals of human genetic diversity, is that it directly describes variation at the level of the genotype: all of the systems listed previously describe characteristics that are essentially phenotypic. Although it is often possible to take environmental effects into account in the analysis of such characters, the underlying genetic diversity may be masked. Even when protein sequence data are used, not all of the underlying DNA sequence variation is detected, owing to the redundancy of the genetic code.

This does not mean that the existing body of non-genetic information must be discarded, or that earlier conclusions based on such data are inaccurate. The application of serological and protein electrophoretic techniques to the study of human variation has revealed facets not visible before. For example, within the Pacific region discussed in this chapter, serum protein data have been used to demonstrate that linguistic affiliation between populations is not a good indicator of biological relatedness (Boyce *et al.*, 1978). These and other results (Kirk, 1980; Serjeantson, Kirk & Booth, 1983) show that geographical distance correlates with biological relatedness better than does linguistic similarity, and that any attempt to explain the relationships of the populations of the Pacific based solely on language may lead to erroneous conclusions.

The use of molecular genetic techniques in the analysis of population relationships and history has been particularly extensive in this region of the world. Substantial amounts of data have now been obtained for several genetic loci in many different regions of the Pacific (Hill & Serjeantson, 1989). In this chapter I shall summarize the data currently available for two loci in particular – α-globin and mitochondrial DNA – and show how these data have supported the consensus view of Pacific colonization, but I also hope to show how the same data have been used to support alternative scenarios.

The debate over Pacific colonization

When anthropologists first studied the populations of the Pacific in an attempt to describe their relationships and histories, the physical differences they observed led to the theory that the region was populated by 'successive migrations of 'pure' and stable races and cultures' (Bellwood, 1989). Admixture between these populations was then said to account for any cultural, linguistic and physical differentiation between races in the region. Subsequent investigations by archaeologists, linguists and biologists have revealed this to be a crude oversimplification. Many researchers believe that population movements have indeed occurred, but that these were not the organized, motivated movements proposed by the first workers in the field (Bellwood, 1978).

The current consensus view begins with the first settlement of what are now the separate continental islands of Australia and New Guinea over 35 000 years ago by Australoid peoples originating in Southeast Asia. At that time, lowering of the sea level caused by glacial expansion produced a land bridge fusing these islands into a larger land mass known as Sahulland (White & O'Connell, 1982). The sea level never fell low enough to join Sahul with the supercontinent of Sunda (Java, Borneo and Bali) to the west; thus this first settlement must have entailed a sea crossing of approximately 70 km (Birdsell, 1977). The land bridge between Australia and New Guinea remained until the final Pleistocene rise in sea level around 9–7000 years ago; the inhabitants of what are now isolated land masses had the opportunity of genetic exchange for approximately 30 000 years.

Archaeological dating of sites found in New Guinea and on the Bismarck and northern Solomon Island groups to the east show signs of human habitation dating back to around 30 000 years ago (Groube *et al.*, 1986; Allen *et al.*, 1988). No land bridges existed between New Guinea and these islands at this time, but there is a continuous chain of island intervisibility from the northern to the south-eastern Solomon Islands (Irwin, 1992) and so it is possible that further evidence of ancient occupation will be seen in the archipelago. Beyond that point, however, inter-island distances tend to increase, and it is likely that further movement eastward did not occur at this time.

Climatic fluctuations since the first settlement of Sahul would have created a harsh environment for the inhabitants prior to 15 000 years ago. The post-glacial thaw which flooded the land bridge between Australia and New Guinea around 9–7000 years ago also improved the climate to the extent where evidence of horticulture can be seen soon afterwards in New Guinea. This has significant implications for the understanding of the genetic constitution of these Melanesian peoples[1]. The land-clearing that accompanies horticulturalism often facilitates the spread of the mosquito vector that transmits malaria in man. The interactions between mosquito, parasite and man in this region probably date back to this time (Groube, 1993) and have resulted in the current distribution of α-thalassaemia (Boyce, Harding & Martinson, 1995).

Most scholars working on Pacific history agree with the above account. The controversies emerge when more recent population movements are considered. Much of the debate centres on the emergence of a characteristic pottery known as 'Lapita' (Green, 1979). Over the relatively short period of 600 years (from 1600 to 1000 BC) this pottery is found to have spread over 4000 miles, from the Admiralty Islands north of Papua New Guinea to Samoa in central Polynesia (Bellwood, 1987). Little evidence of it has been seen in New Guinea itself, although one or two fragments have been recovered (Terrell, personal communication). The majority view is that the Lapita manufacturers were agriculturalists who spoke a different family of languages (referred to as 'Austronesian') to the original inhabitants of New Guinea (who spoke languages referred to as 'Papuan') (Wurm, 1983).

The ancestors of the Lapita manufacturers are thought to have originated in island Southeast Asia and spread from there southwards and eastwards to settle in western Melanesia (primarily in the Bismarck and Solomon archipelagos) and establish long-term interactions with the existing Papuan-speaking populations. The more advanced seafaring skills of the Lapita makers enabled them to move further eastward into the unpopulated islands of Vanuatu, New Caledonia and Fiji[2]. The biological phenotype of these people cannot be determined accurately from the available skeletal material (Pietrusewsky, 1985) but it is believed that it may have resembled that of present-day Polynesians.

Once the easterly bounds of this region had been settled, by around 1500

[1] 'Melanesia', 'Polynesia' and 'Micronesia' are geographical terms; their use to describe populations is a common, but sometimes misleading, shorthand.

[2] Terrell has noted that many scholars describe the Lapita people in terms more usually associated with mythical superheroes endowed with powers far in advance of their more mortal neighbours (Terrell, 1994).

BC, no further successful eastward migration occurred for approximately 800 years, as no sites in eastern Polynesia pre-dating 200 BC have been found (Kirch, 1986). It is yet another controversy whether the settlement of eastern Polynesia involved accidental drift-voyaging or intentional navigation (Levinson, Ward & Webb, 1973; Irwin, 1992). The speed with which the Lapita culture and its descendants spread across the reaches of the Pacific has been likened to an express train (Diamond, 1988), carrying its passengers through congested Melanesia into the vacant sites of Polynesia. The extreme version of this view allows little scope for contact and exchange between the Lapita people and the Melanesians whom they encountered on their journey. At the most, a one-way transfer is allowed whereby the Lapita people leave some of their pottery and language behind in Melanesia, but acquire nothing of use to take with them on the train.

As might be expected, this is not the sole interpretation of the archaeological and linguistic data. A minority view, most consistently argued by Terrell, rejects the 'family tree' approach to the analysis of variation which attempts to relate populations to one another on the basis of shared characteristics, in favour of an 'entangled bank' metaphor in which individuals interact with each other in complex networks of trade, intermarriage and inherited friendships which can result in a heterogeneous mixture of cultural and linguistic characteristics throughout the region (Terrell, 1986, 1988; Welsch, Terrell & Nodalski, 1992). Using this approach, it is not necessary for the characteristics associated with the so-called 'Austronesian'-speaking Lapita culture to have been introduced from outside the Pacific in the recent past: the current pattern of languages and culture could have arisen *in situ* by differentiation following a single ancient settlement of the region.

All of the above controversy is based mainly on linguistic, archaeological and cultural data. Can the molecular genetic data that have been so exhaustively accumulated in the past two decades resolve it to everyone's satisfaction? It should come as no surprise that this happy state of affairs has not yet been achieved. Supporters of one or other view see, in the genetic data, support for their own theories, and should any geneticist attempt to offer their own opinion then their detractors claim that the sophistication of their laboratory techniques is not matched by a similar maturation in their anthropological thinking (Terrell & Fagan, 1975). Nevertheless, I shall take the risk.

Haemoglobin mutations as population markers in Oceania

Haemoglobin is probably the most extensively studied protein in human populations. Its abundance in blood meant that it was one of the few

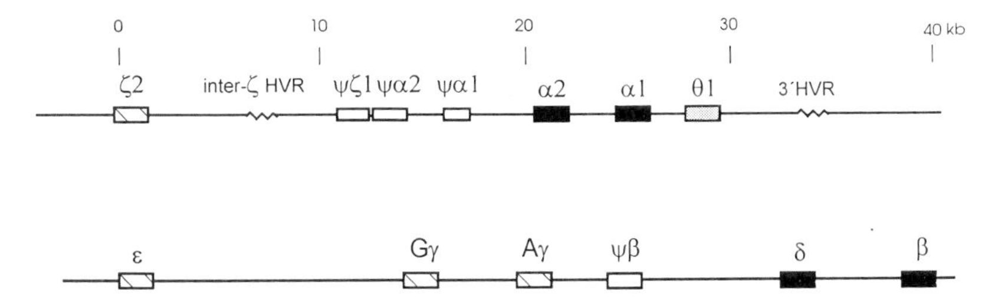

Fig. 13.1. The α and β globin complexes. Functional genes are denoted by shaded boxes: pseudogenes by white boxes. The main adult genes are shown in black, and the foetal genes are striped. The θ gene is of unknown function. The δ gene is represented in $< 5\%$ of the adult haemoglobin.

proteins which could be extracted in sufficient quantities for amino acid sequencing and protein structure determination as these techniques were developed. The ease with which variant forms of haemoglobin could be characterised has revealed many different forms in human populations (Livingstone, 1975). The genetics of the peptide components of haemoglobin have been the subject of molecular analysis since the advent of these techniques in the 1970s and 1980s (Higgs & Weatherall, 1993) and much is now known about the structure, function and regulation of the genes which encode the different types of haemoglobin produced throughout foetal and adult life.

The adult form of haemoglobin is a heterotetramer, composed of two α-globin and two β-globin peptides. These peptides are encoded by two gene complexes located on different chromosomes; the α-globin complex is on chromosome 16, β-globin on chromosome 11 (Fig. 13.1). Globin gene mutations can be classified into two main types:

Biosynthetic mutations, in which the amount of peptide produced is reduced or eliminated. The disease states which result from these mutations are known as thalassaemias: in α-thalassaemia the amount of α-globin is reduced, in β-thalassaemia β-globin levels are affected.

Structural variants, in which nucleotide substitutions produce globin chains whose amino acid sequences differ from normal. The variant globins are produced in normal quantities, but some are unstable and can result in thalassaemic or other disease phenotypes. The best-known example is HbS which causes sickle-cell anaemia.

More complex phenotypes can result from combinations of thalassaemia mutations and haemoglobin variants in the same individual. This can occur when an individual inherits a thalassaemia chromosome from one parent and a variant haemoglobin chromosome from the other. Rarer conditions are known where a variant haemoglobin sequence has arisen on a thalassaemia chromosome. One example of this condition that is found in Oceania is $HbJ^{Tongariki}$, where a variant haemoglobin has been produced on an α-thalassaemia chromosome. This particular mutation is an especially informative marker for population movements and affinities in Oceania, and will be discussed in detail later in this chapter.

Several different thalassaemia mutations are known, ranging in severity from those whose phenotype is so benign that the mutation can only be identified with confidence by molecular genetic analysis, to ones which result in death either *in utero* or in early neonatal life. Thalassaemias can result from several different types of mutation. Chromosomal rearrangements which remove one or more genes from the complex are the commonest cause of α-thalassaemia (Higgs, 1993) and are also responsible for some β-thalassaemias. Point mutations affecting either single nucleotides or short regions of the globin genes are the major causes of β-thalassaemia (Thein, 1993), although some α-thalassaemias are known to be produced by this mechanism also. A final, less well-defined class of thalassaemias are caused by trans-acting mutations in other regions of the genome that affect levels of globin gene expression: these cases are rare, however, and not fully understood.

The thalassaemias are probably the commonest and most widely distributed class of genetic disorders in human populations. Their worldwide distribution closely matches that of endemic malaria in the years prior to eradication programs, and led Haldane to suggest that thalassaemic individuals might be protected in some way against this disease (Haldane, 1949). This association was more readily detected for β-thalassaemia (Siniscalco *et al.*, 1966); haemoglobinopathies arising from molecular lesions in the β-globin complex generally exhibit a more severe phenotype than their counterparts in the α-globin complex as the β-globin gene complex contains only one adult gene (Fig. 13.1). Removal of the function of this gene halves the amount of β-globin in the affected individual. As the α-globin complex contains two identical genes, a more varied range of clinical severity can be produced by the removal of one, two, three or all four adult genes. The milder forms of α-thalassaemia can only be detected reliably using molecular genetic techniques.

The first clear indication that levels of α-thalassaemia correlate with the endemicity of malaria was reported in the populations of western Oceania (Flint *et al.*, 1986). There is a strong geographical and altitudinal cline of malarial endemicity in the region (Fig. 13.2), but earlier studies had shown that other markers already known to be selected by malaria were absent or

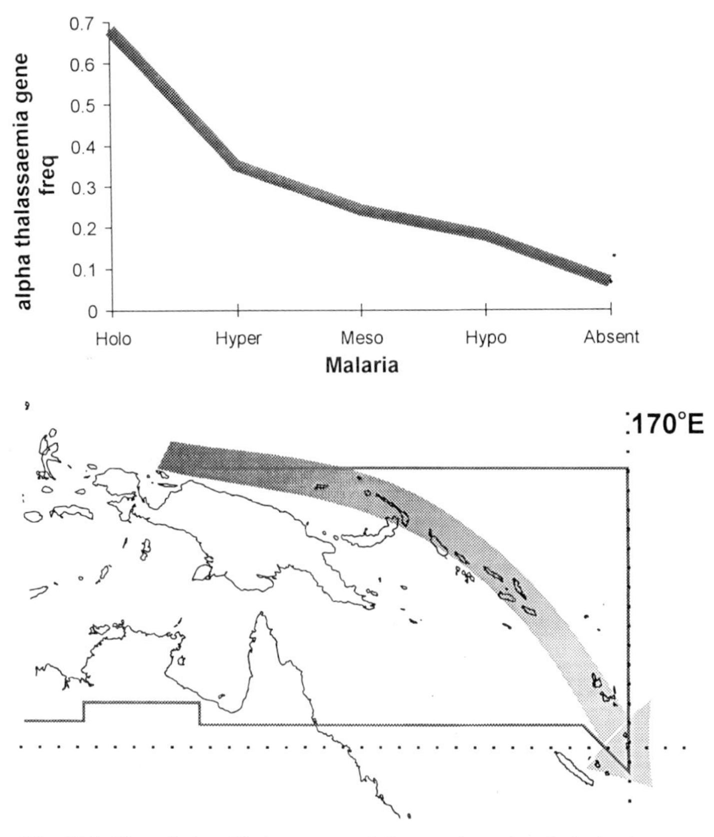

Fig. 13.2. The relationship between α-thalassaemia and malaria in the western Pacific. The correlation between α-thalassaemia frequencies and malarial endemicity is shown in the upper Figure. The lower Figure shows the geographical cline in both α-thalassaemia and malaria in a Southeast direction from Papua New Guinea. A similar reduction is seen with increasing altitude. Outside the latitude and longitude coordinates shown on the map, malaria is absent.

rare (Bowden *et al.*, 1985). Flint *et al.* found high levels of the mild form of α-thalassaemia, caused by the removal of one of the two α-globin genes, on the coast of Papua New Guinea (PNG) where malaria is holoendemic. The frequency of this deletion approached 0.7 in some areas, meaning that over 90% of the population is either heterozygous or homozygous for α-thalassaemia – individuals with the 'normal' phenotype are a small minority! Levels of the deletion decrease with increasing altitude further inland into PNG, and also decrease in the islands to the south-east. In both of these cases the endemicity of malaria shows a corresponding decline. In itself, this finding does not demonstrate the selective effect of malaria; the cline in gene frequency could instead reflect gene flow from a region of high α-thalassaemia into regions where the mutation was absent.

Further analysis of globin and other polymorphisms shows that this is not the case. Flint *et al.* studied polymorphisms at several other loci: no correlation could be detected between the distribution of their variants with either malarial endemicity or α-thalassaemia frequencies. The distribution of α-thalassaemia in this region instead reflects selection by malaria, as Haldane predicted. This conclusion is based solely on epidemiological observations, however, and the physiological advantage conferred by α-thalassaemia has yet to be established. Recent theoretical work has also shown that the population dynamics of malarial selection are more complex than a simple case of a selective advantage conferred by homozygous α-thalassaemia (Boyce, Harding & Martinson, 1995). Clearly, more can be learned by continued study of this system.

The detailed analysis carried out on the α-globin complex has also revealed the extensive variety of genetic lesions that can cause this mild form of α-thalassaemia. Characterisation of the molecular structure of α-thalassaemia deletion chromosomes reveals two common groups of deletion (Embury *et al.*, 1980). In one group, a deletion of 4.2 kb of DNA removes the $\alpha2$ gene (this deletion is denoted as $-\alpha^{4.2})$ in the other common group, a 3.7 kb region of DNA containing the α gene is deleted (the $-\alpha^{3.7}$ chromosome). These deletions can easily be discriminated by RFLP mapping: more extensive RFLP analysis further subdivides the $-\alpha^{3.7}$ chromosomes into three subtypes ($-\alpha^{3.7}$I, $-\alpha^{3.7}$II and $-\alpha^{3.7}$III) according to the precise chromosomal location of the deletion breakpoints (Higgs *et al.*, 1984). The studies of Flint *et al.* and others (Oppenheimer *et al.*, 1984; Yenchitsomanus *et al.*, 1985, 1986) have shown that more than one type of α-thalassaemia deletion is found in the malarious areas of the Pacific. the $-\alpha^{4.2}$ deletion is found mainly on the north coast of PNG and in island Southeast Asia, while the $-\alpha^{3.7}$I predominates on the south coast of PNG and island Southeast Asia. The $-\alpha^{3.7}$III is found mainly in Vanuatu. This shows that α-thalassaemia in this area has had at least three separate origins, and suggests that the different distributions of these mutations may tell us about the histories of the populations. One example makes this clear.

I stated earlier that the geographical distribution of haemoglobinopathies closely matches that of malaria. There is one exception to this observation. In Polynesia the gene frequency of α-thalassaemia is about 15% (and is fairly uniform at that level throughout the whole of Polynesia) despite the total absence of malaria in recorded history. Why should a population sustain such a high level of what is after all a mildly deleterious condition? The α-thalassaemia present in this region is not fatal, even in the homozygous state, but is associated with a greater risk of anaemia, especially in pregnancy (Philippon *et al.*, 1995). Characterization of the types of α-thalassaemia chromosome seen in the different regions of the Pacific provides an explanation. Virtually all of the α-thalassaemia seen in Polynesia is the $-\alpha^{3.7}$III deletion (Hill *et al.*, 1985; Philippon *et al.*, 1995). The genetic rearrangement that produces this deletion can only occur within a very small region of the α-globin complex (Higgs *et al.*, 1984); consequently it is unlikely to be produced very often by mutation. The only other region where the $-\alpha^{3.7}$III chromosome is seen is in the western Pacific, where it is the main type of α-thalassaemia chromosome in Vanuatu, and is also seen at low frequency further west in PNG. The high frequencies of this unusual mutation in Polynesia suggests that the populations who originally colonized the many islands of Polynesia originated from, or had substantial admixture with, the inhabitants of Vanuatu. The persistence of the $-\alpha^{3.7}$III deletion in Polynesia shows that the inhabitants of this region have not lived in the absence of malaria long enough for its frequency to be reduced by selection against the mildly deleterious phenotype. The genetic uniformity of the Polynesians, despite their wide geographical dispersal, is seen with other genetic loci (Hertzberg *et al.*, 1989*a*, *b*) and reflects a recent occupancy of the region involving a small founding population (Flint *et al.*, 1989; Martinson *et al.*, 1993).

Analysis of the distribution of a single variant allele is informative about the recent settlement of the eastern Pacific, but does it help to resolve the controversies concerning the original colonization of the area from the west? The above description of the distribution of α-thalassaemia in Papua New Guinea and island Southeast Asia would appear to suggest either an ancient common ancestry or substantial gene flow between the regions in the past. Such impressions are deceptive, however, as an analysis of the chromosomal background on which the deletions arose reveals.

Detailed studies of the α-globin complex, and its variation between individuals have suggested that, on average, 1 in every 80 bp is polymorphic. Many of these variants are rare and specific to one or another population group. Others, however, reach high frequencies and are found amongst most of the world's peoples. Extensive linkage disequilibrium has been demonstrated between these polymorphic sites, allowing the construction of linkage haplotypes (Higgs *et al.*, 1986). The structure of the α-globin haplotype, with some examples common in the Pacific region, is shown in Fig. 13.3.

Many studies have now been carried out on α-globin haplotype variation in the Pacific, both on normal and on α-thalassaemia chromosomes (O'Shaughnessy *et al.*, 1990; Flint, Clegg & Boyce, 1992). The original studies on non-thalassaemic chromosomes revealed a strong distinction between the present-day inhabitants of island Southeast Asia,

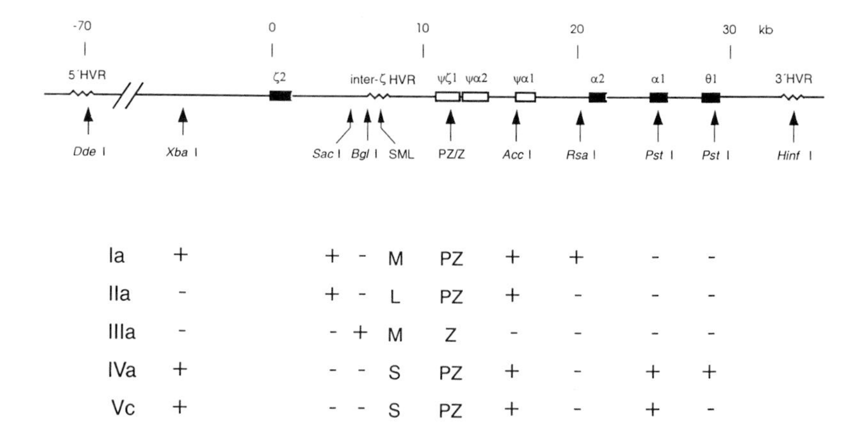

Fig. 13.3. The α-globin RFLP haplotype. The genes are marked as in Fig. 13.1. Zigzag lines mark the position of VNTR loci. The polymorphic loci used to construct the haplotype are indicated, together with the restriction enzymes used to detect the polymorphism. S, M, L refer to the Small, Medium and Large inter-ζ-HVR alleles discernible at the resolution used. Z and PZ refer to the length polymorphism produced by a VNTR within the intron of the ζ-gene, whose size variation is associated with the presence of either a ζ-gene (Z) or a $\psi\zeta$ pseudogene (PZ) at the locus shown. α-globin haplotype nomenclature classifies the major haplotype groups on the basis of shared polymorphisms at the 3′ end of the complex; thus the group I haplotypes end in PZ$++--$, group II in PZ$+---$, etc. Below the map are shown some of the haplotypes common in the populations of Oceania.

Melanesia and Polynesia. The inhabitants of Southeast Asia are characterized by high frequencies of haplotypes Ia and IIa, whereas the inhabitants of Melanesia predominantly exhibit haplotypes IIIa, IVa and Vc (summarized in Table 13.1). In Polynesia, all five haplotypes are seen, in the ratio 70:30 Southeast Asian:Melanesian. Thus the 'Polynesian' gene pool can in some ways be considered as a hybrid between 'Southeast Asian' and 'Melanesian' types (Fig. 13.4).

Haplotype analysis of the α-thalassaemia chromosomes is equally revealing. The apparent sharing of $-\alpha^{4.2}$ and $-\alpha^{3.7}$I chromosomes between the populations of Papua New Guinea and island Southeast Asia is shown to be misleading when the haplotypes on which the deletions occur are discerned. The $-\alpha^{4.2}$ and $-\alpha^{3.7}$I deletions in Southeast Asia occur on haplotypes Ia and IIa, whereas the same deletions in Melanesia are found on IIIa and IVa haplotypes (Flint *et al.*, 1992). Thus, the deletions arose independently in these two populations on chromosomes that are native to those populations – they were not introduced by migration from elsewhere.

The α-globin complex is, of course, only one genetic locus out of the

Table 13.1. *α-globin haplotype percentage frequencies in Oceania*

Haplotype	Southeast Asia	PNG Highlands	Australia	Island Melanesia	Polynesia	Micronesia
Ia	28.6		14.6	8.6	25.5	27.6
Ib	1.9		2.2			
Ia/b			2.2			
Ic	0.4			0.5		2.1
Id	4.5			0.5		2.1
IIa	24.8		11.7		5.9	11.5
IIb	0.4		1.5			
IIc	6.0		0.7		1.5	12.0
IId	1.5			4.3	8.6	12.5
IIe	6.4		4.4		6.0	2.6
IId/e	11.3			3.6	7.8	
IIf	0.8			1.1		2.1
IIg	2.3					0.5
IIf/g					0.7	
IIIa	8.3	57.9	44.5	28.8	13.4	13.0
IIIb		21.1	5.8	0.4		
IIIa/b			2.2	14.2	2.0	
IIIc	1.1				0.4	3.6
IIIe	0.8			1.1	1.3	0.5
IIIh	1.1				0.9	
IVa		5.3	8.0	42.2	12.2	6.8
IVb		5.3	0.7	0.1		0.5
IVa/b				10.1	3.8	
IVc			1.5			
Vb					0.1	1.0
Vc		5.3		11.2	1.8	
Vd		5.3				
chromosomes	266	38	137	369	697	192

The data simplified from Boyce *et al.* (1995). Some haplotypes were not completely determined, and are shown as Ia/b, etc.

hundreds of thousands present in the human genome, and any one locus has limitations as a marker for population variation. A robust consensus based on molecular genetic markers will only be obtained once a variety of different markers has been studied to the same extent. One or two other nuclear loci have been studied in the Pacific (Hill & Serjeantson, 1989), but the only system with a comparable amount of data available is mitochondrial DNA (mtDNA). This is a system with unique benefits, and unique problems.

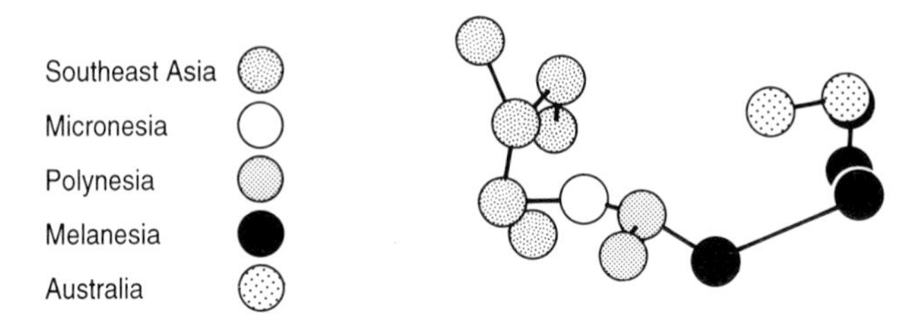

Fig. 13.4. Genetic distances between Pacific populations, based on α-globin haplotypes. This shows a multidimensional scaling analysis of Nei distances among Oceanic populations. The minimum-spanning tree (Gower & Ross, 1969) is superimposed on the plot. A simplified Table of the data used to generate the plot is shown in Table 13.1.

Mitochondrial DNA-variation ancient and modern

More individuals have been characterized in detail for mtDNA variation than for any other molecular genetic marker. The high degree of polymorphism that has been uncovered makes them ideal markers for population screening. The matrilineal haploid inheritance pattern means that the main cause of polymorphic variation is nucleotide substitution. This simplifies the analysis of mtDNA data; the effects of genetic recombination do not have to be taken into account, as they do with studies on most nuclear markers. Another key advantage of mtDNA is that mitochondria are present at very high copy numbers per human cell – a typical nucleated human cell contains around 10 000 mitochondria, but only one pair of each nuclear chromosome. This last feature means that there is a greater chance of recovery of mtDNA than of nuclear DNA from small quantities of poorly preserved material such as plucked hairs, cheek swabs and, most spectacularly, ancient bone from archaeological sites and museum specimens. Other markers can be informative about present-day variation, and permit historical relationships to be inferred on the basis of modern geographical variation, but only mtDNA allows us to look at historical variation directly.

Variation in mtDNA was initially characterized by RFLP analysis on restriction enzyme-digested DNA (Cann, Stoneking & Wilson, 1987). Although highly informative, this approach was superseded by PCR amplification (Wrischnik *et al.*, 1987). the characterization of mtDNA variation now takes two main forms:

> The small size of the mtDNA genome (16 kb) means that the entire genome can be amplified in a small number of fragments. These fragments can then be screened for RFLP

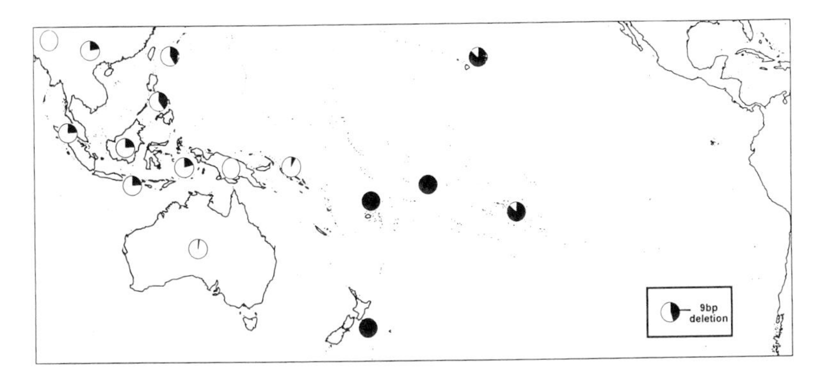

Fig. 13.5. The distribution of the mtDNA 9 bp deletion in the Pacific region. The frequency of the 9 bp deletion is given by the amount of black shading in each individual piechart. The data were collected from the references given in the text.

variation and haplotypes constructed that provide information on the variation in the whole mtDNA genome. One particular region of the mtDNA genome (the D-loop hypervariable region) shows especially high levels of variation which can be detected by PCR and direct DNA sequencing.

Both approaches have been used on samples from the Pacific region, although rather more D-loop region sequence data have been obtained.

The data first obtained from mtDNA seemed to indicate a very strong association between Polynesians and Southeast Asians. The initial study of PCR-amplified DNA found a characteristic deletion of 9 basepairs of DNA in several individuals of Southeast Asian origin (Wrischnik *et al.*, 1987). More extensive studies on populations in the region found that this deletion was found in 93% of the individuals studied from Polynesia, and in 82% of Fijian individuals studied. In contrast, it was absent in Papua New Guinea highlanders and Aboriginal Australians (Hertzberg *et al.*, 1989*a*). Other studies at around the same time showed that PNG highlanders have a greater amount of mtDNA variation – consistent with an ancient history of occupation in that region – than do inhabitants of the coast (Stoneking *et al.*, 1990). The PNG coastal variation was more indicative of recent population movements through the region, and includes some examples of the 9 bp deletion. The distribution of the 9 bp deletion throughout Southeast Asia and the Pacific is shown in Fig. 13.5.

The extensive distribution of the 9 bp deletion throughout Southeast Asia indicates an ancient origin for this mutation (Ballinger *et al.*, 1992).

More recent analyses do not just detect the presence or absence of the 9 bp deletion in the samples studied, but in addition report the sequence backgrounds on which the deletions are found. More detailed lineage analyses can thus be performed on the data, and these show that the deletion has probably arisen more than once in Southeast Asia, and has also been produced independently elsewhere in the world (Redd *et al.*, 1995). Sequence analysis has also shown that the differences between the 'Southeast Asian' mtDNAs and the 'Melanesian' types are very great (Lum *et al.*, 1994; Melton *et al.*, 1995; Sykes *et al.*, 1995). The lineage most strongly associated with the 9 bp deletion in Polynesia is seen as a component of the overall diversity in Southeast Asia, and lineages related to this 'Polynesian motif' are seen with the greatest frequency in the corridor running from Taiwan south and east through the Philippines and eastern Indonesia (Melton *et al.*, 1995). Some studies have been made of the non-deletion mtDNAs found in Polynesia: these are generally related to the complex and diverse set of lineages found in Vanuatu and Papua New Guinea, but only account for around 3.5% of the lineages seen in Polynesia (Sykes *et al.*, 1995).

The mtDNA data provide the strongest set of genetic evidence to support the 'Express Train' hypothesis (Gibbons, 1994), and are the most consistently cited to support the notion that the original settlers of the central Pacific, and hence Polynesia, were Southeast Asian in origin and passed through Melanesia without much interaction with the populations that already inhabited the region. The assumption is also that they spoke Austronesian languages and made Lapita pottery. But this last set of assumptions, which would establish the 'express train' as the only possible explanation of events, is simply not borne out by the totality of the data.

The unique feature of mtDNA that is so interesting to those anthropologists interested in human population history is that it can be extracted and analysed from archaeological material. Once the technical problems, chiefly concerned with avoiding contamination with contemporary material (amplified museum curator DNA will tell us little about the settlement of the Pacific), have been overcome then the potential is immense. The relatively few samples that have so far been studied from the Pacific, however, are already sufficient to derail the 'express train' for some (Hagelberg & Clegg, 1993; Gibbons, 1994). The oldest material studied dates to between 2500–2000 BP and was recovered from New Britain, Vanuatu and Fiji in sites associated with Lapita pottery (including the site generally agreed to be the oldest Lapita-associated burial in the Pacific). None of the samples studied contained the 9 bp deletion. This is only seen in the more recent samples from Polynesia, dating from 700 BP forwards. As one is dealing with negative evidence – the absence of an 'Austronesian' marker rather than the presence of a 'Melanesian' one – one must be cautious, but at the very least Hagelberg & Clegg's evidence gives little support to the 'express train', and at worst implies that the Lapita manufacturers could well have been 'Melanesian'.

A genetic consensus?

Can the genetic data be reconciled with any one of the views of Pacific colonization? It appears that different loci agree with different viewpoints. I have already shown how the α-globin data currently available indicate that the gene pools present in present-day Southeast Asia and Melanesia are clearly different from each other, but that modern-day Polynesians contain both sets of markers (Fig. 13.4). This leads to a view of Pacific settlement that lies somewhere between the two extremes described previously. I find it very difficult to account for the observed distribution in terms of one single ancient settlement followed by *in situ* differentiation: this would require the total loss of the 'Southeast Asian' haplotypes Ia and IIa from the populations of Papua New Guinea, and a corresponding loss of the 'Melanesian' IIIa, IVa and Vc haplotypes from island Southeast Asia. There have undoubtedly been more recent population movements into island Southeast Asia in the past 5-10 000 years (Bellwood, 1985) but it is unlikely that this would have resulted in the total replacement of the indigenous gene pool with that of the incoming population. Similarly, there is evidence of genetic drift within the populations of Melanesia (Giles, Wyber & Walsh, 1970), but I do not think that this has been sufficient to account for the removal of markers that are so common in the populations both to the east and west.

This does not mean either that I am an 'express-trainspotter'. The α-globin data suggest to me that a major component of the Polynesian gene pool was until recently located in island Southeast Asia. That population then expanded or diffused to the south and east into island Melanesia but that stage was sufficiently long-lasting, and with enough exchange with the existing inhabitants of the region, for that population to acquire a 30% component of Melanesian nuclear markers. The peoples who then went on to settle eastern Polynesia had their immediate origin in Vanuatu (as this is where the $-\alpha^{3.7}$III deletion arose) but had spent some considerable time there prior to their eastward migration. I do not feel that the extreme version of the Express Train view is justified by the data currently available: the data suggest to me that there was interaction and gene flow between the ancient inhabitants of the regions we now call Melanesia and Southeast Asia prior to the settlement of the eastern Pacific, but this did not result in a panmictic population. There were two ancient groups, but they communicated with each other.

It is definitely not the case that any population settling Polynesia from a Southeast Asian homeland did so because of any genetic advantage. Haplotype analysis of the different α-thalassaemia mutations present in Melanesia and Southeast Asia clearly shows that the deletions arose independently in the different populations, on chromosomes already native to those populations. This contrasts with the suggestion, based on Gm alleles, that the early settlers of the malarious areas of Melanesia originated from Southeast Asia and were genetically better adapted to the rigours of living in a malarious environment than were the indigenous inhabitants of Melanesia (Clark & Kelly, 1993). The α-globin data show that the oldest inhabitants of Melanesia were perfectly capable of generating their own resistance to malaria.

The expansion eastward from Papua New Guinea through to Vanuatu, Fiji and Polynesia could well have been made by a population that was not entirely 'Southeast Asian'. The settlement of eastern Polynesia undoubtedly involved small founding populations, possibly as small as a canoe-load (Alpers, 1987), with genetic consequences that have not yet been fully established (Flint *et al.*, 1989; Martinson *et al.*, 1993). The predominance of the 9 bp deletion in Polynesia could be a result of such demographic crises; certainly the effects on mtDNA would be different to those for nuclear markers. Also, one must not forget that mtDNA is maternally inherited and may not be giving us the full story. The emerging use of Y-chromosomal DNA as an alternative system (see Chapter 2) will be particularly informative in this regard (Pena *et al.*, 1995). It will also be rewarding in the future to acquire data from other nuclear loci in Oceanic populations to see if the 30% 'Melanesian' component of the gene pool of modern-day Polynesians is a common feature of nuclear loci, or whether the α-globin data themselves are reporting an unusually high degree of 'admixture' and that the 5% suggested by mtDNA is closer to the consensus value.

VNTR lineage analysis – a hierarchy of chromosomes

Variable number of tandem repeat (VNTR) loci (Nakamura *et al.*, 1987) are found widely spread throughout the human genome. Their high polymorphism arises from differences in the number of tandemly reiterated short sequence motifs that comprise the locus. In addition to this variation, a second level of polymorphism exists for most VNTR loci. Few loci are composed of just one type of sequence repeat: most have two or more repeat units of related sequence and their different arrangements within the VNTR array can easily give rise to enormous variation between alleles. It is

Table 13.2. *Structure of the α-globin 3′HVR*

Repeat	Sequence	Length/bp
a	aacagcgacacgggggg	17
*b*1	aacagcgacacgggagg	17
*b*2	aacagcgacacggggagg	18
*b*3	aacagcgacacgggggagg	19
*b*4	aacagcgacacggggggagg	20
c	aacagcgacacgggagggagg	21

The recognition sequence for the restriction enzyme *Mnl* I is shown underlined.

possible for two alleles at one VNTR locus to be similar or identical in size, but yet to be substantially different in internal composition. These internal structure differences can be revealed by internal mapping or sequencing (Armour & Jeffreys, 1992): determination of the internal structure of related VNTR alleles can reveal much about the mutation mechanisms by which VNTR allele diversity is generated (Jeffreys *et al.*, 1994).

The α-globin gene complex is flanked by two VNTR loci, the 5′HVR and the 3′HVR (Jarman *et al.*, 1986; Jarman & Higgs, 1988). Both of these loci are highly polymorphic: the 5′HVR array is composed of copies of a 57 bp repeat unit and has a heterozygosity of around 70% in most populations, whereas the 3′HVR is more variable still, with heterozygosities approaching 100% in some populations. This locus is composed of several related repeat units which differ both in length as well as in sequence (Table 13.2). A unique feature of the α-globin complex is that these VNTRs are separated by 100 kb of DNA, containing the α-globin RFLP haplotype: when allelic variation at all of these loci is considered together, the resulting 'superhaplotype' is a powerful marker for population analysis. The RFLP haplotype can be used to divide the total population of 3′HVR and 5′HVR alleles into subpopulations of alleles that share a linkage relationship with one or another of the haplotypes. This has been done for the populations of eastern Polynesia, and shows that the various haplotypes present are each associated with significantly different size distributions of 3′HVR alleles (Martinson, Boyce & Clegg, 1994).

A still finer subdivision can be made in the populations of Oceania. The $-\alpha^{3.7}$III deletion is present at high frequencies in many Oceanic populations due to malarial selection and genetic drift. Several hundred examples of this deletion have been studied to date, and all of them are on chromosomes bearing a IIIa haplotype (Hill *et al.*, 1985; Martinson, Boyce & Clegg, 1994; Philippon *et al.*, 1995), indicating that this deletion had a single origin arising from a chromosomal rearrangement involving a IIIa

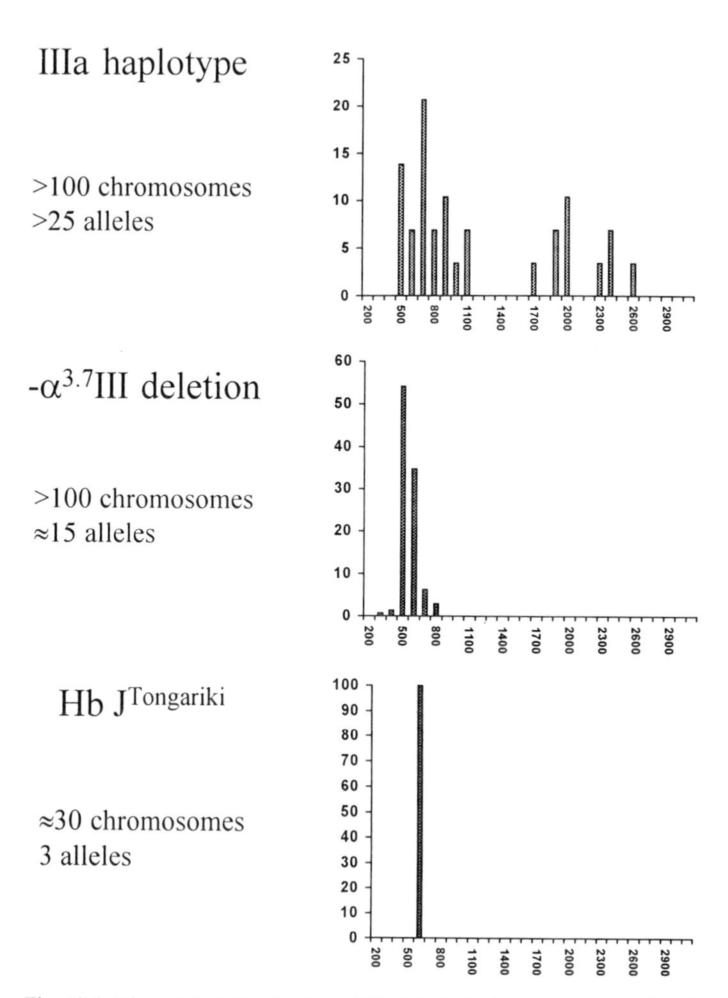

Fig. 13.6. Hierarchical distribution of 3′HVR alleles in western Oceania. The percentage frequencies of each size of 3′HVR allele are shown for the three populations described. The allele size intervals on the horizontal axis are larger than the 3′HVR repeat unit; consequently each bar represents more than one size of 3′HVR allele.

haplotype. All the $-\alpha^{3.7}$III chromosomes in Polynesia are the descendants of this progenitor, which would have carried one size of 3′HVR allele. All of the 3′HVR alleles associated with the $-\alpha^{3.7}$III chromosome in Oceania, therefore, form a haploid lineage descended from this progenitor allele, giving rise to a distinct subset of the wider population of IIIa haplotype-associated alleles[3].

[3] The 3′HVR is so close to the α-globin genes that recombination events between the loci, introducing new 3′HVR alleles, is unlikely.

This hierarchy can be extended by one more level. The variant haemoglobin known as $HbJ^{Tongariki}$ is present in many populations of western Oceania (Gadjusek *et al.*, 1967; Livingstone, 1975). It is always found in association with α-thalassaemia in these populations (Old *et al.*, 1978; Bowden *et al.*, 1980) and is now known to have arisen on a $-\alpha^{3.7}$III deletion chromosome, probably in Vanuatu (Hill, O'Shaughnessy & Clegg, 1989). As no examples of $HbJ^{Tongariki}$ have yet been seen in eastern Polynesia, this variant was probably produced recently, after the initial settlement of central Polynesia (alternatively, it could have arisen earlier, but in a population that had little genetic input into the gene pool of the ancestral Polynesian population).

This three-level hierarchy of chromosome lineages is reflected in the distribution of 3′HVR alleles found associated with each member, as shown in Fig. 13.6. The greatest diversity is seen with the IIIa haplotype chromosomes, reflecting their greater age. The IIIa haplotype chromosome on which the $-\alpha^{3.7}$III mutation arose appears to have carried a 3′HVR allele in the 500–800 bp size range, as this is the range of alleles associated with the deletion. The small range in size indicates that the majority of the mutation events that generate VNTR allele diversity involve the gain or loss of small numbers of repeat units, rather than large-scale changes. This is in agreement with many other studies of VNTR allele mutation processes (Jeffreys *et al.*, 1988).

Although a large number of different alleles is seen associated with the $-\alpha^{3.7}$III chromosome in Oceania, three alleles make up the majority of the distribution, with the remainder present at lower frequencies. Only two of those three alleles are present at high levels in the populations of eastern Polynesia: a higher diversity of alleles is seen in Vanuatu, where malaria is present. It is still not clear whether the majority of those alleles existed at the time of the settlement of Polynesia, but were not carried out (reinforcing the idea of a population bottleneck during the settlement of the region), or whether they have been produced subsequently. If malarial selection was acting to raise the frequency of the $-\alpha^{3.7}$III and other α-thalassaemia mutations in western Oceania over the past few thousand years, then any new 3′HVR alleles produced on α-thalassaemia chromosomes would have a higher chance of being retained in the population, due to the hitchhiking effect of the nearby α-thalassaemia mutation. In the absence of malaria, newly produced 3′HVR alleles on α-thalassaemia chromosomes would be subject to random genetic drift and would be less likely to persist in the gene pool.

A relatively recent origin for the $HbJ^{Tongariki}$ mutation is supported by the 3′HVR allele frequency distribution. Although Fig. 13.6 indicates that three alleles are associated with the $HbJ^{Tongariki}$ mutation, two of these

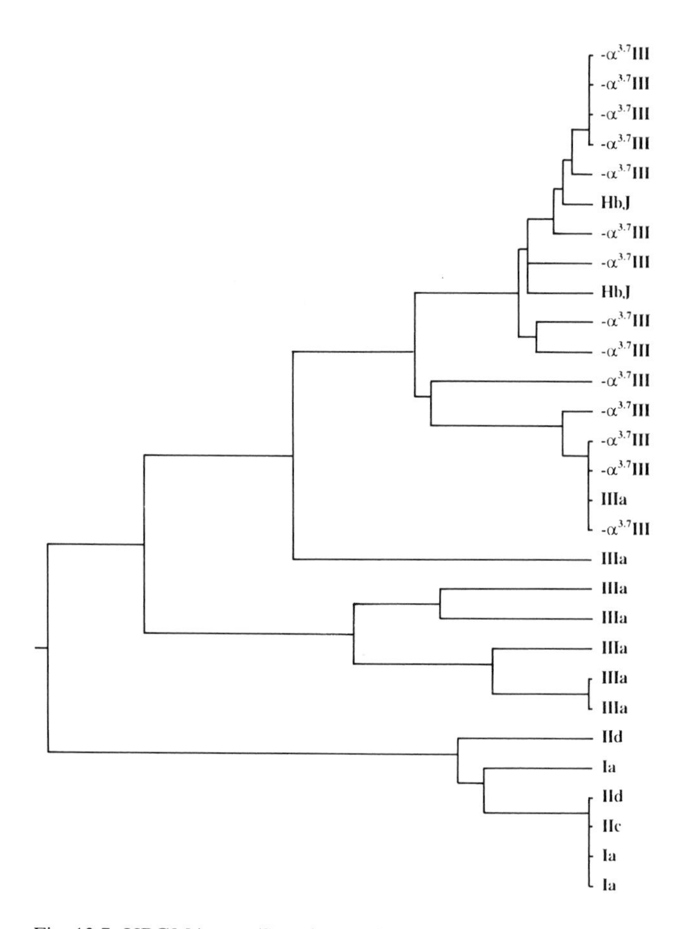

Fig. 13.7. UPGMA tree (Sneath & Sokal, 1973) based on maps for several different haplotypes as well as $-\alpha^{3.7}$III and HbJ^{Tongariki} chromosomes, obtained from different samples in eastern Polynesia, Southeast Asia and Vanuatu.

alleles are singletons – only one copy of each has been detected to date. The great majority of 3′HVR alleles associated with the HbJ^{Tongariki} mutation are of one size – and this allele is the one that is not seen in Polynesian samples. The widespread distribution of the HbJ^{Tongariki} mutation throughout western Oceania indicates that substantial amounts of gene flow have persisted throughout this region in the recent past.

The repeat unit variation within the 3′HVR allele array (Table 13.2) can be used to explore the population relationships in this region further. As many of the repeat units contain the recognition sequence for the restriction enzyme *Mnl* I it is possible to determine the repeat unit structure of PCR-amplified 3′HVR alleles by RFLP mapping (Smith & Birnstiel, 1976;

Martinson *et al.*, 1994). Clear differences emerge when alleles associated with different haplotypes are compared, and the hierarchical relationship between the IIIa haplotype, $-\alpha^{3.7}$III and HbJ^{Tongariki} alleles is clearly represented, as in Fig. 13.7. This shows a UPGMA tree (Sneath & Sokal, 1973) based on maps for several different haplotypes as well as $-\alpha^{3.7}$III and HbJ^{Tongariki} chromosomes, obtained from different samples in eastern Polynesia, Southeast Asia and Vanuatu. The deepest branch is between the cluster containing the Ia, IIc and IId haplotypes and the cluster containing the IIIa and related haplotypes. This reflects the substantial differences that exist between the haplotypes of 'Southeast Asian' origin and those of 'Melanesian' origin – for example, the Ia, IIc and IId haplotype-associated 3′HVR alleles contain c-type repeats, which are absent from the IIIa-related alleles.

Within the cluster of IIIa-related alleles, it can be seen that the next two deepest branches separate off some IIIa haplotype-associated alleles from the cluster of IIIa, $-\alpha^{3.7}$III and HbJ^{Tongariki} 3′HVR alleles. This shows that there is some heterogeneity in allele structure within the IIIa haplotype-associated set, and that the $-\alpha^{3.7}$III mutation arose on one chromosome whose allele has some closely-related alleles, and some more distantly-related ones, in the same population. Given the age of the IIIa haplotype in this region, that is not surprising. Many branches of equal or nearly-equal length separate many of the $-\alpha^{3.7}$III and HbJ^{Tongariki} 3′HVR alleles, indicating that their repeat unit compositions are rather similar, and that short mutational steps are responsible for the allelic variation.

References

Allen, J., Flannery, T., Gosden, C., Jones, R. & White, J. P. (1988). Pleistocene dates for the human occupation of New Ireland, Northern Melanesia. *Nature*, **331**, 707–9.

Alpers, A. (1987). *The World of the Polynesians*. Auckland: Oxford University Press.

Armour, J. A. L. & Jeffreys, A. J. (1992). Biology and applications of human minisatellite loci. *Current Opinion in Genetics and Development*, **2**, 850–6.

Ballinger, S. W., Schurr, T. G., Torroni, A., Gan, Y. Y., Hodge, J. A., Hassan, K., Chen, K. -H. & Wallace, D. C. (1992). Southeast Asian mitochondrial DNA analysis reveals genetic continuity of ancient mongoloid migrations. *Genetics*, **130**, 139–52.

Bellwood, P. S. (1978). *Man's Conquest of the Pacific*. Auckland: Collins.

Bellwood, P. S. (1985). *Prehistory of the Indo-Malaysian Archipelago*. Sydney: Academic Press.

Bellwood, P. S. (1987). *The Polynesians: Prehistory of an Island people*. London: Thames & Hudson.

Bellwood, P. S. (1989). The colonisation of the Pacific: some current hypotheses. In

The Colonization of the Pacific: A Genetic Trail, ed. A. V. S. Hill and S. W. Serjeantson, pp.1–39. Oxford: Clarendon Press.

Birdsell, J. B. (1977). The recalibration of a paradigm for the first peopling of Greater Australia. In *Sunda and Sahul*, ed. J. Allen, J. Golson and R. Jones, pp. 113–67. Academic Press.

Bowden, D. K., Pressley, L., Higgs, D. R., Clegg, J. B. & Weatherall, D. J. (1980). α-Globin gene deletions associated with Hb J Tongariki. *British Journal of Haematology*, **51**, 243–9.

Bowden, D. K., Hill, A. V. S., Higgs, D. R., Weatherall, D. J. & Clegg, J. B. (1985). Relative roles of genetic factors, dietary deficiency, and infection in Vanuatu, south-west Pacific. *Lancet*, **ii**, 1025–8.

Boyce, A. J, Harrison, G. A., Platt, C. M., Hornabrook, R. W., Serjeantson, S. W. & Booth, P. B. (1978). Migration and genetic diversity in an island population. *Proceedings of the Royal Society of London Series B*, **202**, 269–95.

Boyce, A. J., Harding, R. M. & Martinson, J. J. (1995). Population genetics of the α-globin complex in Oceania. In *Human Populations: Diversity and Adaptation*, ed. A. J. Boyce and V. Reynolds, pp. 217–32. Oxford: Oxford University Press.

Cann, R. L., Stoneking, M. & Wilson, A. C. (1987). Mitochondrial DNA and human evolution. *Nature*, **325**, 31–6.

Clark, J. T. & Kelly, K. M. (1993). Human genetics, paleoenvironments, and malaria: relationships and implications for the settlement of Oceania. *American Anthropologist*, **95**, 612–30.

Diamond, J. M. (1988). Express train to Polynesia. *Nature*, **326**, 307–8.

Embury, S. H., Miller, J. A., Dozy, A. M., Kan, Y. W., Chan, V. & Todd, D. (1980). Two different molecular organisations account for the single α-globin gene of the α-thalassemia-2 genotype. *Journal of Clinical Investigation*, **66**, 1319–25.

Flint, J., Hill, A. V. S., Bowden, D. K., Oppenheimer, S. J., Sill, P. R., Serjeantson, S. W., Bana-Koiri, J., Bhatia, K., Alpers, M. A., Boyce, A. J., Weatherall, D. J. & Clegg, J. B. (1986). High frequencies of α-thalassaemia are the result of natural selection by malaria. *Nature*, **321**, 744–9.

Flint, J., Boyce, A. J., Martinson, J. J. & Clegg, J. B. (1989). Population bottlenecks in Polynesia revealed by minisatellites. *Human Genetics*, **83**, 257–63.

Flint, J., Clegg, J. B. & Boyce, A. J. (1992). Molecular genetics of globin genes and human population structure. In *Molecular Applications in Biological Anthropology*, ed. E. J. Devor, pp.103–78. Cambridge: Cambridge University Press.

Gadjusek, D. C., Guiart, J., Kirk, R. L., Carrell, R. W., Irvine, D., Kynoch, P. A. M. & Lehmann, H. (1967). Haemoglobin J Tongariki (α115 alanine→aspartic acid): the first new haemoglobin variant found in a Pacific (Melanesian) population. *Journal of Medical Genetics*, **4**, 1–6.

Gibbons, A. (1994). Genes point to a new identity for Pacific pioneers. *Science*, **263**, 32–3.

Giles, E., Wyber, S. & Walsh, R. J. (1970). Microevolution in New Guinea: additional evidence for genetic drift. *Archaeology and Physical Anthropology in Oceania*, **5**, 60–72.

Gower, J. C., & Ross, G. J. S. (1969). Minimum spanning trees and single linkage cluster analysis. *Applied Statistics*, **18**, 54–64.

Green, R. C. (1979). Lapita. In *The Prehistory of Polynesia*, ed. J. D. Jennings, pp. 27–60. Cambridge, MA: Harvard University Press.

Groube, L. (1993) Contradictions and malaria in Melanesian and Australian prehistory. In *Occasional Papers in Prehistory, No.* 21: *A Community of Culture: the People and Prehistory of the Pacific*, ed. M. Spriggs, D. E. Yen, W. Ambrose, R. Jones, A. Thorne and A. Andrews, pp. 164–86. Canberra: Australian National University.

Groube, L., Chappell, J., Muke, J. & Price, D. (1986). A 40,000 year-old human occupation site at Huon Peninsula, Papua New Guinea. *Nature*, **324**, 453–5.

Hagelberg, E. & Clegg, J. B. (1993). Genetic polymorphisms in prehistoric Pacific Islanders determined by analysis of ancient bone DNA. *Proceedings of the Royal Society of London Series B*, **252**, 163–70.

Haldane, J. B. S. (1949). The rate of mutation of human genes. *Proceedings of the 8th International Congress of Genetics. Hereditas* (supplement), **35**, 267–73.

Hertzberg, M., Mickleson, K. N. P., Serjeantson, S. W., Prior, J. F. & Trent, R. J. (1989*a*). An Asian-specific 9-bp deletion of mitochondrial DNA is frequently found in Polynesians. *American Journal of Human Genetics*, **44**, 504–10.

Hertzberg, M., Jahromi, K., Ferguson, V., Dahl, H. H. M., Mercer, J., Mickleson, K. N. P. & Trent, R. J. (1989*b*). Phenylalanine hydroxylase gene haplotypes in Polynesians: evolutionary relationships and absence of alleles associated with severe phenylketonuria. *American Journal of Human Genetics*, **44**, 382–7.

Higgs, D. R. (1993). α-Thalassaemia. In *The Haemoglobinopathies*, ed. D. R. Higgs and D. J. Weatherall, pp. 117–50. London: Baillière Tindall.

Higgs, D. R. & Weatherall, D. J. (1993). *The Haemoglobinopathies.* Ballière's Clinical Haematology. London: Ballière Tindall.

Higgs, D. R., Hill, A. V. S., Bowden, D. K., Weatherall, D. J. & Clegg, J. B. (1984). Independent recombination events between the duplicated human α-globin genes: implications for their concerted evolution. *Nucleic Acids Research*, **12**, 6965–77.

Higgs, D. R., Wainscoat, J. S., Flint, J., Hill, A. V. S., Thein, S. L., Nicholls, R. D., Teal, H., Ayyub, H., Peto, T. E. A., Falusi, Y., Jarman, A. P., Clegg, J. B. & Weatherall, D. J. (1986). Analysis of the human α-globin gene cluster reveals a highly informative locus. *Proceedings of the National Academy of Sciences, USA*, **83**, 5165–9.

Hill, A. V. S. & Serjeantson, S. W. (eds). (1989). *The Colonisation of the Pacific: A Genetic Trail.* Oxford: Clarendon Press.

Hill, A. V. S., Bowden, D. K., Kent, R. J., Higgs, D. R., Oppenheimer, S. J., Thein, S. L., Mickelson, K. N. P., Weatherall, D. J. & Clegg, J. B. (1985). Melanesians and Polynesians share a unique α-thalassemia mutation. *American Journal of Human Genetics*, **37**, 571–80.

Hill, A. V. S., O'Shaughnessy, D. F. & Clegg, J. B. (1989). Haemoglobin and globin gene variants in the Pacific. In *The Colonisation of the Pacific: A Genetic Trail*, ed. A. V. S. Hill and S. W. Serjeantson, pp. 246–85. Oxford: Clarendon Press.

Irwin, G. (1992). *The Prehistoric Exploration and Colonisation of the Pacific.* Cambridge: Cambridge University Press.

Jarman, A. P. & Higgs, D. R. (1988). A new hypervariable marker for the human α-globin gene cluster. *American Journal of Human Genetics*, **43**, 249–56.

Jarman, A. P., Nicholls, R. D., Weatherall, D. J., Clegg, J. B. & Higgs, D. R. (1986). Molecular characterisation of a hypervariable region downstream of the human α-globin gene cluster. EMBO Journal, **5**, 1857–63.

Jeffreys, A. J., Royle, N. J., Wilson, V. & Wong, Z. (1988). Spontaneous mutation rates to new length alleles at tandem-repetitive hypervariable loci in human

DNA. *Nature*, **332**, 278–81.

Jeffreys, A. J., Tamaki, K., MacLeod, A., Monckton, D. G., Neil, D. L. & Armour, J. A. L. (1994). Complex gene conversion events in germline mutation at human minisatellites. *Nature Genetics*, **6**, 136–45.

Kirch, P. V. (1986). Rethinking east Polynesian prehistory. *Journal of the Polynesian Society*, **95**, 9–40.

Kirk, R. L. (1980). Language, genes and peoples in the Pacific. In *Population Structure and Genetic Disorders*, ed. A. N. Eriksson, H. Forsius, H. R. Neranlinna, P. L. Workman and P. K. Norio, pp. 113–26. London: Academic Press.

Levinson, M., Ward, R. G. & Webb, J. W. (1973). *The Settlement of Polynesia: A Computer Simulation*. Minneapolis: University of Minnesota Press.

Livingstone, F. B. (1975). *Frequencies of Hemoglobin Variants*. Oxford: Oxford University Press.

Lum, J. K., Rickards, O., Ching, C. & Cann, R. L. (1994). Polynesian mitochondrial DNAs reveal three deep maternal lineage clades. *Human Biology*, **66**, 567–90.

Martinson, J. J., Harding, R. M., Philippon, G., Flye Sainte-Marie, F., Roux, J., Boyce, A. J. & Clegg, J. B. (1993). Demographic reductions and genetic bottlenecks in humans: minisatellite allele distributions in Polynesia. *Human Genetics*, **91**, 445–50.

Martinson, J. J., Boyce, A. J. & Clegg, J. B. (1994). VNTR alleles associated with the α-globin locus are haplotype and population related. *American Journal of Human Genetics*, **55**, 513–25.

Melton, T., Peterson, R., Redd, A. J., Saha, N., Sofro, A. S. M., Martinson, J. J. & Stoneking, M. (1995). Polynesian genetic affinities with Southeast Asian populations as identified by mtDNA analysis. *American Journal of Human Genetics*, **57**, 403–14.

Nakamura, Y., Leppert, M., O'Connell, P., Wolff, R., Holm, T., Culver, M., Martin, C., Fujimoto, E., Hoff, M., Kumlin, E. & White, R. (1987). Variable Number of Tandem Repeat (VNTR) markers for human gene mapping. *Science*, **235**, 1616–22.

Old, J. M., Clegg, J. B., Weatherall, D. J. & Booth, P. B. (1978). Haemoglobin J Tongariki is associated with α-thalassaemia. *Nature*, **273**, 319–20.

Oppenheimer, S. J., Higgs, D. R., Weatherall, D. J., Barker, J. & Spark, R. A. (1984). α-thalassemia in Papua New Guinea. *Lancet*, **i**, 424–6.

O'Shaughnessy, D. F., Hill, A. V. S., Bowden, D. K., Weatherall, D. J. & Clegg, J. B. (1990). Globin genes in Micronesia: origins and affinities of Pacific Island peoples. *American Journal of Human Genetics*, **46**, 144–55.

Pena, S. D. J., Santos, F. R., Bianchi, N. O., Bravi, C. M., Carnese, F. R., Rothhammer, F., Gerelsaikhan, T., Munkhtuja, B. & Oyunsuren, T. (1995). A major founder Y-chromosome haplotype in Amerindians. *Nature Genetics*, **11**, 15–16.

Philippon, G., Martinson, J. J., Rugless, M. J., Moulia-Pelat, J. P., Plichart, R., Roux, J. F., Martin, P. M. V. & Clegg, J. B. (1995). α-Thalassaemia and globin gene rearrangements in French Polynesia. *European Journal of Haematology*, **55**, 171–7.

Pietrusewsky, M. (1985). The earliest Lapita skeleton from the Pacific. *Journal of the Polynesian Society*, **94**, 389–414.

Redd, A. J., Takezaki, N., Sherry, S. T., McGarvey, S. T., Sofro, A. S. M. & Stoneking, M. (1995). Evolutionary history of the COII/tRNA^{Lys} intergenic 9 base pair deletion in human mitochondrial DNAs from the Pacific. *Molecular Biology and Evolution*, **12**, 604–15.

Serjeantson, S. W., Kirk, R. L. & Booth, P. B. (1983). Linguistic and genetic differentiation in New Guinea. *Journal of Human Evolution*, **12**, 77–92.

Siniscalco, M., Bernini, L., Filippi, G., Latte, B., Khan, P.M., Piomelli, S. & Rattazzi, M. (1966). Population genetics of haemoglobin variants, thalassaemia and glucose-6-phosphatase dehydrogenase deficiency, with particular reference to the malaria hypothesis. *Bulletin of the World Health Organisation*, **4**, 379–93.

Smith, H. O. & Birnstiel, M. L. (1976). A simple method for DNA restriction site mapping. *Nucleic Acids Research*, **3**, 2387–98.

Sneath, P. H. A. & Sokal, R. R. (1973). *Numerical Taxonomy*. San Francisco: Freeman.

Stoneking, M., Jorde, L. B., Bhatia, K. & Wilson, A. C. (1990). Geographic variation in human mitochondrial DNA from Papua New Guinea. *Genetics*, **124**, 717–33.

Sykes, B., Leiboff, A., Low-Beer, J., Tetzner, S. & Richards, M. (1995). The origins of the Polynesians: an interpretation from mitochondrial lineage analysis. *American Journal of Human Genetics*, **57**, 1463–75.

Terrell, J. (1986). *Prehistory in the Pacific Islands*. Cambridge: Cambridge University Press.

Terrell, J. (1988). History as a family tree, history as an entangled bank: constructing images and interpretations of prehistory in the South Pacific. *Antiquity*, **62**, 642–57.

Terrell, J. (1994). *Lapita as History and Culture Hero*. Unpublished manuscript.

Terrell, J. & Fagan, J. L. (1975). The savage and the innocent: sophisticated techniques and naive theory in the study of human population genetics in Melanesia. *Yearbook of Physical Anthropology*, **19**, 1–18.

Thein, S. L. (1993). β-Thalassaemia. In *The Haemoglobinopathies*, ed. D. R. Higgs and D. J. Weatherall, pp. 151–75. London: Baillière Tindall.

Welsch, R. E., Terreu, J. E. & Nodalski, J. A. (1992). Language and culture on the north west of New Guinea. *American Anthropologist*, **94**, 560–600.

White, J. P. & O'Connell, J. F. (1982). *A Prehistory of Australia, New Guinea and Sahul*. Sydney: Academic Press.

Wrischnik, L. A., Higuchi, R. G., Stoneking, M., Erlich, H. A., Arnheim, N. & Wilson, A. C. (1987). Length mutation in human mitochondrial DNA: direct sequencing of enzymatically amplified DNA. *Nucleic Acids Research*, **15**, 529–42.

Wurm, S. A. (1983). Linguistic prehistory in the New Guinea area. *Journal of Human Evolution*, **12**, 25–35.

Yenchitsomanus, P. T., Summers, K. M., Bhatia, K., Cattani, J. & Board, P. G. (1985). Extremely high frequencies of α-globin gene deletions in Madang and on Kar Kar Island, Papua New Guinea. *American Journal of Human Genetics*, **37**, 778–84.

Yenchitsomanus, P. T., Summers, K. M., Board, P. G., Bhatia, K. K., Jones, G. L., Johnston, K. & Nurse, G. T. (1986). Alpha-thalassaemia in Papua New Guinea. *Human Genetics*, **74**, 432–7.

14 *Population ancestry on Tristan da Cunha – the evidence of the individual*

D. F. ROBERTS AND H. SOODYALL

As with gene data, knowledge of frequencies of molecular variants from human population samples helps in tracing population affinities, histories, diversification. Other applications of practical interest are at the individual level, mostly in a forensic context for identification of individuals. In addition to these, however, knowledge of individual DNA constitution may help to illuminate the more remote genetic history of populations, to show where their genes come from. This is illustrated by the example of Tristan da Cunha. The data are from the full report by Soodyall *et al.* (1995).

The population of Tristan da Cunha, a small island in the middle of the south Atlantic ocean, dates from the arrival there of a British garrison in 1816. During the course of its history, 15 women (Table 14.1, column 1) brought genes to the population, excluding those transient individuals who were shipwrecked and stayed on the island for only a few weeks or months before they could resume their journeys. Many of these genes were lost from the population as the women or their descendants died or left the island (Roberts, 1968), so that to the nuclear gene pool of the population of today (December 31, 1994) only seven women have contributed (Table 14.1, column 3). Altogether, today's gene pool was endowed by 22 ancestors, and there are only seven surnames among the 300 members of the population.

The inbreeding and endogamy that have occurred have produced a population closely knit by kinship (Roberts & Bear, 1980), in which all individuals share a strong physical likeness, with no sign of clustering of morphological features in the surname groups (Brothwell & Harvey, 1965). This striking similarity in appearance, essentially European, masks a heterogeneity of origin. The geographical provenances of all those who introduced genes to the population are known, but not necessarily the origin of their genes. All the male ancestors were of western European

Table 14.1. *Female contributors to the genetic constitution of the Tristan population*

These brought genes to the population	Arrived	Percentage contribution to the 1989 gene pool (nuclear)	*Descendants with SSO type*		
			Observed Number	%	Expected Number
M. L. (Mrs W. G.)	1816	7	11	6.8	24.5
Mrs S. W.	1821	-	-		
Mrs J. M.	1822	-	-		
S. W. (Mrs T. S.)	1827	16	46	28.6	56
M. W. (Mrs A. C.)	1827	4	34	21.1	14
My. W. (Mrs P. G.)	1827	3	-		
Mrs G. P.	1827	-	-		
Mrs P. P.	1827	-	-		
S. K. (Mrs R. R.)	1827	-	-		
F. R. (Mrs C. C.)	1870	-	-		
S. P. (Mrs S. S.)	1862	10	45	28.0	35
R. B. (Mrs T. C.)	1882	-	-		
An. S. (Mrs J. H.)	1908	-	-		
Ag. S. (Mrs J. G.)	1908	6 }	25	15.5	31.5
E. S. (Mrs R. G.)	1908	3 }			
		49	161	100.00	161

origin (all except two were of north-west European derivation), and this presumably accounts for the strongly European appearance. There are, however, in a minority of the population occasional non-European features. These include a slightly darker pigmentation, broader nose, prognathism, appearances which, together with somewhat elevated frequencies of the African Gm haplotypes 1,5,6,14,17 and 1,5,6,17, the Gd^A allele of the G6PD system, Rhesus haplotype cDe, and the ADA^2 allele (Jenkins, Beighton & Steinberg, 1985), suggest some African ancestry. These it seems were introduced by some of the women, for they were of more varied origins than the men. On these, their origins, the mitochondrial DNA (mtDNA), with its unilineal mode of transmission from mother to offspring, may be expected to provide useful information.

The founding mothers

Of the seven ancestresses who contributed to today's gene pool, one arrived in 1816, three in 1827, one in 1862, and two in 1908. With one exception, all the women's names were English and are non-informative as regards origin. For the exception, the wife of William Glass the founder of the settlement, her Christian name (Maria Magdalena) has been variously attributed as Portuguese and Dutch, and her surname (Leanders) is clearly Dutch. Born in the Cape, she married William Glass in 1814 and two years later accompanied him to Tristan when posted there as a member of the garrison. People who met her, e.g. Augustus Earle in 1824 (subsequently the draughtsman on the 'Beagle') and William Taylor (1856), the first resident clergyman on the island, described her as a Cape Creole but it is not clear whether the term at that time signified a hybrid or the offspring born overseas of two European parents. The sketch by Earle of the first two Glass children shows them as fully European in appearance, with no sign of coloured ancestry. Henriksen & Oeding (1946) list her as 'Dutch(?)', and Munch (1971) accepts her as a 'Cape Creole of Dutch extraction' which he interpreted as meaning that she was born at the Cape of European parents whom he concluded were Dutch.

There is less difficulty about the pair of sisters who were the latest to arrive. They came to the island from the Cape, the youngest being born there and her older sister in Ireland, and there is no reason to doubt their designation as Anglo-Irish. They, and their third sister, all married islanders and all accompanied their husbands to Tristan in 1908, though the third sister returned to the Cape and did not contribute to the present gene pool.

Little is known about the women from St Helena. Of the five adults who were brought in April 1827 as wives for the then five unmarried men on Tristan, two of them were recorded in Taylor's 1851 census as sisters: he noted that the older of them had an English father. Elsewhere she is referred to as a Negress and the other four women as mulattoes. The older sister (S.W.) brought her four children with her to Tristan, one of whom (My W.) contributed genes to the present population; the other children did not, nor did the three women other than the sisters. The latest arrival from St. Helena (S.P.) in 1862 married there and accompanied her husband back to Tristan. She too is referred to as coloured.

Materials and methods

Sequence-specific oligonucleotide (SSO) probes were applied which had been designed (Stoneking *et al.*, 1991) to detect specific mutations in the control region of the mtDNA, where there is a great deal of diversity within and between populations. These probes define four regions in hypervariable segment I (located between positions 15 996 and 16 401 in the published sequence (Anderson *et al.*, 1981)) and five in segment II (between positions 29 and 408). MtDNA variation in these nine sequence locations was assessed using 24 specific SSO probes, and then the SSO type for each specimen classified by considering the variant observed in each location in the following order – IA, IB, IE, IC, ID, IIA, IIB, IIC, IID. This procedure was applied to specimens collected in 1982 from 161 Tristan da Cunha islanders and provided their SSO types. For confirmation of the validity of these types and for greater refinement, the two hypervariable regions were sequenced in 75 of the specimens and compared with the reference sequence (Anderson *et al.*, 1981).

Results and discussion

The SSO typing analysis in the 161 specimens showed five SSO types to be present. Polymorphism was observed in eight of the nine locations, but location IE was monomorphic. The most frequent type (31111-2131) was present in 46 subjects, and the second most frequent (12131-2321) in 45. The third type (12112-2231) occurred in 34, and the fourth (11111-1311) in 25. The least frequent (32131-2022) was found in 11. The ancestry of each of the 161 individuals was traced back maternally. The ancestry of all subjects with a particular type converged on one of the founding mothers. The most frequent type traced to S.W., the second most common to S.P., and the third to M.W. The descendants of the pair of Anglo-Irish sisters are all identical in type and their ancestries converge on one or the other sister. The least frequent type traces to M.L. These results confirm that the Anglo-Irish women were sisters born of the same mother. By contrast, they show that the two St Helena women listed as sisters were not full sisters and had different mothers, so they may have been half-sisters with a common father, as their names suggest, or may not have been biologically related at all.

The number of subjects of each SSO type (Table 14.1, column 4) was compared with that expected from the proportional contributions of these six ancestresses to the nuclear gene pool (column 6). There is a highly significant difference ($\chi^2 = 42.0$, d.f. $= 4$). The proportion in the population of those with mtDNA derived from S.P. and particularly M.W. is in excess over that expected from their nuclear gene contribution and that from the other ancestresses is in deficit. This difference may be due to a sampling effect, but this is unlikely since over half the population (161 out of the then 294) were included in the sample. It is more likely to be attributable to the vagaries of variation in the sex ratio of the offspring over the generations. Support for this explanation comes from the absence of mtDNA from the daughter (My W.) of S.W., not of course detectable from the SSO results but from the pedigree information. Of her seven children, four were boys, through whom her genes were transmitted to the present population, but no descendant of any of her three daughters remains on Tristan today. It is therefore not possible to check whether she was the biological daughter of S.W. as is stated.

Table 14.2. *Variant nucleotide sequences in Tristan mtDNA lineages compared with the reference sequence*

Position	Reference Sequence	Founder				
		SW	SP	MW	E/AS	ML
Control region I						
16086	T	C	.	.	.	.
16129	G	A	.	.	.	A
16223	C	.	T	T	.	T
16230	A	.	.	.	.	G
16243	T	.	.	.	.	C
16263	T	.	C	.	.	.
16295	C	.	.	T	.	.
16311	T	.	C	.	.	C
16324	T	C	.	.	.	.
16362	T	.	.	C	.	.
Control region II						
73	A	G	G	G	.	G
146	T	.	.	C	.	C
152	T	.	C	.	C	C
195	T	.	C	.	.	C
199	T	C	.	C	.	.
204	T	.	C	.	.	.
247	G	.	.	.	.	A
263	A	G	G	G	G	.
294	T	.	.	.	.	A

(. = same as reference)
Source: Soodyall *et al.* (1995).

In view of their frequency differences, and consequent usefulness in distinguishing between major ethnic groups (Connor & Stoneking, 1994), these SSO types throw light on the origins of their introducers. Thus, at location IB variant 2 is some eight times as frequent in American Africans as in Europeans and variant I correspondingly more common in Europeans than Africans; at location II2 variant 2 is some 13 times as frequent in Africans as Europeans. Search for other populations in which these types have been found is informative. Of the five types listed (See page 199), the first has not been found in surveys of 1000 Asians and 1000 Africans but has occurred in one Filipino and one European; a type that differs only in the IIC location has been found in one African and one Asian individual (Soodyall *et al.*, 1995). The second type has been observed in two Africans and 13 Malagasy. The third has been seen in two Africans, one Asian and 24 Malagasy, the fourth in five Europeans and the fifth in 24 Khoisan (Soodyall *et al.*, 1995). However, for greater refinement the results of the sequence analysis were used for this purpose.

Sequencing confirmed the distinction between the five SSO types, and no variation was detected among individuals within a given maternal lineage. However, it demonstrated the presence of additional mutations in each type outside and between the locations examined by the SSO probes, so that the most frequent type showed six mutations from the reference sequence (Anderson *et al.*, 1981) and the others (in order of frequency) showed eight, seven, two and eleven (Table 14.2). The least number of deviations from the reference sequence occurred in the mtDNA from the Anglo-lrish sisters and the largest in that from M.L. Thus by comparison with the reference, in the sequence from S.W. in control region I a C nucleotide replaces T at positions 16 086, 16 324, and A replaces G at position 16 129, while in control region II a G replaces A at positions 73 and 263 and there is C instead of T at 199. In M.W. there is T instead of C at positions 16 223, 16 295, C instead of T at positions 16 362, 146, 199, and G instead of A at 73 and 263. The Anglo-Irish sisters have no variant in control region I but C instead of T at 152 and G instead of A at 263 in control region II.

An appreciable amount of information is accumulating on the control region sequences in individuals from Africa, Asia and Europe (Vigilant *et al.*, 1991; Soodyall, 1993). The Tristan sequences were compared with others available from Europeans, Africans and Asians, and the results summarized in a phylogenetic tree using the neighbour-joining method of Saitou & Nei (1987) to indicate genetic affinity (Fig. 14.1). The sequence from the Anglo-Irish sisters clusters clearly with all other European sequences, as expected from their known ancestry. Those from S.W. and M.W. cluster in branches of the tree made up of Chinese and European but quite close to branches that include Yoruba, 'Bantu' and one pygmy sequence. The sequence from S.P. clusters closely with some Europeans but is approached by those from West Africans and Chinese. The three women from St Helena all have the same mutations at positions 73 and 263. The apparent presence of an oriental component in the Tristan gene pool, at first appearing somewhat surprising, is perhaps not unexpected. One or two observers on first approaching Tristan have remarked on a 'Malay' cast of countenance, and at the beginning of the nineteenth century the composition of the St Helena population was reported as 50% African, 25% each European and Chinese (Shine, 1970) The sequence from M.L. falls clearly with those from Khoisan, resolving the problem of the interpretation of Creole when applied to her; it is most likely that her mother's lineage derived from one of the unions, so numerous in the earlier days of settlement and colonization, between a Dutch man and a Hottentot woman.

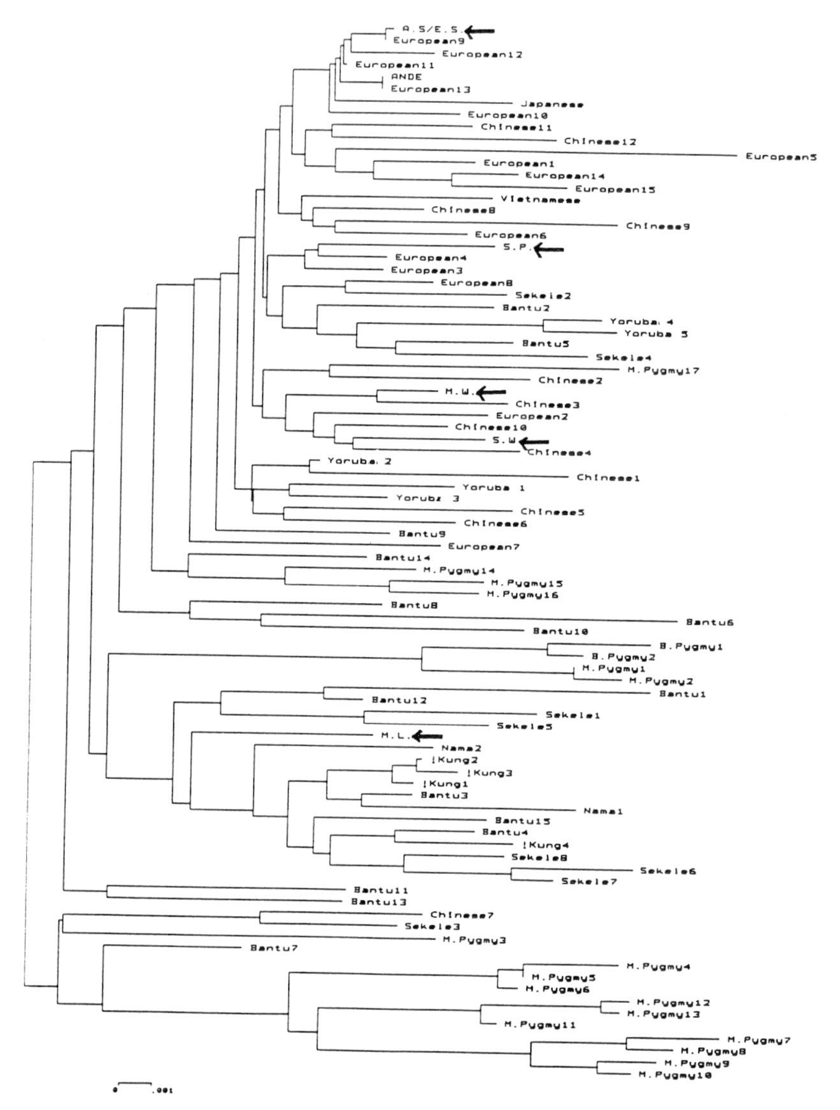

Fig. 14.1. Neighbour-joining tree comparing Tristan control region sequences with others available from Europeans, Africans and Asians.

Conclusions

Analysis of the mtDNA in a sample of the population of Tristan da Cunha confirms the accuracy of the maternal lineages of the pedigree. It has identified the mtDNA types of the individual founding mothers, the original ancestresses who brought their genes to the population where they have stayed, and has resolved questions regarding them and their relationships on which the historical sources were ambiguous. The types and the more detailed mtDNA sequences, compared with those in other populations, limited though the data on them are, clearly confirm the presence of contributions to the Tristan genetic stock from African sources (Khoisan and West/Central Africa) and somewhat unexpectedly suggest a possible Asian component, perhaps Chinese, perhaps Malay, arriving via St Helena.

This identification of individual mtDNA and its use in indicating origins is of more than academic interest. Examining the detailed Tristan pedigree in the light of this knowledge allows assessment of the stability of mtDNA and sets an upper limit to the mtDNA mutation rate (Soodyall *et al.*, 1995). Moreover, in isolated populations there sometimes occur syndromes that are difficult to identify by reference to the European spectrum of inherited disease, and Tristan is no exception. The present findings suggest that, in seeking a match to the so-called Tristan syndrome (Samuels, 1963), it may be worth looking in Oriental or African populations. Once again, the Tristan population, remote and of minimal political or economic significance, seems to have made a contribution to academic biology out of proportion to its numbers.

Acknowledgements

Acknowledgement is gratefully made to our friends and colleagues who have generously allowed us to draw on unpublished results for this discussion and especially Professor Mark Stoneking and Professor Trefor Jenkins.

References

Anderson, S., Bankier, A. T., Barrell, B. G., de Bruijn, M. H. L., Coulson, A. R., Eperon, I. C. *et al.* (1981). Sequence and organisation of the human mitochondrial genome. *Nature*, **290**, 457–65.

Brothwell, D. R. & Harvey, R. G. (1965). Facial variation. *Eugenics Review*, **57**, 167–81.

Connor, A. & Stoneking, M. (1994). Assessing ethnicity from human mitochondrial DNA types determined by hybridization with sequence-specific oligonucleotides. *Journal of Forensic Sciences*, **39**, 1360–71.

Henricksen, S. D. & Oeding, P. (1946). Medical survey of Tristan da Cunha. In *Results of the Norwegian Scientific Expedition to Tristan da Cunha* 1937-38, 1, pp. 1–147. Oslo, Hos Jacob Dybwad.

Jenkins, T., Beighton, P. & Steinberg, A. G. (1985). Serogenetic studies on the inhabitants of Tristan da Cunha. *Annals of Human Biology*, **12**, 363–71.

Munch, P. (1971). *Crisis in Utopia.* London: Longman.

Roberts, D. F. (1968). Genetic effects of population size reduction. *Nature*, **220**, 1084–8.

Roberts, D. F. & Bear J. C. (1980). Measures of genetic change in an evolving population. *Human Biology*, **52**, 773–86.

Saitou, N. & Nei, M. (1987). The neighbour-joining method: a new method for reconstructing phylogenetic trees. *Molecular Biology and Evolution*, **4**, 406–25.

Samuels, N. (1963). Experiences of a medical officer on Tristan da Cunha, June-October 1961. *British Medical Journal*, **ii**, 1013–17.

Shine, I. (1970). *Serendipity in St. Helena*. Oxford: Pergamon Press.

Soodyall, H. (1993). Mitochondrial DNA polymorphisms in Southern African populations. Unpublished PhD. thesis, University of the Witwatersrand.

Soodyall, H., Jenkins, T., Mukherjee, A., du Toit, E, Roberts, D. F. & Stoneking, M. (1995). The founding mitochondrial DNA lineages of Tristan da Cunha islanders. *American Journal of Physical Anthropology*, in press.

Stoneking, M., Hedgecock, D., Higuchi, R. G., Vigilant, L. & Ehrlich, H. A. (1991). Population variation of human mtDNA control sequences. *American Journal of Human Genetics*, **48**, 370–82.

Taylor, W. F. (1856). *Some account of the settlement of Tristan da Cunha*. London: Society for Promoting Christian Knowledge.

Vigilant, L., Stoneking, M., Harpending, H., Hawkes, K. & Wilson, A. C. (1991). African populations and the evolution of human mitochondrial DNA. *Science*, **253**, 1503–7.

15 *Linguistic divergence and genetic evolution: a molecular perspective from the New World*

R. H. WARD

A unique characteristic of our species is that recent evolution has proceeded in two dimensions: biological and cultural. From a paleoanthropological perspective, the past two million years are defined as much in terms of cultural innovations as they are in terms of morphological change. The recognition that humans have been shaped by the intersection of two evolutionary processes has led to much speculation about the degree of interdependence and interaction between cultural and biological evolution. Attempts to understand the interplay between the two processes have ranged from metaphysical speculations (Teilhard de Chardin, 1957) to specific mathematical models (e.g. Cavalli-Sforza & Feldman, 1981; Lumsden & Wilson, 1983). However, apart from some broad generalizations, the parameters of this interaction remain undefined. In particular, we have little concrete knowledge about whether the tempo and mode of evolutionary change are concordant from one realm of evolution to the other. In part, this is because there are no wholly satisfactory measures of the cultural components of evolution: the cultural attributes that have been examined are either too superficial and ephemeral (e.g. incised patterns on pottery), or so fundamental to human behaviour that the interplay between biology and culture is inextricably intertwined (e.g. the transition between Oldowan and Acheulian). Linguistic evolution may represent a cultural parameter that is both amenable and is uniquely suitable for making a careful comparison with biological evolution.

There are several reasons why the development of human languages represents an appropriate foil against which to contrast biological evolution. Language is an integral and characteristic part of cultural identity: one of the most predominant defining features of any human culture is the language spoken by its members. Further, following Darwin's (1871) insight, there is general agreement that the similarity between the processes of linguistic evolution and those of biological evolution is more than superficial. Modern linguistic theory explicitly recognizes that the demographic and ecological factors thought to influence rate of linguistic differentiation are remarkably similar in process and effect to the factors that influence rate of genetic differentiation (Nichols, 1990). Hence, there is a strong case for conducting a detailed investigation of whether the tempo and mode of linguistic change are comparable to the rate and process of genetic evolution. The answers are likely to illuminate the larger issue of how the two dimensions of culture and biology have interacted to influence the course of human evolution.

Initial studies in this area noted a correspondence between the distribution of genetic markers and linguistic traits. This correlation has only recently been explored in depth, with the advent of a large number of informative genetic markers. In the course of a comprehensive analysis of genetic polymorphisms, Cavalli-Sforza and colleagues found a striking degree of correspondence between the hierarchy of linguistic affiliation and the hierarchy of genetic affiliation, as revealed by genetic dendrograms (Cavalli-Sforza *et al.*, 1988; Cavalli-Sforza, Minch & Mountain, 1992) However, while informative and stimulating, these studies say little about the relative rates of linguistic divergence and genetic differentiation.

At the most basic level, the correlation between gene frequency distributions and linguistic affiliation simply indicates that people who speak the same language also tend to share genes. It would be surprising if this were not so. In essence, the concordance between the two hierarchical taxonomies denotes congruence of pattern, rather than congruence of process. For the latter, some comparison of the tempo of change is needed. This requires knowledge of time depth. Without independent estimates of time depth, it is impossible to know whether associations are due to recent constellations of population structure, or reflection of ancient origins. Consequently, it is still a matter of conjecture as to whether the hierarchy of linguistic relationships defined by Ruhlen's (1991) genetic classification of languages bears any relationship to degree of biological ancestry, as measured on an evolutionary scale of mutational change.

The analysis of sequence divergence provides a potential solution to the problem of determining whether linguistic evolution and biological evolution have proceeded in concert. Sequence divergence, unlike allele frequency distributions, has the potential to reveal time depth without recourse to an excessive number of assumptions. To estimate time depth from the difference between allele frequencies requires a number of critical demographic and biological assumptions: stable populations of known size, absence of migration, similar population structure, etc. While such assumptions can be defended for short-term processes, such as the relationship between dialects and genetic differentiation (Smouse & Ward, 1978; Barbujani & Sokal, 1990), these assumptions are unrealistic for the time span required to generate linguistic phyla. The number of nucleotide differences between two sequences affords a more direct estimate of the time depth since they diverged from their common ancestor. Even though mutation rates are only known with imprecision, the distribution of nucleotide differences between sets of sequences can still be used to define relative amounts of temporal divergence. In addition, the phylogenetic analysis of sequences, in conjunction with their geographic location (phylogeography) can be extremely informative (Avise *et al.*, 1987).

Advantages of Amerindian populations

Amerindian populations appear to represent one of the better opportunities to study the relationship between cultural characteristics and biological ancestry. Unlike most major ethnic groups, the temporal and spatial coordinates of the origin of Amerindians can be defined with a fair degree of certainty. The constraints of geography, coupled with high levels of biological resemblance, indicate that Amerindian populations had their origin in North-eastern Asia. In addition, despite some scattered claims to the contrary, there is little doubt that the ancestors of contemporary Amerindians first entered the Americas by way of Beringia, some 12 000 to 15 000 years ago (Hoffecker *et al.*, 1993; Meltzer, 1993; Szathmary, 1993). Hence, Amerindians are a relatively recent offshoot of the human tree in which the time and place of divergence can be specified with a reasonable degree of accuracy.

Of equal importance, Amerindian populations have mostly evolved without experiencing the mass population interactions with other ethnic groups that occurred in Eurasia and Africa. Consequently, Amerindians have a fairly cohesive gene pool. Further, with a few exceptions, the development of agriculture, and the foundation of civilizations, appears to have been much more localised in the Americas than in the Old World. Unlike the spread of farming in Europe, maize appears to have spread more by diffusion than by movements of populations. Similarly, compared to their Old World counterparts, Amerindian empires extended their boundaries more by trade than by invasion and resettling of populations. As a consequence, the impact of Amerindian civilizations appears to have been more localized and have resulted in less intermingling of formerly distinct gene pools than in Eurasia, where the past 5000 years has led to complicated distributions of genetic and linguistic affiliations. Thus the contentious issue of language replacement, which will obscure, or bias, the relationship between genetic divergence and linguistic divergence, appears to have been far less common in the Americas. Hence, excepting the catastrophic disruption of the post-Columbian period, Amerindian populations may represent an almost ideal test bed to examine the degree of concordance between linguistic and genetic evolution.

From a linguistic perspective, the distribution of languages in the New World also favours a comparison between genetic divergence and degree of linguistic affiliation. Taking the entire New World as a whole, the complete range of languages spoken by Amerindian populations have been classified into three language phyla (Greenberg, 1987). However, these three phyla differ markedly in their complexity and extent of geographic range. Two of the phyla, Eskimo–Aleut and Na–Dene, are essentially restricted to populations north of latitude 55. With only 85 000 speakers, Eskimo–Aleut is by far the smallest phylum, containing only two linguistic stocks and nine extant languages (Woodbury, 1984; Nichols, 1990). Na–Dene, also containing only two linguistic stocks, is a somewhat larger phylum with some 202 000 speakers and 34 distinct languages (Ruhlen, 1991). The third phylum, Amerind, is appreciably larger and much more complex: with some 18 million speakers located in North, Central and South America, the 583 extant languages of this phylum are classified into 51 linguistic stocks and 69 language families (Nichols, 1990; Ruhlen, 1991). Thus, Amerindian populations not only have a fairly well-defined temporal origin, and are reasonably genetically homogeneous, but allow a full range of hierarchical comparisons at the linguistic level. This stands in contrast to the situation in Europe where, despite the antiquity and complex genetic origins of the population (Bowcock *et al.*, 1991), there are essentially only two linguistic phyla, each containing some 19 (Uralic) to 50 (Indo-Hittite) languages.

Despite these potential advantages, little consensus has resulted from the few studies that have attempted to examine the relationship between linguistic diversity and genetic differentiation in Amerindian tribal populations. A series of comprehensive studies of Chibchan speaking tribes of Central America found a remarkably high level of congruence between a hierarchical classification of languages and a genetic dendrogram (Barrantes *et al.*, 1990). In large part, this degree of correspondence appears to be related to the geographic stability of these Chibchan populations over several millennia. More recent studies of mtDNA variability support the conclusion that the Central American Isthmus has been a region of evolutionary stasis for a long period (Torroni *et al.*, 1994; Kolman *et al.*, 1995), and also indicate that the distribution of mtDNA lineages tend to correspond to linguistic affiliation (Santos, Ward & Barrantes, 1994; Kolman *et al.*, 1995). By contrast, two studies from South America failed to find such an association: for both the relatively large tribal populations of the Andean region (Chakraborty *et al.*, 1976), and the small isolated tribes of the Amazon region (Black *et al.*, 1983), linguistic affiliation proved to be a poor predictor of genetic similarity. Compilation of North American data gave variable results: in some instances the genetic distance between tribes appeared to be a reflection of their linguistic affiliations, whereas in other instances, this was not the case (Spuhler, 1979). Thus, the available data neither support, nor refute, the basic premise of Cavalli-Sforza *et al.* (1988, 1992). Clearly, more focused studies need to be carried out in this potentially informative segment of the human population.

Variability of Amerindian mtDNA at the tribal level

A potential handicap in evaluating linguistic affiliation and molecular divergence in the New World is that, due to its relatively recent origin and cohesive nature, the Amerindian gene pool may contain insufficient variability at the mtDNA sequence level. Preliminary studies, using restriction enzyme data, found minimal numbers of mitochondrial haplotypes, and led to the conclusion that the Amerindian gene pool had passed through a severe bottleneck (Wallace, Garrison & Knowler, 1985; Schurr *et al.*, 1990). However, mtDNA restriction types exhibited relatively little variability at the local level, in contrast to the marked genetic differentiation observed with classical genetic markers. Consequently, it seemed likely that restriction site polymorphisms would not have sufficient resolution to identify local evolutionary processes.

Following Vigilant *et al*'s. (1991) demonstration that sequence data from the mitochondrial control region provided a significant increase in evolutionary resolution, it seemed that this kind of data would equal or exceed the informativeness of studying a series of classical polymorphisms. Earlier work with classical polymorphisms indicated that genetic differentiation at the tribal level provided valuable clues to the evolutionary processes of traditional human populations. In particular, evaluation of genetic diversity at the tribal level indicated the extent to which socio-demographic processes such as population fissioning and migration, influenced the distribution of genetic diversity within and among subpopulations (Ward & Neel, 1970; Ward, 1972). Furthermore, tribal populations contained an appreciable proportion of the genetic diversity seen at the continental level (Neel & Ward, 1970).

To evaluate the degree of mtDNA diversity within a single Amerindian tribal population, we carried out a comprehensive study of control region sequence diversity within the Nuu–Chah–Nulth, a Wakashan speaking tribe from the West coast of Vancouver Island (Ward *et al.*, 1991). The contemporary population consists of 14 distinct bands, scattered along 250 miles of rugged coastline. Since geography and the existence of three dialects, within the Nuu–Chah–Nulth, suggested significant differentiation among bands, we sampled 63 maternally unrelated individuals, with representation from as many bands as possible. Sequencing the first 360 nucleotides of the control region identified 24 variable sites, which defined 28 distinct mitochondrial lineages. Phylogenetic analysis indicated that the majority of lineages clustered into four clades, plus a small number of isolated lineages. Subsequent work (Valencia, 1992) indicated that each of these clades corresponded to one of the four major mitochondrial haplotypes that had been defined on the basis of restriction site polymorphisms (Ballinger *et al.*, 1992).

The distribution of sequence divergence within the Nuu–Chah–Nulth sample indicated the divergence within clades was relatively small, amounting to no more than 4000–8000 years of evolution. By contrast, the divergence between clades was more substantial, being equivalent to 13 000–25 000 years. This suggests that the molecular diversity within this tribal population had its origins well before human populations first migrated to the New World. It also suggests that the colonizing populations which first came to the New World by way of Siberia, were genetically heterogeneous, and probably of appreciable size.

With an average pairwise sequence difference of 1.5%, the overall degree of sequence diversity within the Nuu–Chah–Nulth is approximately 80% of the mean pairwise sequence difference observed in a large sample of Japanese (Horai *et al.*, 1993), and is 60% of the value observed by Vigilant *et al.* (1991) in a heterogeneous sample of sub-Saharan Africans. Hence, as was true for the situation with classical genetic markers (Neel & Ward, 1970), the magnitude of sequence diversity within a single tribal population is an appreciable proportion of the diversity defined at the continental scale. Furthermore, the bulk of this diversity is due to sequence divergence between clades, rather than to the divergence within clades. This indicates that the molecular diversity within tribes is due to evolutionary events that antedate the existence of the tribe by a considerable margin. However, there is no evidence that Amerindian tribal populations are characterized by containing a specific clade of mitochondrial lineages.

By extending this initial study to a larger sample of 128 individuals, selected to represent a random sample of maternally unrelated individuals from the six major band coalitions, Valencia (1992) was able to examine the relationship between language dialects and the distribution of mitochondrial lineages. She found, unlike the situation for classical polymorphisms, mitochondrial lineages show no tendency to cluster within bands, or within dialect areas. Further, the average pairwise sequence difference within bands was as great as the average sequence difference among bands. The overall distribution of mitochondrial lineages within the Nuu–Chah–Nulth population was essentially random, exhibiting no correspondence with geography, linguistic dialect, or socio-political affiliation. Given the time scale required to generate new mtDNA lineages by mutation, the amount of migration between bands appears sufficient to obliterate the effects of population subdivision. Although much larger samples might show differences in lineage frequency between bands, the task of generating sequence data for 400–500 individuals is beyond the scope of most laboratories. Thus at the level of mtDNA sequence divergence, the tribe, rather than the band, is the appropriate sampling unit.

Linguistic affiliation and mtDNA in the Pacific Northwest

Traditional Amerindian communities of the Pacific Northwest provide an ideal test case for determining whether the tempo of linguistic divergence is equivalent to that of biological evolution. Besides being one of the most linguistically diverse areas in the New World (Nichols, 1990), the Pacific Northwest contains representatives of two of the major linguistic phyla that occur in the Americas: Na–Dene and Amerind. Thus, adjacent populations may speak languages belonging to different phyla – a fairly rare situation which facilitates the comparison between genes and language. With relatively small geographic distances between linguistically disparate populations, contrasts among Pacific Northwest populations will not be unduly influenced by the otherwise nearly universal correlation between linguistic distinctiveness and magnitude of geographic separation. Moreover, most of the traditional communities in this region have similar subsistence patterns and, consequently, similar population sizes. Accordingly, the demographic factors that influence rate of genetic and linguistic differentiation will be similar in populations that belong to different linguistic phyla.

Our initial study of the Nuu–Chah–Nulth indicated the amount of mtDNA diversity in an Amerind speaking tribe of the Pacific Northwest region. To maximize the contrast between linguistic groups in this region, we studied a second Amerind speaking tribe (Bella Coola), plus the Haida, a Na–Dene speaking tribe (Ward *et al.*, 1993). The traditional Bella Coola number some 600 individuals, who live along the Bella Coola river in British Columbia. Their language is regarded as an outlier within the Salishan language family, which is composed of 16 languages and is linguistically very close to the Wakashan language family. Unlike the much more numerous Nuu–Chah–Nulth, there is no marked geographic structuring of communities within the Bella Coola. The approximately 1200 Haida, who speak an isolated language within the Na–Dene phylum, are distributed in two bands located on the Queen Charlotte Islands, plus a small population in Alaska. If the rate of linguistic differentiation proceeded at the same pace as the rate of biological divergence, then the molecular divergence between the two Amerind tribes, belonging to closely related language families, should be much smaller than their molecular distance from the Na–Dene speaking Haida.

Comparable samples sizes of 40 and 41, respectively, were taken from the Bella Coola, and the first 360 nucleotides of the mtDNA control region were sequenced exactly as was done for the Nuu–Chah–Nulth sample. The Bella Coola sample exhibited 20 variable positions, which defined 11 mtDNA lineages of which four occurred in the Nuu–Chah–Nulth. Despite the smaller sample size of the Bella Coola, compared to the Nuu–Chah–Nulth, the average diversity value, $h = 90.4\%$, was essentially the same as the diversity value calculated for the sample of 63 Nuu–Chah–Nulth, $h = 95.4\%$. The mean pairwise sequence difference within the Bella Coola, 1.42%, was virtually identical to that observed in the Nuu–Chah–Nulth, 1.47%. The distribution of mean pairwise sequence differences was also very similar between the two tribes (Ward *et al.*, 1993). Thus, both Amerind speaking tribes appear to have remarkably similar distributions of molecular diversity indicating that their long-term evolution, and their short-term population history, have been roughly equivalent. This supposition is strengthened by the fact that the relative evolutionary effective size of the two populations (estimated by applying Ewens sampling formula to estimate $\theta = 2N_f\mu$, then taking the ratio of the two estimates) was almost exactly the same as the relative size of the two populations (Ward *et al.*, 1993).

Superficially, the Haida sample appeared similar. It identified 17 variable positions which defined ten lineages, three of which were found in the Nuu–Chah–Nulth and one in the Bella Coola. However, in terms of sequence diversity, the Haida were much less variable than the two Amerind tribes: the diversity value was much less, $h = 70.9\%$, the mean pairwise sequence difference was only 0.69%, and the estimate of $\theta = 3.9$ was 40% of the value predicted by the relative size of the Haida population. Further, 75% of the Haida sequences are contained in two lineages that differ by only a single nucleotide. Hence, 29% of pairwise comparisons within the Haida involve identical lineages, and 34% involve lineages differing by only a single nucleotide. This is in marked contrast to the figures for the two Amerind tribes, 6.8% and 6.2%, respectively. These data suggest that the Na–Dene speaking Haida have had a much more recent origin than either of the two Amerind speaking groups. Other explanations, such as an extremely severe bottleneck, a much smaller population throughout the past 8000 years, or more extensive population substructure, are not supported by the available archaeological and historical record

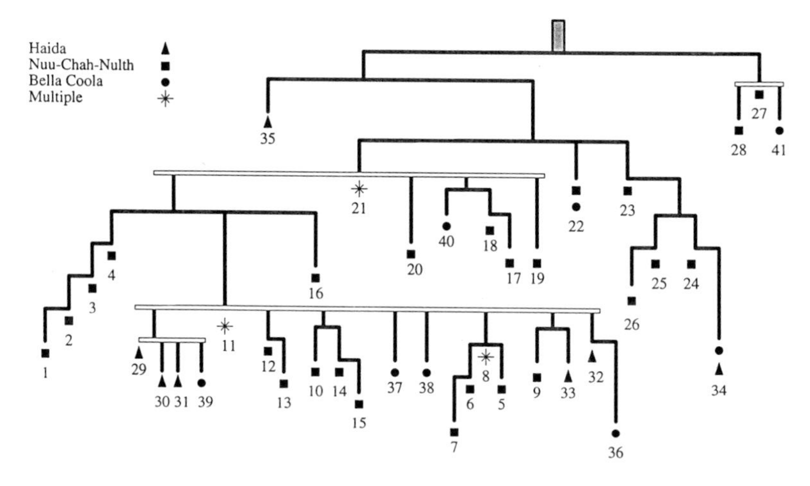

Fig. 15.1. Depiction of the molecular phylogeny for the 41 mtDNA lineages observed in a sample of 144 Amerindians from the Pacific Northwest, belonging to two linguistic phyla: Amerind ($N = 103$) and Na-Dene ($N = 41$). The specific tribes in the sample, and the symbols used to identify in which tribe the lineages were observed, are: Nuu-Chah-Nulth ($N = 63$, squares), Bella Coola ($N = 40$, circles), Haida ($N = 41$, triangles). Lineages that occur in all three tribes are indicated by an asterisk. Lineages numbered 1–29 were first observed in the Nuu-Chah-Nulth, those numbered 29–35 were first observed in the Haida, and those numbered 36–41 observed only in the Bella Coola. The phylogeny was estimated by the DNAML 'maximum likelihood' procedure of the PHYLIP phylogeny package. Nodes that failed to give statistically reliable estimates of branching order are indicated by open boxes. Branch lengths in the tree are scaled with respect to the amount of sequence divergence between lineages.

(Ward *et al.*, 1993). Hence, if the hierarchy of linguistic classification were to reflect time depth, the Haida appear much younger than predicted by the linguistic distinctiveness of the Na–Dene phylum.

Evaluation of the molecular divergence between the three tribal populations, and a phylogenetic analysis of the lineages is also at variance with the simple predictions based on linguistic affiliation. While the mean pairwise difference between Bella Coola lineages and Nuu–Chah–Nulth lineages is 5.5 nucleotides, the mean difference between Haida lineages and those of the Amerind tribes is slightly smaller (4.1 and 4.7 nucleotides, respectively), rather than larger. Further, when all 41 lineages found in these three Pacific Northwest tribes are placed in a maximum likelihood phylogeny (Fig. 15.1), the Haida lineages fail to form a separate clade. They also have a much more restricted distribution than the lineages found in the two Amerind populations.

Clade I (lineages 1–3) and clade IV (lineages 27,28,41) contain no Haida lineages, while clade III (lineages 23–26, 34) has only a single Haida lineage

(34). Since lineage 34 is also found in the Bella Coola it is likely that this is an Amerind lineage that has been introduced into the Haida, following an intertribal raid. Lineage 35 is an aberrant lineage which may be due to non-Amerindian admixture (Ward *et al.*, 1993). Seven of the remaining eight Haida lineages are distributed in clade II, while lineage 21, is found in all three tribes. This lineage occupies a nodal position suggesting that it is an ancestral lineage that predates the divergence of Na–Dene and Amerind. None of the seven Haida lineages that fall into clade II, form a 'Na–Dene specific clade': two lineages appear to be ancestral lineages within this clade (lineages 8 and 11), while the other five lineages are scattered randomly throughout the clade.

These phylogenetic data imply that the mtDNA lineages found in the contemporary Haida originated no earlier than the population separation that led to clade II. With an average sequence divergence of 0.8%, the age of clade II is estimated to be no more than 7500 years. Since this represents an upper limit for population divergence, one interpretation of these data is that the ancestry of the Na–Dene speaking Haida is not appreciably greater than 7500 years. By contrast, the average sequence divergence within the two Amerind speaking tribes implies a time depth of around 13 500 years, nearly twice that for the Haida. The restriction of Haida mtDNA lineages to only one of the four clades that are otherwise found throughout Amerind speaking populations of the new World, implies that the Haida have a much more recent biological origin than that of Amerind groups. Consequently, the distinctiveness of the Na–Dene language may be due to a rapid burst of linguistic differentiation, rather than the steady accumulation of language changes over time.

Linguistic affiliation and mtDNA divergence in Beringia

While the preceding analysis of three tribal groups in the Pacific Northwest suggests that language and genes may evolve at very different rates, it is hardly definitive: three populations is a minimal number of comparisons, the Haida may not be representative of other Na–Dene groups, or the entire Na–Dene phylum could represent a special case. Clearly, a more extensive contrast, involving more populations and more linguistic phyla, is needed. Accordingly, we extended the comparison between molecular divergence and linguistic divergence to an additional five populations located in the Beringian area, or the Circumarctic of North America (Shields *et al.*, 1993). The Beringian area holds a special importance for studies of New World origins. This is the region through which all early migrants must have passed. Consequently, it is likely that the evolutionary affinities of the contemporary populations in the region will shed some light on the biological composition of the groups that entered the New World by this route. Although initial archaeological data suggested a shallow time depth (Dumond, 1980), more recent work indicates that Beringia has a complex series of cultural assemblages that date back to around 12 000 years ago (Dumond, 1987; Laughlin, 1987; Hoffecker, Powers & Goebel, 1993), and may provide a record of occupation that predates the colonization of the New World.

More important for the present comparison, is the fact that the Beringian area contains the third language phylum found in Amerindian people, the Eskimo–Aleut phylum. It also contains representatives of two other distinct language phyla: Chukchi, belonging to the Chukchi–Kamchatka phylum, and Yukaghir, belonging to the Uralic language phylum. In addition, by extending the study to the Beringian area, we were able to use members of a sixth language phylum, the Altai, as an outgroup for both the biological and linguistic contrasts. Since there is no evidence that contemporary Altaic speaking populations have any close biological relationship to Amerindian groups, the use of the Altai as an outgroup gives an independent perspective on the relative evolutionary divergence within and among Amerind and the other linguistic phyla.

To provide a more comprehensive assessment of the degree of molecular divergence in Amerind tribes, we sampled 42 individuals from a third Amerind speaking group from the Pacific Northwest – the Yakima, a Penutian-speaking group of Washington state. We also added a second Na–Dene speaking group to the analysis, by adding a sample of 21 interior Athapaskans from Alaska. This sample of indisputable Na–Dene speaking Athapaskans provides a comparison with the Haida. Two widely geographically separated populations were sampled as representatives of the Eskimo–Aleut phylum: a set of 47 Yup'ik speakers from Western Alaska and Eastern Siberia, plus a group of 22 Greenlandic Inuit. We also examined a small sample of 7 Inupiaq speakers from Alaska. Three exclusively Siberian populations were also sampled: a sample of 80 Chukchi, plus two small samples of 16 Yukaghir and 17 Altai.

The initial analysis of these samples (Shields *et al*., 1993), which excluded the Yukaghir and only included a reduced sample set for the Chukchi and Yup'ik, amply confirmed the conclusions of the Pacific Northwest study. Both Na-Dene populations had mean sequence differences of 2.5 nucleotides, approximately half the value observed in each of the three Amerind tribes. Hence, there is no evidence that the Haida are somehow nonrepresentative of Na–Dene populations. The Greenlandic Inuit exhibited an even smaller mean pairwise sequence difference (2.0), while the values in the small samples of Chukchi and Yup'ik were similar to the Na–Dene values. By contrast, the Altai sample had a mean pairwise sequence difference that was slightly greater (5.52) than the values for the three Amerind tribes (Shields *et al.*, 1993).

Thus, the samples representing the Na–Dene, Eskimo–Aleut and Chukchi language phyla all had amounts of sequence divergence that were consistent with a time depth of 5000–7100 years. This is substantially smaller than the time depth of 12 100–13 200 years estimated for the Amerind tribes. For a more stringent comparison, we calculated the distribution of pairwise sequence differences for all 90 sequences belonging to the combined set of 'Circumarctic' populations. The mean pairwise difference for this sample, which incorporates all three language phyla (Na–Dene, Eskimo–Aleut, Chukchi), was 2.8 ± 2.5, which is significantly smaller than the mean value of 6.1 ± 1.8, calculated for an 'average' Amerind tribe. Moreover, the majority of pairwise comparisons within an 'average' Amerind tribe involved 5–13 nucleotide differences, whereas most of the pairwise comparisons within the total set of Circumarctic lineages involved zero to three nucleotide differences (Shields *et al.*, 1993).

Since the previous analysis had indicated that the mtDNA lineages in the Haida clustered in only a single clade, rather than being distributed among the four clades observed in Amerind populations (Schurr *et al.*, 1990; Ward *et al.*, 1991; Torroni *et al.*, 1992; Horai *et al.*, 1993), we carried out a phylogenetic analysis of the 33 'Circumarctic' lineages (Shields *et al.*, 1993). The results, presented in Fig. 15.2, mimic those from the analysis of the 41 Pacific Northwest lineages (Fig. 15.1). Within the 'Circumarctic' lineages, there is no evidence of clades equivalent to those seen in Amerind populations. Further, in an even more marked manner than Fig. 15.1, lineages belonging to different populations are interspersed throughout the phylogenetic tree. This intermingling of lineages is all the more notable when it is realised that, unlike the Pacific Northwest analysis, samples are distributed across three linguistic phyla and a wide geographic zone (Greenland to Siberia). This argues against any long-term separation between the Circumarctic populations. This conclusion is reinforced by the fact that the mean pairwise sequence difference between Circumarctic populations is always less than 3.0, and less than half the mean pairwise sequence difference between Amerind tribes. Both pieces of data imply that the three different language phyla spoken by the Circumarctic populations (Na–Dene, Eskimo–Aleut, and Chukchi) are all of relatively recent origin.

As was true for Fig. 15.1, lineages that occur in multiple populations (starred lineages) tend to occupy nodal positions, consistent with the interpretation that they are ancestral lineages that became incorporated into more than one descendent population. The exception is lineage 58, which is shared between two Eskimo groups, probably as a result of recent migration. Lineage 11, which occupies a nodal position in both figures,

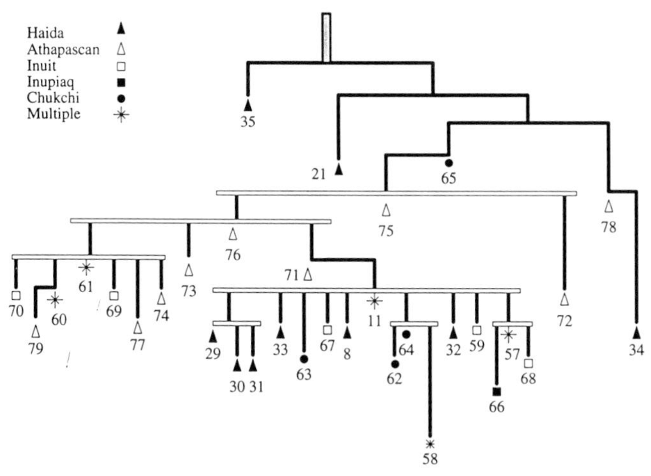

Fig. 15.2. Depiction of the molecular phylogeny for the 33 mtDNA lineages that were observed in a sample of 90 individuals, drawn from the following linguistic groups: Na-Dene–Haida ($N = 41$, filled triangles), Interior Athapascans ($N = 21$, filled triangles); Eskimo–Aleut–Greenlandic Inuit ($N = 17$, open squares), Alaskan Inupiaq ($N = 5$, filled squares); Chukchi–Kamchatka–Chukchi ($N = 7$, filled circles). Lineages that occur in more than one population are indicated by asterisks. The phylogeny was estimated by DNAML, as in Fig. 15.1, and is scaled in proportion to sequence divergence.

accounts for half the Haida sample and one-quarter of all Circumarctic lineages. Since this lineage is also frequent in Amerind tribes (Fig. 15.1), lineage 11 may be ancestral to many of the mtDNA lineages belonging to clade II that occur in North America. Comparison of Fig. 15.2 with Fig. 15.1 serves to emphasize the relative youth of the 'Circumarctic' phylogeny. The bulk of the 'Circumarctic' lineages (excepting lineages 35, 21, 65, 78, 34) fall within the branch leading to clade II in Fig. 15.1. However, Fig. 15.2 indicates that there is a subclade within clade II (lineages emanating from the node occupied by lineage 61). This subclade is absent from Amerind populations to the south. This suggests that, compared to Amerind populations, Na–Dene and Eskimo–Aleut speakers shared a more recent common ancestor. It also suggests that the common ancestral population of these two groups was similar, but not identical, to the population that had previously given rise to Amerind populations.

The final analysis consisted of defining a population phylogeny, based on the mtDNA sequences for the full data set (Ward *et al.*, submitted). This phylogeny, displayed in Fig. 15.3, was constructed by first using pairwise sequence differences between populations to define a distance measure, then constructing the phylogeny according to the least squares criteria of

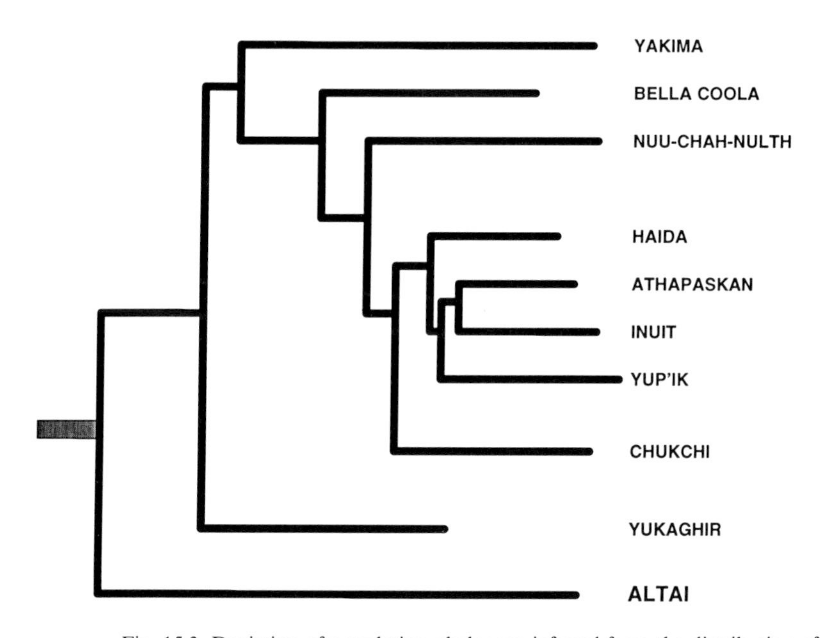

Fig. 15.3. Depiction of population phylogeny inferred from the distribution of mean pairwise sequence differences observed between mtDNA lineages sampled from populations in the following linguistic phyla: Altaic–Siberian Altai ($N = 17$); Uralic–Siberian Yukaghir ($N = 16$); Chukchi–Kamchatka–Chukchi ($N = 80$); Eskimo–Aleut–Siberian Yup k ($N = 47$), Inuit from Greenland ($N = 17$) and Alaska ($N = 5$); Na-Dene–Interior Athapascan from Alaska ($N = 21$), Haida from British Columbia ($N = 41$); Amerind–Nuu-Chah-Nulth from British Columbia ($N = 63$), Bella Coola from British Columbia ($N = 40$), Yakima from Washington Sate ($N = 42$). The tree was estimated by a least squares analysis of the mean pairwise differences by using the program FITCH in the PHYLIP package. Branch lengths are proportional to sequence divergence.

the FITCH algorithm in the PHYLIP package. The topology of this phylogeny serves to confirm the conclusions based on the lineage-specific phylogenies (Figs. 15.1 and 15.2), and agrees with the distribution of sequence divergence within and among populations. As expected, the Altai form a marked outlier from the other nine populations and clearly represent a biological outgroup, as well as a linguistic one. Also, despite their small sample size, the Yukaghir form a distinct branch at a level that is consistent with their affiliation in the Uralic language phylum. Since the initial divergence of the three Amerind tribes occurs at roughly the same level as the split leading to the Yukaghir, this also agrees with the consensus classification of Ruhlen (1991, 1994), and most other linguists (Nichols, 1990). Hence, up to this point the mtDNA population tree is roughly concordant with the pattern of linguistic divergence, in agreement with the data from classical markers (Cavalli-Sforza *et al.*, 1988, 1992; Renfrew, 1992, 1994). However, from this point on, the topology of the tree becomes increasingly discordant with linguistic affiliation.

There are essentially three areas in which the molecular phylogeny of the populations disagrees with the consensus linguistic hierarchy. First, the very long branch lengths leading to the three Amerind speaking tribes, plus the internode lengths that are significantly greater than zero, conflicts with the close similarity of the languages spoken by these populations. However, this is no basis for resurrecting the criticism of Greenberg's (1987) proposal that all Amerind languages should be grouped into a single phylum. There is a palpable similarity between the Wakashan language family (Nuu–Chah–Nulth) and the Salishan language (Bella Coola) that cannot be misinterpreted. While the Penutian language family (Yakima) is somewhat more distinct, there is no serious quarrel with the assessment that all three languages should be grouped together. Hence, the languages are close, but the genes are not.

The second conflict concerns the fact that the populations representing the three distinct language phyla of Na–Dene, Eskimo–Aleut, and Chukchi–Kamchatka have substantially smaller amounts of molecular divergence, and hence time depth, than is the case for the other language phyla. In particular, these molecular data provide no support for the contention that the linguistic distinctiveness of Na–Dene indicates it is an ancient division, that is, an early offshoot of the Nostratic super-group (Ruhlen, 1994). Thirdly, is the fact that, while the Chukchi appear somewhat distinct, there is clearly no support for any biological differentiation between Na–Dene and Eskimo–Aleut that reflects the linguistic hierarchy of these groups. With no statistical support for the internodes separating them, the four innermost populations (Haida, Athapaskan, Inuit, Yup'ik) are essentially a single multifurcation. Hence, in the more critical details, the population phylogeny fails to reveal concordance between the languages and genes of Amerindian populations. The magnitude of the discordance suggests that the Na–Dene and Eskimo–Aleut groups have had a vastly different evolutionary history to that of the Amerind linguistic group.

Conclusions

These studies indicate that mitochondrial divergence within the four Beringian linguistic phyla is substantially less than that observed within the

Amerind phylum. Even more dramatic, the mitochondrial divergence within a single Amerind-speaking tribe can exceed the divergence observed within the entire set of the four 'Beringian' linguistic phyla. This is in total conflict with the prediction that the relative degree of molecular divergence will be concordant with the relative degree of linguistic affiliation. Moreover, the distribution of pairwise sequence differences within and among these five linguistic phyla indicates that a single language family can have a larger time depth than the time depth of an entire phylum or a set of phyla. It is difficult to imagine how such a result could obtain if the tempo of linguistic evolution were consistently uniform. The data suggest that the Na–Dene and Eskimo–Aleut phyla have evolved in less time than was required for the differentiation of a single Amerind language family. These observations argue that, in the New World, linguistic evolution has proceeded at a dramatically different rate from the more steady rate of biological differentiation. Instead, linguistic evolution of the New World languages appears to have been characterized by periods of stasis and periods of spurts that presumably reflect changes in social and cultural factors, as well as in demographic factors.

Since the molecular data allow an initial assessment of the time depth separating populations, some impression can be gained of the extent to which the evolution of linguistic attributes has proceeded in fits and starts. In making this comparison, we recognize that the molecular estimates of divergence are subject to potential error in calibration, so that the relative degree of temporal divergence between populations should be emphasized rather than the absolute time depths. We also note that since molecular divergence tends to antedate the population divergence (Ward *et al.*, 1991, 1993), the temporal depth for linguistic evolution will be even smaller than the numbers given here.

Using the statistically derived calibration for this region of the mitochondrial genome, sequence divergence is estimated to occur at a rate of 1% per 8,900 years (Lundstrom, Tavaré & Ward, 1992). This implies that the ancestry of each Amerind group extends back approximately 14 500 years, while the Eskimo–Aleut and Na–Dene groups have more modest time depths of only 9000 and 6600 years, respectively. Further, combining the Na–Dene and Eskimo–Aleut populations yields an average time depth of no more than 8000 years. This suggests that there is a little biological differentiation between these two linguistic groups. However, the Chukchi and Uralic language phyla have intermediate time depths of 12 000 and 8000 years, respectively. Notably, when the Na–Dene, Eskimo–Aleut, Chukchi and Yukaghir samples are combined in a single group, their combined time depth of 10 600 years is still appreciably less than the time depth for the Amerind populations combined.

The pairwise sequence divergence between populations suggests that the differentiation leading to the establishment of the Na–Dene and Eskimo–Aleut-speaking groups occurred during a time span of no more than 9000 years, while 16 000 years were required for the differentiation of the three Amerind populations. If linguistic evolution proceeds by differentiation of linguistic elements within culturally cohesive communities, then the development of these three linguistic phyla has to be contained within these time frames. Since our estimated time depth for the Amerind populations coincides with the most commonly accepted date of entry of humans into the New World, it is likely that the estimated time depth reflects biological reality. Since humans are believed to have occupied the Beringean region for 40 000 years (Hoffecker *et al.*, 1993), the much younger time depth for the four Beringian language phyla suggests that the development of these languages did not occur in situ. Rather, linguistic diversification appears to have followed an expansion into this marginal ecological zone, or penetration of a previously uninhabited zone, along the lines envisaged by Renfrew (1992). However, while cultural evolution may have proceeded apace, the degree of biological differentiation was relatively slight.

References

Avise, J. C., Arnold, J., Ball, R. M., Bermingham, E., Lamb, T., Neigel, J. E., Reeb, C. A. *et al.* (1987). Intraspecific phylogeography, the mitochondrial DNA bridge between population genetics and systematics. *Annual Review of Ecology and Systematics*, **18**, 489–522.

Ballinger, S. W., Schurr, T. G., Torroni, A., Gan, Y. Y., Hodge, J. A., Hassan, K., Chen, K-H. & Wallace, D. C. (1992). Southeast Asian mitochondrial DNA analysis reveals continuity of ancient Mongoloid migrations. *Genetics*, **130**, 139–52.

Barbujani, G. & Sokal, R. R. (1990). Zones of sharp genetic change in Europe are also linguistic boundaries. *Proceedings of the National Academy of Sciences, USA*, **87**, 1816–19.

Barrantes, R., Smouse, P. E., Mohrenweiser, H. W., Gershowitz, H., Azofeifa, J., Arias, T. D. & Neel, J. V. (1990). Microevolution in lower Central America, genetic characterization of the Chibcha speaking groups of Costa Rica and Panama, and a consensus taxonomy based on genetic and linguistic affinity. *American Journal of Human Genetics*, **46**, 63–84.

Black, F. L., Salzano, F. M., Berman, L. L., Gabbay, Y., Weimer, T. A., Franco, M. H. L. P. & Pandey, J. P. (1983). Failure of linguistic relationships to predict genetic distances between the Waiapi and other tribes of the lower Amazon. *American Journal of Physical Anthropology*, **60**, 327–35.

Bowcock, A. M., Kidd, J. R., Mountain, J. L., Hebert, J. M., Carotenuto, L., Kidd, K. K. & Cavalli-Sforza, L. L. (1991). Drift, admixture and selection in human evolution, a study with DNA polymorphisms. *Proceedings of the National*

Academy of Sciences, USA, **88**, 839–43.
Cavalli-Sforza, L. L. & Feldman, M. W. (1981). *Cultural Transmission and Evolution, A Quantitative Approach*. Princeton: Princeton University Press.
Cavalli-Sforza, L. L., Minch, E. & Mountain, J. L. (1992). Co-evolution of genes and languages revisited. *Proceedings of the National Academy of Sciences, USA*, **89**, 5620–4.
Cavalli-Sforza, L. L., Piazza, A., Menozzi, P. & Mountain, J. (1988). Reconstruction of human evolution: bringing together genetic, archeological and linguistic data. *Proceedings of the National Academy of Sciences, USA*, **85**, 6002–6.
Chakraborty, R., Blanco, R., Rothhammer, F. & Llop, E. (1976). Genetic variability in Chilean Indian populations and its association with geography, language and culture. *Social Biology*, **23**, 73–81.
Darwin, C. (1871). *The Descent of Man*. London: J. Murray.
Dumond, D. E. (1980). The archeology of Alaska and the peopling of America. *Science*, **209**, 984–91.
Dumond, D. E. (1987). A re-examination of Eskimo-Aleut prehistory. *American Anthropologist*, **89**, 32–56.
Greenberg, J. H. (1987). *Language in the Americas*. Stanford: Stanford University Press.
Hoffecker, J. F., Powers, W. R. & Goebel, T. (1993). The colonization of Beringia and the peopling of the New World. *Science*, **259**, 46–53.
Horai, S., Kondo, R., Nakagawa-Hattori, Y., Hayashi, S., Sonoda, S. & Tajima, K. (1993). Peopling of the Americas founded by four major lineages of mitochondrial DNA. *Molecular Biology and Evolution*, **10**, 23–47.
Kolman, C. J., Bermingham, E., Cooke, R., Ward, R. H., Arias, T. D. & Guionneau-Sinclair, F. (1995). Reduced mtDNA diversity in the Ngöbé Amerinds of Panamá. *Genetics*, **140**, 275–83.
Laughlin, W. S. (1987). From Ammassalik to Attu, 10,000 years of divergent evolution. *Objets et Monde*, **25**, 141–8.
Lumsden, C. J. & Wilson, E. O. (1983). *Promethean Fire, Reflections on the Origin of Mind*, Cambridge, Massachusetts: Harvard University Press.
Lundstrom, R., Tavaré, S. & Ward, R. H. (1992). Estimating substitution rates from molecular data using the coalescent. *Proceedings of the National Academy of Sciences, USA*, **89**, 5961–5.
Meltzer, D. J. (1993). Pleistocene peopling of the Americas. *Evolutionary Anthropology*, **1**, 157–69.
Neel, J. V. & Ward, R. H. (1970). Village and tribal genetic distances among American Indians and the possible implications for human evolution. *Proceedings of the National Academy of Sciences, USA*, **65**, 323–30.
Nichols, J. (1990). Linguistic diversity and the first settlement of the new world. *Language*, **66**, 475–521.
Renfrew, C. (1992). Archeology, genetics and linguistic diversity. *Man*, **27**, 445–78.
Renfrew, C. (1994). World linguistic diversity. *Scientific American*, **270**, 104–10.
Ruhlen, M. (1991). *A Guide to the World Languages, Classification*. Stanford: Stanford University Press.
Ruhlen, M. (1994). *The Origin of Language, Tracing the Evolution of the Mother Tongue*. New York: Wiley.
Santos, M., Ward, R. H. & Barrantes, R. (1994). mtDNA variation in the Chibcha

Amerindian Huetar from Costa Rica. *Human Biology*, **66**, 963–77.
Schurr, T. G., Ballinger, S. W., Gan, Y. Y., Hodge, J. A., Merriwether, D. A., Lawrence, D. N., Knowler, W. C., Weiss, K. M. & Wallace, D. C. (1990). Amerindian mitochondrial DNAs have rare Asian mutations at high frequencies, suggesting they derived from four primary maternal lineages. *American Journal of Human Genetics*, **46**, 613–23.
Shields, G. F., Schmiechen, A. M., Frazier, B. L., Redd, A., Voevoda, M. I., Reed, J. K. & Ward, R. H. (1993). Mitochondrial DNA sequences suggest a recent evolutionary divergence for Beringian and Northern North American populations. *American Journal of Human Genetics*, **53**, 549–62.
Smouse, P. E. & Ward, R. H. (1978). A comparison of the genetic infrastructure of the Ye'cuana and the Yanomama: A likelihood analysis of genotypic variation among populations. *Genetics*, **88**, 611–31.
Spuhler, J. S. (1979). Genetic distances, trees and maps of North American Indians. In *The First Americas, Origins, Affinities and Adaptation*, ed. W. S. Laughlin and A. B. Harper, pp.135–83. New York: Fischer.
Szathmary, E. J. E. (1993). Genetics of aboriginal North Americans. *Evolutionary Anthropology*, **1**, 202–20.
Teilhard de Chardin, P. (1957). *Oeuvres de Pierre Teilhard de Chardin*, 3. *La Vision du Passé*. Paris: Éditions du Seuil.
Torroni, A., Neel, J. V., Barrantes, R., Schurr, T. G. & Wallace, D. C. (1994). Mitochondrial DNA 'clock' for the Amerinds and its implications for timing their entry into North America. *Proceedings of the National Academy of Sciences, USA*, **91**, 1158–62.
Torroni, A., Schurr, T. G., Yang, C-C., Szathmary, E. J. E., Williams, R. C., Schanfield, M. S., Troup, G. A., Knowler, W. C., Lawrence, D. N., Weiss, K. M. & Wallace, D. C. (1992). Native American mitochondrial DNA analysis indicates the Amerind and Na–Dene populations were founded by two independant migrations. *Genetics*, **130**, 153–62.
Valencia, D. (1992). *Mitochondrial DNA Evolution in the Nuu–Chah–Nulth Population*. Unpublished MS thesis, University of Utah.
Vigilant, L., Stoneking, M., Harpending, H., Hawkes, K. & Wilson, A. C. (1991). African populations and the evolution of human mitochondrial DNA. *Science*, **253**, 1503–7.
Wallace, D. C., Garrison, K. & Knowler, W. C. (1985). Dramatic founder effects in Amerindian mitochondrial DNAs. *American Journal of Physical Anthropology*, **68**, 149–55.
Ward, R. H. (1972). The genetic structure of a tribal population, the Yanomama Indians. V: Comparison of a series of genetic networks. *Annals of Human Genetics*, **36**, 21–43.
Ward, R. H., Frazier, B. L., Dew-Jaeger, K. & Pääbo, S. (1991). Extensive mitochondrial diversity within a single Amerindian tribe. *Proceedings of the National Academy of Sciences, USA*, **88**, 8720–4.
Ward, R. H. & Neel, J. V. (1970). Gene frequencies and micro-differentiation among the Makiritare Indians. IV: a comparison of a genetic network with ethno-history and migration matrices; a new index of genetic isolation, *American Journal of Human Genetics*, **22**, 538–61.
Ward, R. H., Redd, A., Valencia, D., Frazier, B. L. & Pääbo, S. (1993). Genetic and linguistic differentiation in the Americas. *Proceedings of the National Academy*

of Sciences, USA, **90**, 10663–7.

Ward, R. H., Voevoda, M. I., Shields, G. F., Frazier, B. L., Bauer, K., Osipova, L. P. & Pääbo, S. (1995). Discordant evolution of languages and genes. *Science*, submitted.

Woodbury, A. C. (1984). Eskimos and Aleut languages. *Handbook of North American Indians*, **5**, 49–63.

16 *Allelic sequence diversity at the human β-globin locus*

S. M. FULLERTON

Despite nearly 20 years of molecular genetic investigation of human polymorphism, relatively little is known about the nature of variation in the human genome. Only a handful of loci have been investigated in any detail, and most studies have relied on indirect methods for the detection of DNA-level variation, most notably surveys using infrequently-cutting restriction enzymes. DNA sequence analysis (Sanger, Nicklen & Coulson, 1977), the only method with the potential for identifying all nucleotide and length variation present in a given genomic region has been applied to the analysis of the maternally inherited, haploid mitochondrial (mt) DNA genome (e.g. Vigilant *et al.*, 1991; Ruvolo *et al.*, 1993) but not nuclear DNA loci. As a consequence, high resolution investigation of genetic variation has been confined to a small, and largely atypical, portion of the human genome.

There are many reasons why DNA sequence investigation of human nuclear loci has lagged behind equivalent analysis of mtDNA. Mitochondrial DNA is far easier to isolate and analyse than nuclear DNA and phylogenetic interpretation of the allelic variation found there is simplified by the absence of interallelic recombination and gene conversion. Nuclear DNA is not only subject to the (potentially confounding) effects of recombination, it also contains fewer variable nucleotide sites than mtDNA due to its approximately ten-fold lower rate of point mutation (Cann, Stoneking & Wilson, 1987). Thus, much more nuclear DNA must be sequenced to identify an equivalent number of 'informative' polymorphic sites. Despite these apparent shortcomings, it is clear that DNA sequencing studies of nuclear DNA are long overdue. Without information about fine-scale genetic variation at the nuclear loci, we are unlikely to understand fully the nature of human genetic variation at the sequence level, or indeed identify the way in which random genetic drift and gene flow, natural selection, mutation and most especially interallelic recom-

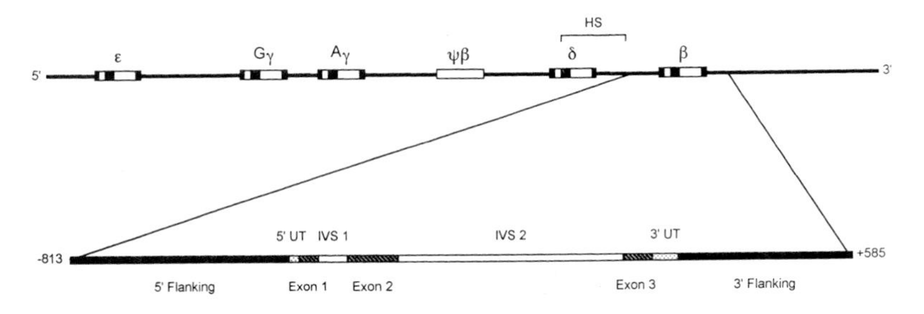

Fig. 16.1. The 3 kb genomic region sequenced and its relationship to the human β-globin gene cluster. Exons, introns (abbreviated IVS for intervening sequence), and untranslated (UT) regions of the human β-globin gene are indicated. HS marks the approximate boundaries of a previously identified hotspot for recombination.

bination, have interacted to determine the level and distribution of variation present in the majority of the human genome.

The human β-globin gene was chosen to initiate the investigation of human nuclear DNA at the sequence-level in a population context. The β-globin locus is one of a group of five functional loci (and a closely related, but non-expressed, pseudogene) on chromosome 11 (Fig. 16.1) which code for the β chains of the haemoglobin molecule during different stages of human development (Karisson & Neinhuis, 1985). The adult-expressed β-globin gene is the most extensively studied locus in the cluster – known best as the gene implicated in the inherited disorders sickle cell anaemia and β thalassaemia (Antonarakis, Kazazian Jr. & Orkin, 1985). This might, at first glance, appear an inappropriate locus for the investigation of population genetic variation, given the demonstrably protective effect these haemoglobinopathies afford heterozygous carriers in regions of high malarial endemicity (Flint *et al.*, 1993). For a pioneering investigation of this kind, however, it would seem ideal to focus on a locus about which the effects of selection are already well known, so that we are better placed to understand and interpret the variation identified.

DNA sequence variation in a 3 kilobase-long genomic region including the β-globin gene and 5′ and 3′ flanking DNA (Fig. 16.1) was surveyed in samples drawn from four separate geographic locales in Melanesia, Indonesia, and West Africa (Fig. 16.2). While investigation of variation in a single population would have gone some way toward addressing the question of 'what does human nuclear DNA variation look like at the sequence level?', it would not have begun to approach the question of 'why does the variation look the way it does?'. By identifying variation in several different populations, patterns arising as a consequence of a common molecular environment (i.e. the effects of mutation and recombination)

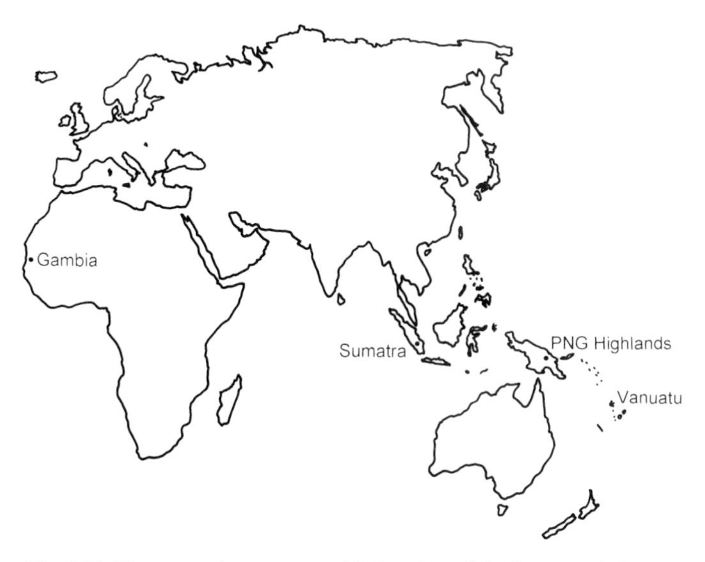

Fig. 16.2. The approximate geographic location of the four populations surveyed. PNG = Papua New Guinea.

could be distinguished from those due to population-specific phenomena such as migration, drift, and natural selection. In addition, because of differences in the level of malarial prevalence in the geographic regions chosen for investigation, the study had the potential to partition or otherwise distinguish the effects of natural selection from those of gene flow and drift, at least in the populations investigated. Thus, although the study described here samples only a portion of β-globin sequence variation in the human species, it does so in such a way as to maximize insights into the nature of forces acting on the locus.

In total, 144 chromosomes were investigated and 53 distinct alleles were identified at the human β-globin locus. The nature of this allelic sequence variation and its distribution within and between the populations surveyed, as well as within and between particular genomic subregions, is described here. Consideration of observed polymorphism in the light of neutral theory predictions (Kimura, 1983) highlights the importance of recombination in determining the nature of variation present in the β-globin gene region.

DNA sequence variation at the β-globin locus

Traditionally, variation at a genetic locus is defined in terms of the number of alleles present and the frequency of those alleles in the sample under

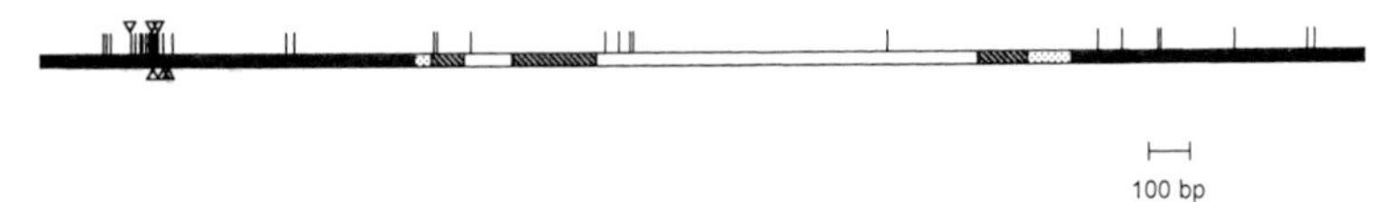

Fig. 16.3. The 3 kb sequenced region, showing the location of the 34 polymorphic positions identified. Lines identify nucleotide substitutions while triangles mark the position of insertion (downward pointing) or deletion (upward pointing) events.

investigation. DNA sequence diversity may be described in similar terms, with alleles being identified according to the 'sequence haplotype' in the genomic region under investigation, i.e. the particular combination of nucleotides found on a single chromosome. Within a given genomic region, many different nucleotide sites may vary, so that DNA sequence variation can also be described in terms of the number and distribution of variable sites. In general, these two ways of describing sequence variation yield similar estimates of diversity, because the introduction of a new polymorphic site at a locus will usually concomitantly increase allelic variation. Sometimes, however, there are discrepancies, and these (depending on the nature of the difference identified) may be related to the influence of demographic, selective, or molecular processes. For this reason, it is important to consider both kinds of information in an investigation of DNA sequence variation.

DNA sequence diversity at the human β-globin locus was characterized by polymerase chain reaction (PCR) amplification (Saiki *et al.*, 1985) and direct sequence analysis of the amplified region. Standard PCR amplification of diploid nuclear gene regions, while sufficient for the identification of polymorphic nucleotide sites, results in sequence data uninterpretable in terms of allelic sequence variation. This is because DNA from both alleles is simultaneously amplified and the resulting sequence ladder is a composite of the nucleotides present on both chromosomes. In order to identify sequence haplotype polymorphism at the β-globin locus, therefore, allele-specific PCR amplification and sequencing was performed. This method (described in detail in Fullerton *et al.*, 1994) proved a rapid and reliable way of unambiguously identifying linkage relationships at the locus, so that both nucleotide and sequence haplotype polymorphism could be assessed.

Twenty-eight of the 3000 nucleotide positions surveyed in the β-globin gene region varied polymorphically in the populations surveyed. These polymorphic positions were non-randomly distributed across the genomic region investigated (Fig. 16.3) : 13 of the 28 dimorphic sites were found in the 5′ flanking DNA (approximately 800 basepairs long), eight in the

Table 16.1. *The number of sequence haplotypes of each type observed in the four geographic locales surveyed, as well as in the total sample*

Type of sequence haplotype	*Number observed (frequency)*				
	Vanuatu ($n=61$)	PNG ($n=24$)	Sumatra ($n=28$)	The Gambia ($n=31$)	Total ($n=144$)
A	26 (42.6)	1 (4.2)	3 (10.7)	10 (32.3)	40 (27.7)
B	19 (31.1)	12 (50.0)	9 (32.1)	18 (58.1)	58 (40.3)
C	11 (18.0)	11 (45.8)	16 (57.1)	2 (6.4)	40 (27.7)
D	5 (8.2)	0 (0.0)	0 (0.0)	1 (3.2)	6 (4.2)

The percentage of overall sample size is given in parentheses.

β-globin gene itself (1600 bp in length), and seven in the 3′ flanking region (600 bp long). All but two of these polymorphic sites (one of the two observed in Exon 1 and another found in Intron 1) were phenotypically 'silent', resulting in changes with no known effect on β-globin protein expression or function. In accord with neutral theory predictions, the vast majority of the polymorphic sites identified fall in non-coding flanking and intervening sequences. In addition to the 28 point polymorphisms identified, three insertions (one 1 bp and two 2 bp in length) and three 1 bp deletions were observed (Fig. 16.3). All of these length polymorphisms occur in, or just adjacent to, a region of sequence characterized by a 52 bp alternating purine–pyrimidine repeat region.

The 34 polymorphic sites segregated as 53 distinct sequence haplotypes in the four geographic locales investigated. These sequence haplotypes ranged in frequency in the total sample from 22 (15.2%) to 1 (0.7%). In general, the most common haplotypes were also those which were observed in the greatest number of populations. Only one sequence haplotype was observed in all four populations. Among the 53 sequence haplotypes identified, four major types of haplotype, designated A, B, C, and D, were apparent (a distinction based on shared sequence similarities identified by visual inspection and confirmed by UPGMA cluster analysis). All four types of sequences were observed in most of the geographic locales surveyed, with B-type sequence haplotypes predominating in the total sample (Table 16.1). The frequencies of the four types of sequence haplotype varied considerably from one population to the next, however, with A-type sequences being most common in Island Melanesia (Vanuatu), C-type sequences most frequent in Sumatra, Indonesia, and B-type sequences most common in the highlands of Papua New Guinea (PNG) and The Gambia.

Estimating diversity from the observed nucleotide polymorphism

As mentioned above, one way in which genetic variation identified by DNA sequencing may by described is in terms of the number of polymorphic nucleotide sites present in the genomic region under investigation. In practice, this is straightforward: if the infinite sites model of mutation (Kimura, 1969) is assumed, diversity, $\hat{\theta}$, may be estimated from the observed number of silent polymorphic sites, s, present in a sample of n DNA sequences (Watterson, 1975):

$$\hat{\theta} = s \Big/ \sum_{i=1}^{n-1} 1/i$$

At the β-globin locus, where 32 silent sites vary (counting the small insertion/deletion polymorphisms but not the non-silent haemoglobin mutants) among 144 alleles, $\hat{\theta}$ is 5.78, an estimate which can be thought of as a measure of the average number of heterozygous sites in the 3 kb region. This estimate is, in turn, equivalent to a level of heterozygosity per nucleotide of 0.22%, assuming the presence of 2658 effectively silent sites (Kreitman, 1983) at the locus (Fullerton, 1994). In other words, approximately 1 in 500 silent sites varies on average between any two alleles drawn at random from the total sample. This level of nucleotide diversity, which is the same as that calculated for the β-globin gene cluster on the basis of restriction fragment length polymorphism data (0.2%, Kazazian Jr. *et al.*, 1983), and somewhat higher than an equivalent estimate calculated for a collection of human nuclear genomic and cDNA sequences (0.08%, Li & Sadler, 1991), suggests that variation has been accumulating at nuclear loci in human populations for a considerable period of time. A comparison of the calculated diversity with estimates of interspecific (human–chimpanzee) divergence in the same 3 kb region (Savatier *et al.*, 1985 and J. Bond, personal communication) gives an *average* age of β-globin sequence divergence (approximately half the total time to sequence coalescence) of 640 000 years, assuming that human and chimpanzee lineages began diverging 5 million years ago (Saitou & Ueda, 1994).

When similar estimates of nucleotide diversity are calculated for the 5′ flanking DNA, gene, and 3′ flanking DNA subregions of the 3 kb sequence, it is clear that per-nucleotide estimates of heterozygosity are highest in the 5′ flanking region and lowest in the gene in the total sample, as well as within each geographic locale surveyed (Fig. 16.4). Furthermore, estimates of diversity in the three Southeast Asian populations (Vanuatu, PNG Highlands, and Sumatra) are roughly equivalent to one another in all three genomic subregions, whereas nucleotide polymorphism in the West

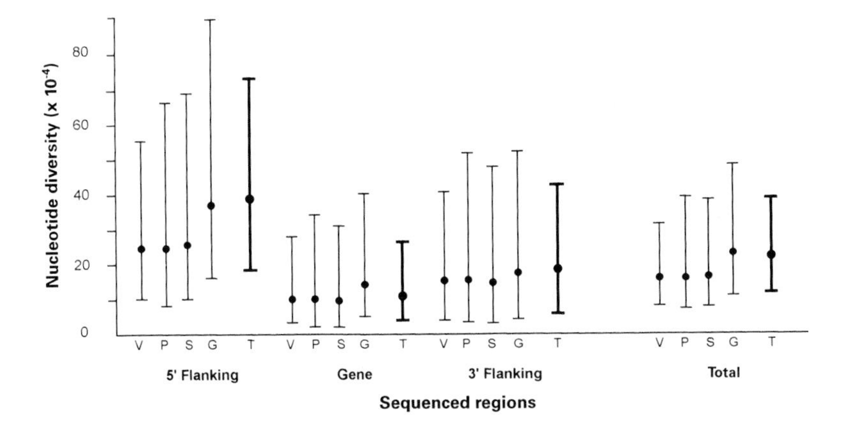

Fig. 16.4. Estimates of nucleotide diversity, θ, by genomic region, for the four populations surveyed (V = Vanuatu, P = PNG Highlands, S = Sumatra, and G = Gambia), as well as for the total sample (T, bold brackets). The brackets represent the 95% confidence interval for each estimate.

African Gambian population is greater, particularly in the 5′ flanking DNA. None of these differences is statistically significant. The consistently higher nucleotide diversity in the 5′ flanking region of all locales surveyed is, nevertheless, striking, and may reflect the importance of regionally specific molecular phenomena. A higher underlying rate of mutation to the 5′ flanking region, for example, could explain the slightly higher level of diversity observed 5′ to the gene.

Differences in sequence diversity between populations

If a higher rate of neutral mutation in the 5′ flanking region of the human β-globin locus is responsible for the higher level of nucleotide diversity observed there *within* geographic locales, we should also expect to see a greater degree of sequence divergence *between* locales in the same genomic region. This is because new, geographically specific, polymorphisms are more likely to arise in the 5′ flanking DNA following population differentiation. One way in which sequence differences within and between populations can be summarized is using the population fixation index, F_{ST} (Wright, 1951). When DNA sequence data are available, F_{ST} is estimated from average pairwise sequence differences within and between populations, and designated N_{ST} (because the estimate is based on genetic divergence at the nucleotide level). N_{ST} is calculated as the ratio of the average genetic distance between genes from different populations, $\hat{v}_b$, relative to that among genes in the population at large (i.e. the sum of

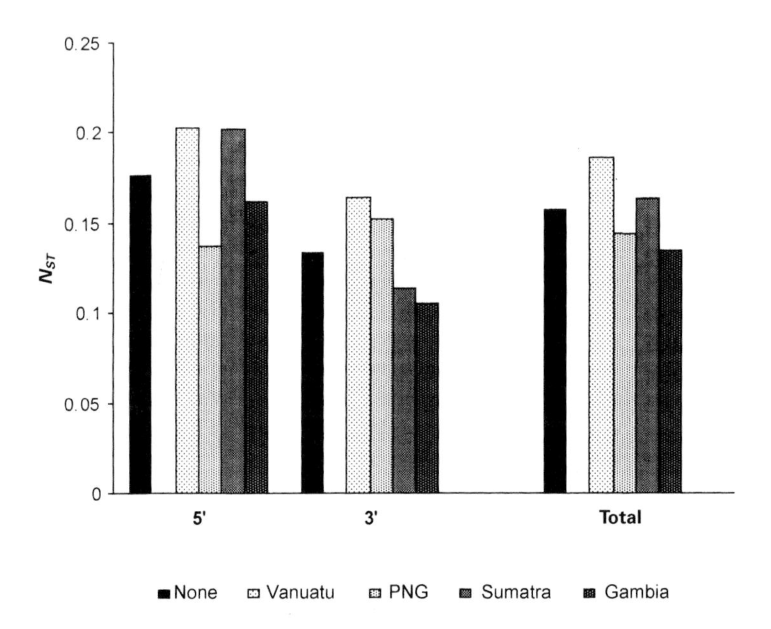

Fig. 16.5. Estimates of N_{ST} for 5′ and 3′ (gene + 3′ flanking DNA) subregions of the 3 kb sequence. In each case, estimates of N_{ST} for the total sample are followed by estimates for four subsets of the total sample, generated by removing data from one of the four populations surveyed.

within- , $\hat{v}_w$, and between-population differences) (Lynch & Crease, 1990):

$$N_{ST} = \frac{\hat{v}_b}{\hat{v}_w + \hat{v}_b}$$

N_{ST} estimates range from 0, if there are no differences between populations, to 1, for the case of complete genetic differentiation.

As shown in Fig. 16.5, when the entire 3 kb region is considered, N_{ST} for the total sample is 0.157. This suggests that 16% of the sequence variation observed at the β-globin locus occurs between particular geographic locales while the remaining 84% is found within populations. When estimates of N_{ST} for the 5′ flanking and gene + 3′ flanking region are compared, genetic differentiation among the four populations sampled is, in fact, greater in the 5′ flanking DNA. This finding appears to confirm the suggestion that a higher rate of mutation in the 5′ flanking region is responsible for the higher level of nucleotide polymorphism observed there within populations. Unexpectedly, however, the pattern of population divergence also differs markedly between the 5′ and 3′ β-globin subregions. This can be seen in the estimates of N_{ST} calculated for four subsets of the total sample (Fig. 16.5), generated by removing each of the main populations in turn (after Slatkin, 1985). These subset estimates, which provide an indication of the degree of genetic isolation of particular populations with respect to the total sample, demonstrate that the distribution of average pairwise sequence differences within and between populations in the 5′ flanking DNA is not the same as in the gene + 3′ flanking region. Such pronounced differences in the structure of sequence variation between adjacent genomic regions are remarkable and clearly cannot be attributed to the effects of mutation alone.

Explaining the observed sequence polymorphism

DNA sequence investigation of the 3 kb region encompassing the β-globin gene has revealed a polymorphic locus with extensive variation within and between the four geographic locales investigated. The challenge is to identify the processes responsible for determining the sequence diversity present at the locus, in particular that factor or factors which have contributed to differences in the level and geographic distribution of polymorphism present in the 5′ flanking DNA.

The effects of natural selection

Given the extensive evidence to suggest that β haemoglobinopathies, such as sickle cell anaemia and β thalassaemia, are maintained in human populations by balancing selection, it is important to consider first what impact natural selection has had on allelic sequence diversity at the locus. It is well known that silent variation may be indirectly subject to the effects of selection via the phenomenon of 'genetic hitchhiking' (Maynard Smith & Haigh, 1974), in which linkage disequilibrium between an advantageous mutation and surrounding genomic regions extends the effects of selection to nearby neutral polymorphism. When a nucleotide substitution is subject to balancing selection, linked neutral mutations are less susceptible to loss by random genetic drift and, consequently, an excess number of silent polymorphisms are expected to accumulate in the vicinity of the selected site over time.

As the majority of β haemoglobinopathies affecting β chain structure and synthesis fall within the coding and intervening sequences of the β-globin gene, it is not unreasonable to expect that 'excess' nucleotide polymorphism associated with effects of genetic hitchhiking would be concentrated in that subregion of the 3 kb sequence. As described above, however, not only are there no significant differences in nucleotide diversity between the 3 subregions investigated (such as would be expected if genetic hitchhiking was acting), in each geographic locale surveyed it is the gene which displays the lowest estimate of nucleotide heterozygosity per effectively silent site (Fig. 16.4). These results suggest that genetic hitchhiking to one or more balanced polymorphisms within the β-globin gene is unlikely to account for the observed pattern of polymorphism. Whether balancing selection acting on sites in the 5′ flanking DNA may explain the higher-than-average nucleotide diversity in that subregion is less clear, and will probably not be resolved until more is known about the functional significance, if any, of sequences in that genomic region.

The lack of evidence for the effects of hitchhiking on silent polymorphism at what is clearly a selected locus are perplexing until we stop to consider the length of time malarial selection has been operating in human populations. It is generally believed that falciparum malaria has only been an important selective force, in terms of its effects on human morbidity and mortality, in the last 2000 to 5000 years (Livingstone, 1989; Flint *et al.*, 1993). It is therefore possible that selection has not been acting on the β-globin locus long enough to affect the number of polymorphic sites found in particular genomic regions.

Another way in which natural selection may disturb the distribution of neutral polymorphism at a focus is by altering rapidly the *frequency* of a selected allele in a population. When a selectively advantageous mutation arises, the frequency of the mutant (and any linked neutral polymorphism) in the population will increase rapidly, often in only a small number of generations. The end result is a few sites of high to intermediate frequency (those that were linked to the selected mutation) and a proportionately greater number of low frequency polymorphisms, a pattern which erodes once the selected mutant ceases to be advantageous, or is fixed in the population. By comparing observed and expected frequencies of nucleotide sites using the Tajima test of neutral mutation (Tajima, 1989), the recent impact of selection may possibly be detected.

When estimates of the Tajima test statistic, D, are compared by genomic subregion for the four main geographic locales investigated, only polymorphism identified in the population from Sumatra, Indonesia displays a significant deviation from neutral theory expectations (Fig. 16.6). This positive deviation is due to a significantly larger number of intermediate (as opposed to low) frequency sites in the gene subregion in this population, a result consistent with the effects of balancing selection on β haemoglobinopathies present in that geographic locale. The lack of a similar pattern of site frequency deviation in other geographic locales subject to malarial selection, however (in The Gambia, non-significant but negative values of D for the gene were obtained) means that this result must be regarded with caution. A possible alternative explanation for the Sumatran findings, for

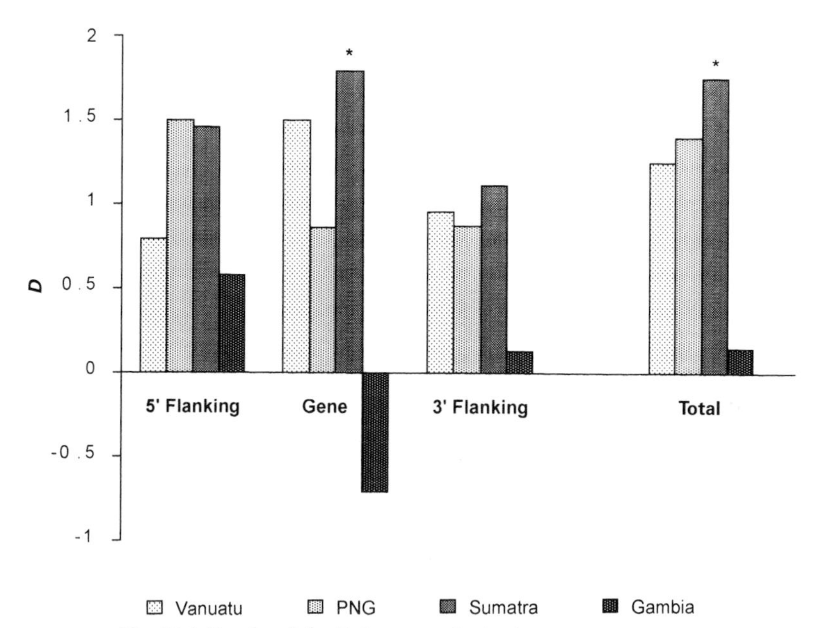

Fig. 16.6. Results of the Tajima test of selective neutrality (Tajima, 1989), by genomic region, for each of the surveyed populations and the total sample. *D* will equal 0 when there is no difference between observed and expected site frequencies.

example, is population admixture. However they are interpreted, these results suggest that the selection of advantageous haemoglobin mutants at the human globin gene does not render the overall pattern of silent polymorphism different from that expected of a neutrally evolving locus. Natural selection is therefore unlikely to account for the unusual pattern of diversity present in the 5′ flanking DNA.

The effects of recombination

A comparison of estimates of diversity based on nucleotide and sequence haplotype polymorphism (Strobeck, 1987) at the β-globin gene does suggest that the influence of recombination at the locus may vary from one genomic subregion to the next. As described above, diversity θ can be estimated from the number of polymorphic sites present in a particular genomic region using the infinite sites model of mutation. θ can also be estimated from the observed number of alleles, or sequence haplotypes, present at the locus using a slightly different model, the infinite alleles model (Kimura & Crow, 1964), which assumes that a new sequence haplotype is created each time a mutation occurs. In the absence of recombination, the

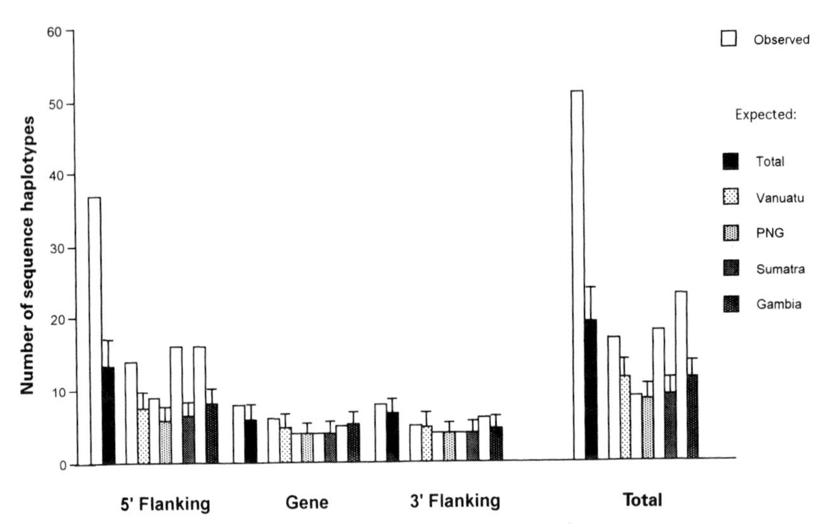

Fig. 16.7. Observed and expected numbers of sequence haplotypes, by genomic region, in each of the populations surveyed as well as the total sample.

infinite alleles estimate will equal the infinite sites estimate. By 'shuffling' existing nucleotide polymorphisms between chromosomes, however, recombination creates new sequence haplotype variation in a manner unrelated to the rate of point mutation, so that the relationship of θ to the number of sequence haplotypes is no longer defined by the infinite alleles model. In this case, many more than the number of sequence haplotypes expected on the basis of the mutation-only estimate of θ are observed (Strobeck & Morgan, 1978). Such greater-than-expected sequence haplotype diversity is found in the region 5′ of the β-globin locus (Fig. 16.7).

The presence of excess haplotype diversity in the 5′ flanking region, but not the gene or 3′ flanking DNA, in each population surveyed is unexpected and suggests that a significantly higher rate of interallelic exchange and gene conversion occurs there. Given the profoundly different molecular environment experienced by the 5′ flanking DNA, which is probably related to the presence of a previously identified recombination hotspot several kb 5′ of the β-globin gene (see Fig. 16.1), it is not surprising that sequence polymorphism in this region differs considerably from that observed in the other genomic subregions investigated. What is not immediately obvious, however, is the way in which recombination has contributed to the observed patterns. For example, it may be that processes involved in strand breakage and repair are inherently mutagenic, so that the point and length polymorphisms found clustered in the 5′ flanking DNA are a direct consequence of interallelic exchange at preferential sites in the region.

In practice, it is extremely difficult to test this hypothesis directly. However, a number of lines of indirect evidence suggest that mutagenicity is not the most plausible explanation of the effect of recombination on the 5′ flanking DNA. First, studies of the β-globin recombination hotspot region in yeast suggest that the site of initiation of recombination lies considerably 5′ of the genomic region sequenced (Treco, Thomas & Arnheim, 1985), so that strand breakage associated with the inferred interallelic exchanges is unlikely to be responsible for the sequence differences observed. Secondly, given the observed pattern of exchanges reflected in the sequence haplotypes identified, it is likely that exchange events resolve at variable distances from the site of initiation. In other words, there is no specific region in the 5′ flanking DNA where sequence exchanges preferentially end, and therefore no region which might be especially susceptible to mutation associated with sequence mismatch repair.

Instead, the influence of recombination on the 5′ flanking region may well be indirect. A similar relationship between levels of recombination and neutral polymorphism was observed in an investigation of variation in the species *Drosophila melanogaster* (Begun & Aquadro, 1992). Although explained initially in terms of the effects of genetic hitchhiking on linked genomic regions (interallelic exchanges disrupt 'selective sweeps', resulting in higher levels of diversity in more recombinogenic regions), this finding has been recently reinterpreted in terms of the effects of purifying selection. The alternative hypothesis, which suggests that the continuous elimination of deleterious mutations from a selectively constrained genomic region can cause a concomitant reduction in the level of linked neutral polymorphism (Charlesworth, Morgan & Charlesworth, 1993; Hudson, 1994), could also explain the relative lack of diversity observed in the β-globin gene and linked 3′ flanking region when compared to the recombinogenic 5′ flanking DNA (i.e. sites effectively 'uncoupled' from the selectively constrained locus). Firm support for this idea awaits further analysis of DNA sequence variation in the remainder of the hotspot region.

Irrespective of its precise mode of action, the impact of interallelic recombination in one portion of the genomic region sequenced provides a unique opportunity to evaluate patterns of geographic differentiation in regions of the genome subject to quite different molecular forces – an ability which can, in certain circumstances, aid in the interpretation of the observed differences. For example, when variation in the entire 3 kb region is considered, only the PNG Highlands sample is genetically differentiated with respect to other Southeast Asian locales investigated (Fig. 16.5). This differentiation, however, derives solely from the distinctiveness of variation found in the 5′ flanking DNA and therefore cannot be attributed to the effects of extended geographic isolation (which we expect would affect variation in the 5′ flanking and gene + 3′ flanking regions similarly). Instead, it is likely that a reduction in population size, sufficient to reduce the extent of recombination in the 5′ flanking region and render variation there more susceptible to the effects of purifying selection (described above), is responsible for the observed patterns. As patterns of differentiation in the gene + 3′ flanking DNA region indicate, this reduction in population size would not have been identified without data from both recombinogenic and non-recombinogenic genomic regions.

Conclusions

Together, these analyses suggest that it is molecular and demographic, rather than selective, processes which best explain the allelic sequence diversity identified at the human β-globin locus. Significant differences in observed and expected levels of sequence haplotype diversity in the 5′ flanking DNA reflect the pronounced impact of interallelic recombination and gene conversion in that genomic subregion. The higher level of nucleotide polymorphism (and interpopulation genetic differentiation) in the 5′ flanking DNA may derive partly from the diversity-preserving effects of frequent interallelic exchanges, which disrupt the effects of purifying selection on the β-globin gene, and partly from a higher underlying rate of neutral mutation unrelated to recombinational activity. Sequence variation in the 5′ flanking region may also be more susceptible to changes in population size than either the gene or 3′ flanking DNA, resulting in a distinct pattern of geographic differentiation in that genomic region. Where recombination is not acting, genetic differences among populations reflect the degree of geographic isolation of each locale, suggesting that mutation and drift, rather than natural selection, determine population differences. Indeed, there is little evidence that selection of advantageous β haemoglobinopathies has had any effect on silent polymorphism at the locus – the variation identified is consistent with that expected of a neutrally evolving human nuclear locus.

These insights into the nature of the forces responsible for determining the level and distribution of variation at the human β-globin locus were, in many cases, unexpected and demonstrate the considerable power that DNA sequence analysis brings to population genetic investigation of human nuclear DNA. Sequence-level investigation identifies all variation in a given genomic region and provides information about the distribution of polymorphic sites among different genomic subregions, as well as among alleles. As we have seen, such data, when considered in the light of neutral theory predictions, can reveal the impact of molecular, selective, and demographic processes undetected by other approaches. In particular, the pronounced impact of interallelic recombination 5′ of the β-globin gene would not have been identified without information about sequence haplotype diversity provided by allele-specific sequence analysis.

Whether or not the observed localised influence of interallelic recombination 5′ of the β-globin gene is typical of the majority of human nuclear loci, the importance of recombination as an evolutionary mechanism is clear. Full understanding of the role of molecular and other forces in the determination of human genetic variation awaits sequence-level investigation of other populations and other loci.

Acknowledgements

The work presented here was conducted at the Institute of Molecular Medicine (IMM) in Oxford and supported by grants from the Rhodes and Wellcome Trusts. DNA samples were provided by Mary Ganczakowski, Doug Higgs, and Adrian Hill of the IMM and ASM Sofro, of the Inter University Centre for Biotechnology, Gadjah Mada University, Yogyakarta. Technical assistance was provided by Jacquie Bond. I am indebted to my academic supervisors at the University of Oxford, John Clegg and Anthony Boyce, as well as to Rosalind Harding, for their support and guidance throughout the period of research.

References

Antonarakis, S. E., Kazazian Jr., H. H. & Orkin, S. H. (1985). DNA polymorphism and molecular pathology of the human globin gene clusters. *Human Genetics*, **69**, 1–14.

Begun, D. J. & Aquadro, C. F. (1992). Levels of naturally occurring DNA polymorphism correlate with recombination rates in *D. melanogaster*. *Nature*, **356**, 519–20.

Cann, R. L., Stoneking, M. & Wilson, A. C. (1987). Mitochondrial DNA and human evolution. *Nature*, **325**, 31–6.

Charlesworth, B., Morgan, M. T. & Charlesworth, D. (1993). The effect of deleterious mutations on neutral molecular variation. *Genetics*, **134**, 1289–303.

Flint, J., Harding, R. M., Clegg, J. B. & Boyce, A. J. (1993). Why are some genetic diseases common?: Distinguishing selection from other processes by molecular analysis of globin gene variants. *Human Genetics*, **91**, 91–117.

Fullerton, S. M. (1994). Allelic Sequence Diversity at the Human β-Globin Locus. Unpublished DPhil thesis, University of Oxford.

Fullerton, S. M., Harding, R. M., Boyce, A. J. & Clegg, J. B. (1994). Molecular and population genetic analysis of allelic sequence diversity at the human β-globin locus. *Proceedings of the National Academy of Sciences, USA*, **91**, 1805–9.

Hudson, R. R. (1994). How can the low levels of DNA sequence variation in regions of the *Drosophila* genome with low recombination rates be explained? *Proceedings of the National Academy of Sciences, USA*, **91**, 6815–18.

Karlsson, S. & Nienhuis, A. W. (1985). Developmental regulation of human globin genes. *Annual Review of Biochemistry*, **54**, 1071–108.

Kazazian Jr., H. H., Chakravarti, A., Orkin, S. H. & Antonarakis, S. E. (1983). DNA polymorphisms in the human β globin gene cluster. In *Evolution of Genes and Proteins*, ed. M. Nei and R. K. Koehn, pp. 137–46. Sunderland, Massachusetts: Sinauer Associates, Inc.

Kimura, M. (1969). The number of heterozygous nucleotide sites maintained in a finite population due to steady flux of mutations. *Genetics*, **61**, 893–903.

Kimura, M. (1983). *The Neutral Theory of Molecular Evolution*. Cambridge: Cambridge University Press.

Kimura, M. & Crow, J. F. (1964). The number of alleles that can be maintained in a finite population. *Genetics*, **49**, 725–38.

Kreitman, M. (1983). Nucleotide polymorphism at the alcohol dehydrogenase locus of *Drosophila melanogaster*. *Nature*, **304**, 412–17.

Li, W. -H. & Sadler, L. A. (1991). Low nucleotide diversity in man. *Genetics*, **129**, 513–23.

Livingstone, F. B. (1989). Simulation of the diffusion of the β-globin variants in the Old World. *Human Biology*, **61**, 297–309.

Lynch, M. & Crease, T. J. (1990). The analysis of population survey data on DNA sequence variation. *Molecular Biology and Evolution*, **7**, 377–94.

Maynard Smith, J. & Haigh, J. (1974). The hitchhiking effect of a favourable gene. *Genetical Research, Cambridge*. **23**, 23–35.

Ruvolo, M., Zehr, S., von Dornum, M., Pan, D., Chang, B. & Lin, J. (1993). Mitochondrial COII sequences and modern human origins. *Molecular Biology and Evolution*, **10**, 1115–35.

Saiki, R. K., Scharf, S. J., Faloona, F. A., Mullis, K. B., Horn, G. T., Erlich, H. A. & Arnheim, N. (1985). Enzymatic amplification of beta-globin sequences and restriction site analysis for diagnosis of sickle cell anemia. *Science*, **230**, 1350–4.

Saitou, N. & Ueda, S. (1994). Evolutionary rates of insertion and deletion in noncoding nucleotide sequences of primates. *Molecular Biology and Evolution*, **11**, 504–12.

Sanger, F., Nicklen, S. & Coulson, A. R. (1977). DNA sequencing with chain-terminating inhibitors. *Proceedings of the National Academy of Sciences, USA*, **74**, 5463–7.

Savatier, P., Trabuchet, G., Faure, C., Chebloune, Y., Gouy, M., Verdier, G. & Nigon, V. M. (1985). Evolution of the primate β-globin gene region: high rate of variation in CpG dinucleotides and in short repeated sequences between man and chimpanzee. *Journal of Molecular Biology*, **182**, 21–9.

Slatkin, M. (1985). Rare alleles as indicators of gene flow. *Evolution*, **39**, 53–65.

Strobeck, C. (1987). Average number of nucleotide differences in a sample from a single subpopulation: a test for population subdivision. *Genetics*, **117**, 149–55.

Strobeck, C. & Morgan, K. (1978). The effect of intragenic recombination on the number of alleles in a finite population. *Genetics*, **88**, 829–44.

Tajima, F. (1989). Statistical method for testing the neutral mutation hypothesis by DNA polymorphism. *Genetics*, **123**, 585–95.

Treco, D., Thomas, B. & Arnheim, N. (1985). Recombination hot spot in the human β-globin gene cluster: meiotic recombination of human DNA fragments in *Saccharomyces cerevisiae*. *Molecular and Cellular Biology*, **5**, 2029–38.

Vigilant, L., Stoneking, M., Harpending, H., Hawkes, K. & Wilson, A. C. (1991). African populations and the evolution of human mitochondrial DNA. *Science*, **253**, 1503–7.
Watterson, G. A. (1975). On the number of segregating sites in genetical models without recombination. *Theoretical Population Biology*, **7**, 256–76.
Wright, S. (1951). The genetical structure of populations. *Annals of Eugenics*, **15**, 323–54.

17 *A nuclear perspective on human evolution*

K. K. KIDD AND J. R. KIDD

Introduction

Human genome diversity and recent human evolution are intimately related. The distribution of DNA sequence variation within and among human populations, however one chooses to define those populations, is the result of numerous evolutionary factors (migration, selection, mutation, and random genetic drift) operating throughout the history of our species and its recent ancestors. The emphasis of human diversity studies is shifting from simple description of which populations are 'more closely related' to identification of the evolutionary forces and historical events that are responsible for the extant diversity. The patterns of variation seen in mitochondrial, sex chromosome, and autosomal DNA are both the result of and reflect those factors, but differently.

Neither mitochondrial nor Y chromosome DNA is sufficient

Mitochondrial DNA (mtDNA) is relatively easy to study and has been exceedingly valuable in addressing a large number of evolutionary and taxonomic issues in a variety of species. However, it has many limitations and cannot provide information on many very relevant questions – especially those related to the amount of genomic diversity within and among human populations. Also, it can provide only partial information relevant to explaining the causes of the genetic diversity found among human populations. Two related characteristics of mtDNA are the primary reasons for the limitations: the absence of recombination and the maternal pattern of inheritance. As a consequence, the mtDNA behaves in evolution as a single locus and reflects only the genetic contributions of females to succeeding generations.

Because the entire mtDNA molecule behaves as a single locus, the pattern of relationships shown by extant mtDNAs, even if that pattern could be perfectly known, is only a single realization of the stochastic process of inheritance from parent to child through generations. As but a single realization, mtDNA will provide a very poor estimate of the parameters of that stochastic process. In addition, any selection that may have operated will have affected all sites simultaneously; with no comparison data the existence of that selection is undetectable against the background stochastic process.

Y chromosome DNA variation, with its paternal transmission and lack of recombination, theoretically should complement the maternal transmission of mtDNA. Unfortunately, the Y chromosome has the disadvantage of containing very little variation (e.g. Jakubicza *et al.*, 1989; Malaspina *et al.*, 1990; Dorit, Akashi & Gilbert, 1995). Furthermore, without recombination, the entire Y chromosome (excepting the pseudoautosomal region, PAR) is transmitted intact as a single locus. Thus, Y chromosome variation also represents a single realization of stochastic population genetic processes. Furthermore, if the relative paucity of polymorphisms on the Y chromosome is due to periodic homogenizing selective sweeps, the time depth of its variation would extend only as far back as the last sweep.

Gene flow has undoubtedly been a significant factor in human population diversification. Migration of individuals between groups is characteristic of many of our closest primate relatives and is certainly seen in a large number of human groups. In many cases, that migration will involve predominantly males or predominantly females as a consequence of various social patterns around the world. Because it is maternally inherited, mtDNA cannot 'see' male migration; analogously, Y chromosome DNA cannot 'see' female migration. Thus, populations that are virtually identical in the nuclear genome can have very different mtDNAs if male migration is responsible for keeping the nuclear genomes identical or very different Y chromosome variation if it is female migration that is responsible for keeping the nuclear genomes identical. Finally, because of their uniparental inheritance and lack of recombination, mtDNA and Y chromosome DNA truly evolve in a lineal manner and their analysis generates gene trees. There are as yet no good methods for estimating the relative similarities of populations based on the particular numbers and relationships of the mtDNA or Y-specific DNA variation found within them.

For analytical purposes, X chromosome and pseudoautosomal DNA variation are generally analogous to autosomal DNA in terms of their utility for population history and affinity studies. As there are only three-quarters the number of X chromosomes, and one-quarter the number of Y chromosomes as there are of any autosome, there are different effective population sizes for loci on the X, the Y, and the autosomal chromosomes.

Eventually, when sufficient data are available, the different levels of variation observed may be usable for more accurate inferences on the effective population sizes in different parts of the world.

Nuclear, autosomal DNA

Nuclear, autosomal DNA exists in many different forms, variation is transmitted biparentally, alleles at different loci can recombine, and different loci can independently assort at each generation. As with the particular characteristics of mtDNA and sex chromosome DNA, these various characteristics are both the strengths and drawbacks of nuclear DNA for studies of human population histories.

Types of nuclear markers and their value

Within nuclear DNA there is a wide variety of different kinds of polymorphic markers, e.g. short tandem repeat polymorphisms (STRPs), variable number of tandem repeats (VNTRs), single nucleotide polymorphisms (SNPs), insertion/deletion polymorphisms, and haplotypes. The VNTRs tend to have higher mutation rates (Jeffreys *et al.*, 1988) than do the STRPs (Weber & Wong, 1993; Weissenbach *et al.*, 1992; Bowcock *et al.*, 1994) and both have higher mutation rates than SNPs and insertion/deletion polymorphisms. This variation in mutation rates allows one to choose the type of marker that will provide the most information for the specific time depth being considered. Restriction fragment length polymorphisms (RFLPs) are defined by a method and are usually, but not always, SNPs involving a restriction site.

Short tandem repeat polymorphisms (STRPs) and variable number tandem repeats (VNTRs)

Recently, short tandem repeat polymorphisms (STRPs) have been used for population studies (e.g. Bowcock *et al.*, 1994; Deka *et al.*, 1995). These are generally loci with a sequence of two to five nucleotides tandemly repeated up to a few dozen times. The polymorphism usually exists as different numbers of repeats; occasionally some alleles involve an irregular sequence. STRPs are being used for population studies because they are highly informative – they have many alleles, some of which are geographically restricted; they are PCR based; and they are generally easy to type and interpret.

Likewise, variable number of tandem repeat (VNTR) polymorphisms have been used for population studies (e.g. Deka *et al.*, 1992). These are loci with seven, eight, or more nucleotides tandemly repeated from a few dozen up to hundreds of times at the same loci. Detection can be based on sequence, PCR, or RFLP methods. This type of polymorphism tends to have many more alleles than STRPs and considerably more than the 2–3 alleles of SNPs. We have had reservations about use of VNTRs, and STRPs also, for population studies because we have been uncertain about implying identity by descent (IBD) of alleles identical in size. In comparing allele frequencies across populations, one of the basic assumptions is IBD of alleles. If alleles are not IBD from a common ancestor, studies of phylogeny, migration, and admixture can all reach wrong conclusions. Clearly, by their very nature (i.e. multiple alleles), the mutation rates at these loci are elevated. Mutation rate estimates have ranged upwards to 7% for some VNTR loci (Jeffreys *et al.*, 1988) and up to 1.2×10^{-3} for STRPs (Weber & Wong, 1993; Weissenbach *et al.*, 1992). In forensics, high mutation rates at some VNTR loci have even confounded paternity testing. How much greater may be the confounding between alleles in populations distantly – and not so distantly – related? Data are beginning to accumulate on the stability of some of the STRP markers (Bowcock *et al.*, 1994; Tishkoff *et al.*, 1996*a*, *b*; Castiglione *et al.*, 1995), but usefulness for population studies will be determined on a marker-by-marker basis.

The accumulating evidence suggests that STRPs generally mutate by 'slippage' with a single-repeat difference being the most likely size of the new mutant allele. Based on that model, new genetic distance measures are being developed (Slatkin, 1995; Goldstein *et al.*, 1995). Just as any other measures of genetic distance, these measures make assumptions that will often be violated by the data on real populations. We are uncertain how robust these measures will be to variation in mutation rates among loci and deviations from the stepwise model of mutation.

Single nucleotide polymorphisms (SNPs) and insertion/deletion polymorphism

Originally, the application of random genetic drift theory to estimate genetic distances did not incorporate mutation; a basic assumption was that alleles identical in type are identical by descent (IBD). It seems likely that most SNPs that occur with moderate heterozygosities in several populations fulfil this criterion. RFLPs are usually examples of such single basepair polymorphisms that create/destroy a restriction site and, along with insertion/deletion polymorphisms, come closest to meeting the criterion of IBD. (The *Eco* RI RFLP at the PLAT locus is an *Alu* insertion polymorphism (Tishkoff *et al.*, 1996*b*).) Mutation rates at single nucleotides must be exceedingly small given the small number of fixed changes in DNA sequence between human, chimpanzee, and gorilla. Most insertion/deletion mutations are unique events with no recurrence. Unfortunately, their bi-allelic nature makes these polymorphisms less informative than multiallelic markers. Most of the two-allele RFLP systems identified as having at least moderate heterozygosities in Europeans are polymorphic in all or virtually all human populations studied to date (Bowcock *et al.*, 1987, 1991*a*, *b*; Kidd *et al.*, 1991). This leads us to the seemingly inescapable conclusion that such ubiquitous polymorphisms existed, and the 'new' allele had already reached polymorphic levels, prior to the dispersal of modern humans around the world. Some alleles, unique to a single or very small number of populations, have been found using the RFLP technique. These are most readily explained by a different site mutating and altering the restriction fragment length and thus are likely to be haplotype systems involving variation at two distinct restriction sites, although nearby insertions or deletions cannot be excluded. Unfortunately, for most of these systems, sufficiently little is known about their precise molecular nature that IBD cannot be considered a proven conclusion. We note that the use of PCR to identify a restriction site polymorphism by amplifying the small encompassing region and then digesting with the relevant enzyme greatly restricts the opportunity to identify additional mutations in the nearby DNA. As scientists work to develop a more uniform set of markers for studying human populations, an important consideration will be the degree to which the marker has already been characterized both molecularly and on a diverse set of human populations.

The allele frequencies of SNPs and insertion/deletion markers, whether detected as RFLPs or by other techniques, are the outcome of the genetic history of both sexes. There are so many markers in the autosomal chromosomes that individual markers or haplotypes for population studies can be selected to be transmitted independently. Thus, with several markers, one can observe several independent realizations of the stochastic transmission of alleles from one generation to the next. Simulation studies, such as that of Astolfi, Piazza & Kidd (1979), have shown that many such independent markers are necessary for correct (or nearly correct) estimation of relative genetic similarities for a set of several populations.

Critically, however, the very richness of autosomal DNA markers can also be a disadvantage in the study of population histories and dynamics. Because there are so many markers identified at the DNA level, the *a priori* probability of overlap in the choice of markers between laboratories is remote. If different markers are typed in different populations in different studies, the overall data will never be cumulative.

Bias in identifying markers

Having concluded that nuclear DNA would carry more information on population and genetic history with more fidelity, we were somewhat disappointed when, as mentioned above, we found that, in initial studies with RFLPs, nearly all populations share nearly all the observed alleles. We had expected that we would see 'semi-private' alleles – those found in only a few populations or regions. Such 'semi-private' alleles would be very informative for migration, admixture, mutation, and population histories. But, in fact, such 'semi-private' or regional alleles are rare. In our recent studies (Kidd, Pakstis & Kidd, 1993), we found five private or semi-private alleles out of 79 observed in 35 loci across 12 populations; Bowcock *et al.* (1991*a*) found 10 private or semi-private alleles out of 226 observed in 100 loci across 5 populations. Adding another African population to those of the Bowcock *et al.* (1991*a*) studies, Poloni *et al.* (1995) found five more private alleles.

In our initial studies in collaboration with L. L. Cavalli-Sforza and his laboratory (Bowcock *et al.*, 1987, 1991*a*, *b*), not only did we find that virtually all RFLP alleles identified were present in all populations, we also found that Europeans had the highest average heterozygosity across 100 mostly two-allele systems. Both the higher heterozygosity and the ubiquity of alleles may be due to an ascertainment bias – virtually all of the polymorphisms studied were previously published because they had been identified as high frequency polymorphisms in samples of individuals of European ancestry. If allele frequencies at polymorphisms were distributed among the loci in a more random fashion among populations, one would expect to find polymorphisms at low heterozygosity in Europe that reached high heterozygosity in some other populations. However, by the criteria operating during the golden years of RFLPs, a low frequency polymorphism identified in an initial screen would never be followed up or published. Therefore, most studies making use of these markers probably suffer to some extent from a European ascertainment bias.

Our discovery of the *Hind* III RFLP at the thymidine kinase locus serves as an example of this European Bias. We reported a *Taq* I polymorphism (Murphy *et al.*, 1986) and later a *BstE* II polymorphism (Murphy *et al.*, 1987) in the cytosolic thymidine kinase (TK1) locus with allele frequencies for the common allele at 60% and 64%, respectively, in large samples of mixed Europeans. In both of those reports, a *Hind* III variant was mentioned only in passing without any details beyond 'rare variant'. As it happens, our screening panel and test families (over 150 unrelated individuals) were all of 'European descent' and we deemed in those early days of RFLPs that a 3% frequency did not warrant description or further study. The *Hind* III polymorphism may prove to be unimportant, indeed, but that will not be because of its low frequency in Europeans as was made clear to us in some of our recent population studies.

Much of our recent population work has focused on 12 populations that have now been typed for at least 35 markers (Kidd *et al.*, 1993) with some

Table 17.1. *Allele frequencies, standard errors, and heterozygosities at the Hind III RFLP at the TK1 locus for selected populations*

Alleles	Populations											
in kb	Kari	Suru	Ticu	Maya	Jeme	Japa	Chin	Camb	Nasioi	Euro	Biaka	Mbuti
4.5	0.15	0.64	0.75	0.48	0.63	0.17	0.44	0.66	0.02	0.03	0.18	0.21
4.0	0.85	0.36	0.25	0.52	0.38	0.83	0.56	0.34	0.98	0.97	0.82	0.79
Std.err.	0.04	0.05	0.04	0.05	0.05	0.04	0.05	0.07	0.02	0.02	0.03	0.05
Obs.chrs.	86	90	122	100	88	88	96	50	44	74	130	76
Het.	0.26	0.46	0.37	0.50	0.47	0.28	0.49	0.45	0.04	0.06	0.30	0.33

Populations are more fully named in Table 17.3. This previously unpublished polymorphism is detected on Southern blots of *Hind* III-digested DNA using the probe pHTK9, which is available from the American Type Culture Collection (ATCC), Rockville, MD 20852. There are two constant bands of 7 and 16 kb in addition to the allelic bands of 4.5 and 4.0 kb.

typed for more than 100 markers (Bowcock *et al.*, 1991*b*). When typing one of the published TK RFLPs on families of European ancestry as part of a linkage study, we inadvertently included a *Hind* III filter of one of the Pygmy populations. When both alleles were seen to be common for that sample of Pygmies, we typed the samples of all eleven non-European populations. As shown in Table 17.1, the 'rarer' allele of this 2-allele system ranges from 2% in Nasioi and 3% Europeans, respectively, to 75% in Tiauna Surui. At an $F_{ST} = .278$ across a sample of 12 populations, the TK1/*Hind* III polymorphism has the highest F_{ST} of 37 two-allele systems, six three-allele systems, one five-allele system, one six-allele system, and seven haplotyped systems ranging in number of haplotypes from 4-17, all typed on the same 12 populations (Kidd *et al.*, 1993, and unpublished).

Although the existence of the European Bias seems obvious, it is difficult to document beyond anecdotes such as our TK1 story; it is impossible to quantify with existing data. The best evidence for its existence probably comes from surveys of STRPs which find that sub-Saharan African populations have more alleles and higher heterozygosities than any other populations (Bowcock *et al.*, 1994). On classical markers and RFLPs, the European Bias operates at the level of which markers are selected and 'new' alleles at those markers are unusual. Thus, markers selected in Europeans are already near their maximum possible heterozygosity and no non-European population can have higher average heterozygosity for those markers. Selection of loci may also be a bias with STRPs but they do not have an upper limit on the number of alleles occurring *at the same locus*. In other words, the threshold for 'identification' of an STRP marker is much lower in its range of possible heterozygosity than has been the general case for classical and RFLP markers and there is 'room' for African populations to have higher heterozygosities. Haplotyped loci could theoretically 'escape' the European Bias, as seen for the globin cluster haplotypes (Wainscoat *et al.*, 1986; Flint *et al.*, 1993), since they would generally be identified through the individual sites without regard to linkage phase. Relevant data we know best are from the haplotype system at the CD4 locus that we have studied on 42 populations (Tishkoff *et al.* (1996*a*). Fig. 17.1 presents graphically some of the CD4 haplotype frequency data from Tishkoff *et al.* (1996*a*) to illustrate the finding that sub-Saharan African populations (illustrated by the Woloff and Biaka) have quite different patterns of disequilibrium from all non-African populations (illustrated by the Finns). All non-African populations that have the *Alu*-deletion allele (*Alu*(-)) have that allele in complete or virtually complete disequilibrium with the 90 bp allele at the pentanucleotide STRP located about 9.8 kb away, as seen in the Finns. Non-African populations have only two other common haplotypes, the 85 and 110 bp STRP alleles in combination with

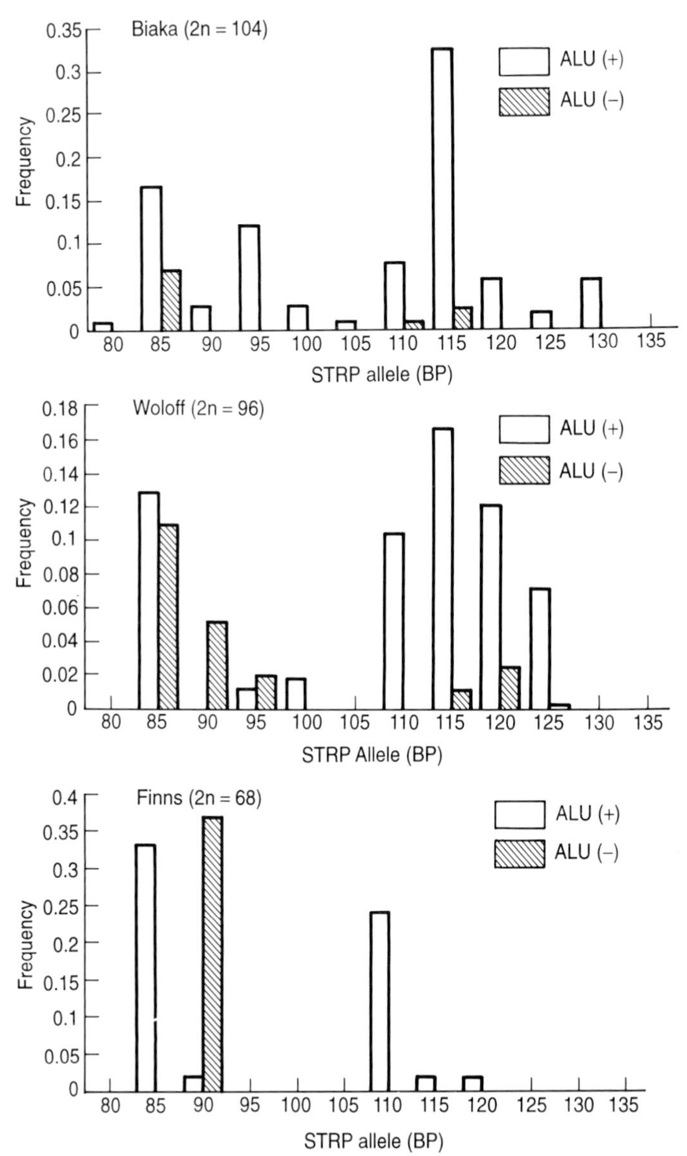

Fig. 17.1. Haplotype frequencies at the CD4 locus. These examples demonstrate the variation that can exist among populations. Alleles ranging from 85 bp to 135 bp have been observed at the pentanucleotide STRP; these can theoretically occur on the same chromosome with either the full-length *Alu* sequence (*Alu* (+) allele) or the partially deleted *Alu* (*Alu* (–) allele) at a site 9.8 kb away. Neither site is in coding sequence or any known regulatory region. Fewer STRP alleles occur in non-African populations and, as illustrated here, disequilibrium is stronger in all non-African populations that have the *Alu*(–) allele. The Woloff sample was originally collected by Dr S. Santachiara (Scozzari *et al.*, 1988) who kindly provided the DNA; the Finnish sample was collected and provided by Dr L. Peltonen. The histograms are drawn from data in Tishkoff *et al.* (1996*a*). See text for discussion.

the full length *Alu*. In contrast, in all sub-Saharan populations haplotypes exist with the *Alu*-deletion allele in combination with several different STRP alleles, there are more STRP alleles altogether, and there is low to absent disequilibrium. We have concluded (Tishkoff *et al.*, 1996*a*) from analyses of the entire set of data on 42 populations that there was only a single early migration of modern humans from Africa, that it occurred less than 90 000 BP, and descendants of that group populated the rest of the world.

Eventually, when more STRP markers and haplotypes are studied and when unbiased assays of heterozygosity in arbitrarily defined segments of DNA are conducted, we should be able to quantify the European Bias, or, better yet, have data not subject to the bias. In the meantime, we need to be concerned about whether the European Bias affects our perception of the relative amounts of evolution for different populations. The scenarios compatible with existing data seem most parsimoniously explained by loss of heterozygosity associated with the exodus from Africa including the failure of many of the alleles present in African populations to persist through the migration of modern humans from Africa and their subsequent expansion into the rest of the world. The failure to identify and hence test for those African-specific alleles will result in African populations appearing more similar to Europeans than they might otherwise.

Unfortunately, almost any screening procedure for SNPs or insertion/deletion polymorphisms (whether RFLPs or not) will be subject to some sort of ascertainment bias. We are attempting to minimize that bias in our search for new polymorphisms through the use of a population tube screening procedure (Ruano *et al.*, 1994). DNA aliquots from five members of a population are combined in a single tube and we generate as many such tubes ('poptubes') of mixed DNA as there are populations to be screened. An aliquot from each tube is PCR amplified (*en masse*), the opportunity for the formation of heteroduplexes is provided, and the PCR products are run on denaturing gradient gels so we can visualize the sequence mismatches at sites of allelic variation. We also generate United Nations tubes ('UN tubes') containing DNA of one male individual from each of five different and geographically widely dispersed populations. The UN tubes protect against absence of a heteroduplex in a poptube as would happen if a single allele, rare and undetected in other populations, were fixed in that poptube sample. The current embodiment of the poptube strategy is not perfect since it does not give a high probability of detecting uncommon alleles, those in the range of 1–5%, even if they occur in many populations. In addition, a decision is required on which of the detected variants are pursued – variants found only in a single population? variants found in at least two (contiguous or non-contiguous) populations? variants found in most populations? Each of those follow-up strategies imposes its own subtle to not-so-subtle biases.

Haplotypes

Is there a perfect marker for population studies? Probably not, but we have begun to look at nuclear DNA haplotypes and are convinced that they can be extremely valuable for elucidating not only the existing distribution of variation but also the more important causes of that distribution. All possible combinations of different types of polymorphisms can be considered as haplotypic markers if they are physically close enough on the chromosome that they are generally transmitted as a unit. Consideration of markers as haplotypes has the advantage of increasing the variability of each locus and additionally increasing the information carried at each marker by measuring disequilibrium. The possibility for more effective 'alleles' (i.e. haplotypes) is realized for most loci that have been studied. Moreover, these alleles are not all present in all populations. In the same study in which we found only five private or semi-private alleles in 35 loci (Kidd *et al.*, 1993), we have found 16 private or semi-private haplotypes out of 58 haplotypes observed at only seven loci in 12 populations. These private or semi-private haplotypes were all observed in the four haplotyped loci that are composed of three to four polymorphic systems; none was observed in three two-system haplotyped loci.

We are by no means the first to recognize the potential value of haplotypes in populations studies. The value of haplotypes at the α- and β-globin and HLA loci for studies of both disease and evolution has been obvious for years. Wainscoat *et al* (1986), Pagnier *et al.* (1984), Flint *et al.* (1993), and many others have described population dynamics in terms of haplotypes. Some of the problems encountered with the various markers discussed above may be eliminated or reduced through the use of haplotypes. (1) The number of independent haplotypes is potentially huge; (2) Patterns of haplotype frequencies reflect both maternal and paternal transmission; (3) Recombination increases the number of haplotypes observed, and the rate of recombination can be controlled by choosing markers to be included in the haplotype with different molecular distances between them; (4) The mutation rate of accompanying component STRP and VNTRs can be titrated against the IBD of single site polymorphisms for the time depth under study; and (5) To some extent the European Bias in the component systems may be diluted in the haplotype combination.

Table 17.2. *Haplotype systems*

Locus name and symbol	Map location	Clone and enzyme	Allele sizes in kb
D4S10	4p16.3	pKO-82/*Pst*I	5.6, 2.4, 2.2
		pKO-82/*Taq*I	3.0, 1.7
		pKO-82/*Hind*IIIa	4.9, 3.7
		pKO-82/*Hind*IIIb	17.5, 15.0
Human leukocyte Antigen, HLA	6p21.3	pRS5.1.0/*Pst*I	2.8, 2.5
		p5.4SH3/*Hind*III	5.4, 4.5
Retinol-binding protein 3, RBP3	10q11.2	H4/*Msp*I	3.0, 2.5, 2.4
		H4/*Bg*lII	6.3, 4.3
D10S20	10q21–q26	OS2/*Taq*I	11.0, 6.0, 2.4
		OS2/*Hind*III	7.4, 3.8
		OS3/*Taq*I	6.2, 5.5, 4.8
D13S2	13q22	p9D11/*Msp*I	15.0, 10.5, 17.0
		p9D11/*Taq*I	7.6, 5.6
		p9D11/*Pst*I	2.1, 1.1
Homeobox B cluster HOXB	17q21–q22	BS2/*Sac*I	2.0, 1.1
		H2-1-3/*Msp*I	2.0, 1.8, 3.0
		H2-1-5/*Taq*I	3.8, 3.2
D20S5	20p12	pRI2.21/*Pvu*lI	2.7, 1.9
		pRI2.21/*Msp*I	3.8, 3.0

Nineteen RFLPs are combined into seven haplotype systems as defined by their clone/restriction enzyme combination and alternative allele fragment lengths as obtained on Southern blots.

A preliminary view of human diversity based on haplotypes: A twelve-population, seven-haplotype study

The 12 population samples included in this study are described in Kidd *et al.*, (1993). The sample sizes ranged from 23 (Nasioi Melanesians) to 68 (Biaka Pygmies) with an average of 49 individuals per population and a total sample size of 591 people distributed across the 12 populations. References to various other previously published studies utilizing these same samples and describing the populations more fully are included in Kidd *et al.* (1996). All DNA was purified from Epstein-Barr virus-transformed cell lines using standard procedures (e.g. Sambrook, Fritch & Maniatis, 1989). Typing was either by Southern blot, PCR with restriction digestion and agarose electrophoresis, or by ^{32}P labelled PCR with polyacrylamide electrophoresis and autoradiography.

The population samples were typed for a total of 19 polymorphic systems at seven haplotyped loci. Table 17.2 presents a brief description of each haplotype and system. Four of the haplotypes are from anonymous segments of DNA (D4S10, D10S20, D13S2, and D20S5) and three are associated with known genes (RBP3, HLA, and HOXB). We have converted several of the systems originally typed by Southern blots to PCR and plan to continue working to convert the remainder (unpublished data).

The number of alleles (haplotypes) possible in a haplotyped locus is the product of the number of alleles at each polymorphic system comprising the haplotype. We actually observed fewer haplotypes in any specific population. The apparent absence of specific haplotypes may be due either to low frequency haplotypes not present in our sample, or to linkage disequilibrium limiting the number of haplotypes present in a population. It was necessary to estimate haplotype frequencies for ambiguous phenotypes (multiple heterozygotes or missing data). The FORTRAN program HAPLO (Hawley, Pakstis & Kidd, 1994; Hawley & Kidd, 1995) combines direct counting of haplotypes with the *expectation-maximization* algorithm (EM) of Dempster, Laird & Rubin (1977) to generate maximum likelihood frequency estimates. Allele frequencies for the individual sites and the haplotypes are presented elsewhere (Kidd, Pakstis & Kidd, 1993; Kidd, 1993). Overall, considering only haplotypes that reach a frequency of at least 5% in some population, there are nearly 50 haplotypes at these seven loci, giving the equivalent of about 40 independently varying alleles. Viewed in this way, the seven haplotyped systems provide the informational equivalent of 40 independent two-allele polymorphic markers, but only 19 such markers needed to be typed.

We have estimated the heterozygosity from the haplotype frequency estimates and averaged across loci to obtain a *mean* heterozygosity for each population at these loci. Table 17.3 presents the heterozygosities (and standard errors) of the twelve populations averaged across all seven haplotyped loci. We see for this study of seven haplotypes that heterozygosity is higher in mixed Europeans than in Africans. While not highly significant, this gives pause since it does not run counter to the European Bias picture. This may be more a function of the specific European and African samples we are currently studying than a general characteristic of haplotypes. Alternatively, there could be something about European history that has operated to preserve large numbers of haplotypes, on average.

On this dataset, most genetic distance measures based on estimates of F_{ST} are highly correlated $r = 0.909\text{-}0.995$ (Kidd, 1993). The Tau (τ) distance measure ($\tau = -\ln(1 - F_{ST})$) is based on the angular transformation and should give an additive distance in units of $t/2N$, (Kidd & Cavalli-Sforza, 1974). We have chosen to analyse the haplotype frequency data using this τ measure as it should have particular biological/historical relevance to population studies. Principal components analysis (PCA) of the τ distance matrix allows the multidimensional display of the informa-

Table 17.3. *Average heterozygosities across seven haplotyped loci for twelve populations*

Population	Average heterozygosity (standard error)
Africa	
Biaka	0.59 (0.05)
Mbuti	0.51 (0.09)
Asia	
Cambodians	0.47 (0.06)
Chinese	0.50 (0.06)
Japanese	0.48 (0.04)
Nasioi	0.46 (0.07)
Europe	
Mixed Europeans	0.66 (0.05)
Americas	
North	
Jemez Pueblo	0.48 (0.06)
Maya	0.54 (0.08)
South	
Ticuna	0.34 (0.09)
R. Surui	0.36 (0.09)
Karitiana	0.35 (0.10)

The average heterozygosity (and standard error) of each haplotyped locus defined in Table 17.2 is presented for a global selection of populations.

tion in the distance matrix without imposing any assumptions about phylogeny. Trees can also be constructed from the genetic distances (Kidd & Sgaramella-Zonta, 1971). Such a tree will be a representation of historical relationships if various assumptions hold but be simply another type of graphical representation of the relative genetic similarities – with no historical/evolutionary implications – if those assumptions do not hold.

In the PCA of the τ distance matrix for these twelve populations and seven haplotyped loci, the first three principal components accounted for 92% of the total variance in the matrix. The Nasioi (Melanesians) accounted for most of the variance in the second principal component. Because the Nasioi are so divergent from the other analysed populations, those other eleven populations tend to be compressed into the first and third principal components. Fig. 17.2 presents the graphical representation of the first two principal components, recalculated after omitting the Nasioi. In this analysis, the first two principal components account for 80% of the total variance. The distribution of populations shown in Fig. 17.2 suggests a genetic cline from Europe through East Asia into North and then South America; the two African populations are not on that cline. We also

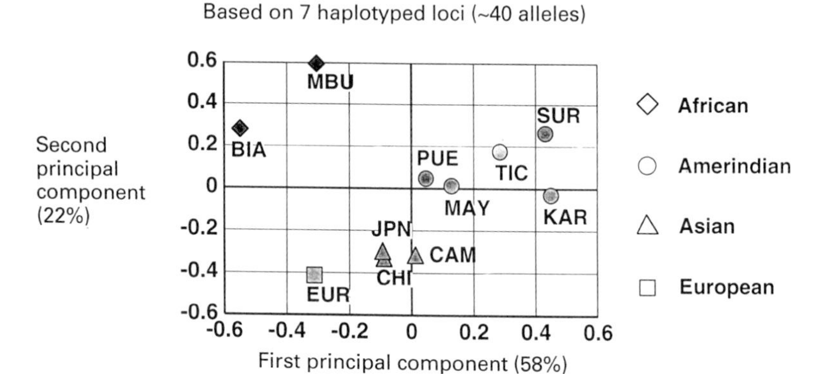

Fig. 17.2. Principal components analysis. The first two principal components of the τ distance matrix of eleven populations suggest a genetic cline from Europe eastward into North America then southward into South America. The two African populations are off that cline. The underlying haplotype data for seven loci used to calculate the genetic distances provide the equivalent of about 40 statistically independent variables (see text for discussion).

note how similar the three East Asian populations are to each other, especially relative to the Amerindian populations. Since similar geographic distances are involved, the relevant histories of the two groups must be different. These results agree with the finding by Cavalli-Sforza, Menozzi & Piazza (1994) that, with the classical markers (blood groups, etc.), the increase in F_{ST} with distance was greater in the New World than in any other continent.

A somewhat different perspective on the relationships among these populations is given by the additive tree structure in Fig. 17.3. We do not strongly argue that this tree represents the evolutionary relationships among the populations, but rather note the general increase in 'amount of evolution' (represented by distance from the 'root', at the left, to the tip of each branch) as one moves from Europe to Asia to the New World. This is exactly the 'cline' seen for the same populations in the PCA (Fig. 17.2). This suggests to us that as modern humans moved to colonize the world outside of our African homeland, genetic distance accumulated along the advancing front from Asia Minor to East Asia to North America and finally to South America. This increasing genetic distance accumulated not by new mutation but rather primarily by genetic drift with concomitant loss of heterozygosity. The signature of that expansion to populate the world is still preserved in the modern populations from the various regions of the world.

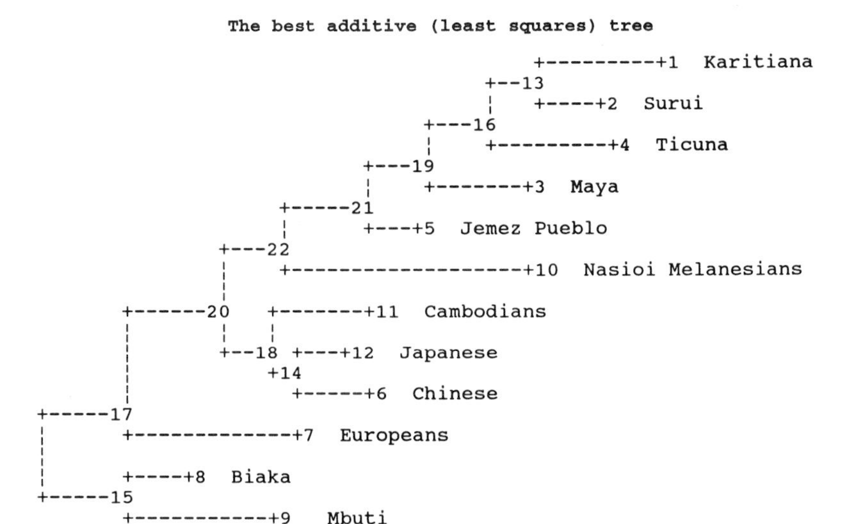

Fig. 17.3. The best additive (least squares) tree based on genetic distances calculated for 23 loci including the seven haplotypes in Table 17.2 and Fig. 17.2 plus 16 unlinked RFLP loci (see Kidd, Pakstis & Kidd, 1993). The 'root' has been arbitrarily placed between African and non-African populations based on the convincing genetic evidence for an African origin of modern humans. This location of the root emphasizes the clinal pattern of increasing genetic distance from Africa as one moves from West (Europe) to East (East Asia) in the Old World and then from North to South in the New World. The segment lengths are roughly proportional to the lengths produced by both of two similar methods: least squares analysis (Cavalli-Sforza & Edwards, 1967; Kidd & Sgaramella-Zonta, 1971) of Tau genetic distances (Kidd & Cavalli-Sforza, 1974) and the FITCH (Fitch & Margoliash, 1967) analysis on chord distances from PHYLIP (Felsenstein, 1993).

The present and the future

New understandings

Fairly extensive data on nuclear DNA markers are now available and they have begun to give us new understandings of recent human evolution. Also, as more is being learned about the ways different types of markers and different specific markers reflect different aspects of population genetics (evolution, history, affinities, etc.), we are becoming able both to tailor our choice of markers to address the specific questions we are considering, and to ask questions not previously answerable genetically. Thus, VNTRs with their high mutation rates appear most useful in studies of very recent population histories; STRPs (at least those with lower mutation rates) seem particularly useful in studies of regional population affinities and somewhat deeper population histories; haplotypes appear to most accurately record migrations (admixture); and, finally, SNPs and insertion/deletion markers seem most useful in studies of variation in the entire species and for titration of mutation rates at other loci. Of course, use of *combinations* of different types of markers may be the most valuable way of examining some population phenomena such as linkage disequilibrium.

Among some of the emerging understandings are the following. First, the existence of a European Bias for both classical markers and RFLPs seems fairly well established though its magnitude and consequences for various analyses remain to be determined. Secondly, the nuclear DNA data strongly support the African origin of modern humans. Haplotype data such as our CD4 study as well as surveys of STRPs show greater genetic diversity within modern African populations, explainable only by the African populations being older and having larger effective population sizes over the relevant time. This last point brings us to a third understanding – effective population sizes (N_e) for human populations are significantly different when averaged over many thousands of years. Those differences exist not only between sub-Saharan and non-African populations, as noted above, but are also relevant to understanding the histories of European, Asian, and New World populations. Gone are the days when population genetic analyses could assume the same N_e for all human groups. Fourthly, the small but growing body of haplotype data shows that the amounts and pattern of linkage disequilibrium can add information on population relationships and histories that is not readily revealed, if at all, by allele frequency data at single sites. Finally, we see that genetic variation, both among chromosomes and among individuals, exists on a background of pervasive similarity. Though each individual is unique, we are all one species. This is becoming one of the major messages for the world at large from modern DNA-based studies of human diversity: we are one human family and the differences between us are 'shallow'.

Marker standardization

The potential for understanding the history of modern human populations is now enormous. Twenty-five years ago it was possible to study on a uniform, global basis fewer than two dozen specific populations typed uniformly for fewer than two dozen specific classical polymorphisms. At that time, about 100 markers were defined as polymorphic in humans but many barely met the formal definition of a polymorphism and consisted of a common allele and a rare allele at 1–5% in a few populations. Moreover, the diversity of techniques required to study more than a dozen or so of the known markers imposed a major impediment to a large number of those markers being uniformly studied. Today, there are probably close to 20 000 polymorphic markers identified at the DNA level in at least one human population, usually Europeans, and the number continues to increase. Unless there is some standardization or convention on which markers should be studied on any population, there is a potential for virtually no overlap in markers across populations and what overlap will exist across a few populations will be the result of studies in a single laboratory. We already see that different laboratories are studying completely different sets of markers. While it is currently desirable for many different markers to be examined for their potential value in human evolutionary studies, for the current virtual absence of commonality to persist is a completely unacceptable outcome scientifically.

The Human Genome Diversity Project (Cavalli-Sforza *et al.*, 1991; Weiss, Kidd & Kidd, 1992) hopes to provide some coordination and standardization so that the results of many different researchers will be synergistic and not disjointed. The ultimate selection of a recommended set of markers must be based on consensus. *A priori*, there is little to recommend any one polymorphism over another. Criteria must include the ease of typing and prior knowledge on the molecular nature and the frequency variation of the marker. Until such time as the Human Genome Diversity Project recommends a panel of markers, the several markers summarized in Table 17.4 can be recommended on our experience, which shows that they can convey substantial information on population relationships. This is not an exhaustive list but represents our personal recommendations based on information we have collected. We base our recommendations on a variety of considerations including molecular characterization, conversion to PCR, ease of typing, sufficient prior data to demonstrate utility for population studies, and a sufficient prior data base to allow ready comparison of a single new population with many other populations. Along with our recommendations are references to the PCR primers, where already published, and to comparative allele frequencies on multiple populations, where already published. Many of the markers that we have converted to PCR and/or have typed on a large number of populations (at least the 12 in Table 17.3 and Fig. 17.3) are not yet published but are either submitted or are manuscripts in preparation; we would be happy to make primer sequences and allele frequency data available to interested researchers prior to publication.

We encourage other researchers to evaluate their data and propose similar lists of markers. For example, we have included none of the polymorphisms in the α- and β-globin clusters; others who have studied them extensively are better qualified to recommend a hierarchy of importance among the numerous sites that could guide researchers who are unable to study all of the sites in each haplotypable region.

Table 17.4. *Markers recommended for broader use in population studies*

Locus symbol	Recommended DNA markers						
	Chromosomal location	Number of Alleles known	Nature of polymorphism	Direction of mutation known	References to PCR primers	Refs. to freqs. in multiple pops.	F_{ST} in 12 pops.
PLAT	8p12–q11.2	2	Alu insertion	yes	Batzer *et al.*, 1994 Tishkoff *et al.*, 1996*b*	Batzer *et al.*, 1994 Tishkoff *et al.*, 1996*b*	.211
HOXB	17q21	2	Sac I restitution site	yes	Kidd & Ruano, 1995	Bowcock *et al.*, 1991*b*	.086
		6	STRP	–	Deinard & Kidd, 1995 Kidd *et al.* unpub.	Kidd *et al.* in prep.	–
DRD2	11q23.1	2	Taq 'A' restitution site	yes	Grandy, Zhang & Civelli, 1993 Castiglione *et al.*, 1995	Barr & Kidd, 1993 Lu *et al.*, 1993	.110
		2	Taq I 'B' restitution site	yes	Castiglione *et al.*, 1995	Lu *et al.*, 1996 Castiglione *et al.*, 1995	.182
		6	STRP	–	Castiglione *et al.*, 1995	Lu *et al.*, 1996 Castiglione *et al.*, 1995	.144
CD4	12pter-12p12	2	deletion	yes	Edwards & Gibbs, 1992 Tishkoff *et al.*, 1996*a*	Edwards & Gibbs, 1992 Tishkoff *et al.*, 1996*a*	.174
		11	STRP	–	Edwards & Gibbs, 1992 Tishkoff *et al.*, 1996*a*	Edwards & Gibbs, 1992 Tishkoff *et al.*, 1996*a*	.165

All of these markers have been typed on the 12 populations reported elsewhere in this study, all but the PLAT insertion polymorphism are components of a known haplotype system, as indicated; and all are already (or in the process of being) molecularly characterized and converted to PCR-based typing.

The future

The ease of communication and the recognition of the need to coordinate among laboratories their studies of various human populations offers great promise for the future. Our history as a species is written in our DNA sequence and in the distribution of variation in that sequence within and among human populations. However, we need a concerted, coordinated effort to study enough of that message that we truly understand it and are not led astray by fragments read out of context. The organization of the Human Genome Diversity Project promises to provide that integrative context and we are excited by the new and more complete understandings of our origins and recent histories that will emerge in the next decade and beyond.

Acknowledgement

This research was supported in part by grants from the National Institute of Alcoholism and Alcohol Abuse (AA09379), the Sloan Foundation (to K.K. Kidd), and the National Science Foundation (SBR9408934 to J.R. Kidd). We wish to thank our many colleagues who have made samples and clones available; and to thank Dr Andrew J. Pakstis for his helpful suggestions throughout the preparation of this manuscript.

References

Astolfi, P., Piazza, A. & Kidd, K.K. (1979). Testing of evolutionary independence in simulated phylogenetic trees. *Systematic Zoology*, **27**, 391–400.

Barr, C. & Kidd, K. K. (1993). Population frequencies of the A1 allele at the dopamine D2 receptor locus. *Biological Psychiatry*, **34**, 204–9.

Batzer, M. A., Stoneking, M., Alegria-Hartman, M., Bazan, H., Kass, D. H., Shaikh, T. M., Novick, G. E., Ioannow, P. A., Scheer, W. D., Herrera, R. J. & Deininger, P. L. (1994). African origin of human-specific polymorphic *Alu* insertions. *Proceedings of the National Academy of Sciences, USA*, **91**, 12288–92.

Bowcock, A. M., Bucci, C., Hebert, J. M., Kidd, J. R., Kidd, K. K., Friedlaender, J. S. & Cavalli-Sforza, L. L. (1987). Study of 47 DNA markers in five populations from four continents. *Gene Geography*, **1**, 47–64.

Bowcock, A. M., Hebert, J. M., Mountain, J. L., Kidd, J. R., Rogers, J., Kidd, K. K. & Cavalli-Sforza, L. L. (1991*a*). Study of an additional 58 DNA markers in five human populations from four continents. *Gene Geography*, **5**, 151–73.

Bowcock, A. M., Kidd, J. R., Mountain, J. L., Hebert, J. M., Carotenuto, L., Kidd, K. K. & Cavalli-Sforza, L. L. (1991*b*). Drift, admixture and selection in human evolution: A study with DNA polymorphisms. *Proceedings of the National Academy of Sciences, USA*, **88**, 839–43.

Bowcock, A. M., Ruiz-Linares, A., Tomfohrde, J., Minch, E., Kidd, J. R. &

Cavalli-Sforza, L. L. (1994). High resolution of human evolutionary trees with polymorphic microsatellites. *Nature*, **368**, 455–7.

Castiglione, C. M., Deinard, A. S., Speed, W. C., Sirugo, G., Pakstis, A. J., Zhang, Y., Grandy, D. K., Bonne-Tamir, B., Kidd, J. R. & Kidd, K. K. (1995). Physical mapping and haplotype frequencies of three polymorphisms at the DRD2 locus. *American Journal of Human Genetics*, **57**, 1445–56.

Cavalli-Sforza, L. L. & Edwards, A. W. F. (1967). Phylogenetic analysis: models and estimation procedures. *Evolution*, **32**, 550–70 (also appears in *American Journal of Human Genetics*, **19**, 233–57).

Cavalli-Sforza, L. L., Menozzi, P. & Piazza, A. (1994). *The History and Geography of Human Genes*. Princeton: Princeton University Press.

Cavalli-Sforza, L. L., Wilson, A. C., Cantor, C. R., Cook-Degan, R. M. & King, M. C. (1991). Call for a worldwide survey of human genetic diversity: a vanishing opportunity for the Human Genome Project. *Genomics*, **11**, 490–1.

Deinard, A. S. & Kidd, K. K. (1996). Levels of DNA polymorphism in extant and extinct hominoids. In: *The Origin and Past of Modern Humans as Viewed from DNA*, Eds. S. Brenner and K. Hanihara, pp. 149–70. New Jersey: World Scientific.

Deka, R., Chakraborty, R., DeCroo, S., Rothhammer, F., Barton, S. A. & Ferrell, R. E. (1992). Characteristics of polymorphism at a VNTR locus 3′ to the Apolipoprotein B gene in five human populations. *American Journal of Human Genetics*, **51**, 1325–33.

Deka, R., Jin, L., Shriver, M., Ling, M. Y., DeCroo, S., Hundreiser, J., Bunker, C. H., Ferrell, R. E. & Chakraborty, R. (1995). Population genetics of Dinucleotide $(dC\text{-}dA)_n \cdot (dG\text{-}dT)_n$ polymorphisms in world populations. *American Journal of Human Genetics*, **56**, 461–74.

Dempster, A. P., Laird, N. M. & Rubin, D. B. (1977). Maximum likelihood from incomplete data via the EM algorithm. *Proceedings of the Royal Statistical Society*, **39**, 1–38.

Dorit, R. L., Akashi, H. & Gilbert, W. (1995). Absence of polymorphism at the ZFY locus on the human Y chromosome. *Science*, **268**, 1183–5.

Edwards, M. C. & Gibbs, R. A. (1992). A human dimorphism resulting from loss of an *Alu*. *Genomics*, **14**, 590–7.

Felsenstein, J. (1993). *PHYLIP (Phylogeny Inference Package) version* 3.5*p*. Distributed by the author. Department of Genetics, University of Washington, Seattle.

Fitch, W. M. & Margoliash, E. (1967). Construction of phylogenetic trees. *Science*, **155**, 279–84.

Flint, J., Harding, R. M., Clegg, J. B. & Boyce, A. J. (1993). Why are some genetic diseases common? Distinguishing selection from other processes by molecular analysis of globin gene variants. *Human Genetics*, **91**, 91–117.

Goldstein, D. B., Ruiz-Linares, A., Cavalli-Sforza, L. L. & Feldman, M. W. (1995). An evaluation of genetic distances for use with microsatellite loci. *Genetics*, **139**, 463–71.

Grandy, D. K., Zhang, Y. & Civelli, O. (1993). PCR detection of the TaqA RFLP at the DRD2 locus. *Human Molecular Genetics*, **2**, 2197.

Hawley, M. E., Pakstis, A. J. & Kidd, K. K. (1994). A computer program implementing the EM algorithm for haplotype frequency estimation. *American Journal of Physical Anthropology*, Supplement **18**, 104.

Hawley, M. E. & Kidd, K. K. (1995). HAPLO: A program using the EM algorithm to estimate the frequencies of multi-site haplotypes. *Journal of Heredity*, **86**, 409–11.

Jakubicza, S., Amemann, J., Cooke, H., Krawczak, M. & Schmidtke, J. (1989). A search for restriction fragment length polymorphism on the human Y chromosome. *Human Genetics*, **84**, 86–8.

Jeffreys, A. J., Royle, N. J., Wilson, V. & Wong, Z. (1988). Spontaneous mutation rates to new length alleles at tandem repetitive hypervariable loci in human DNA. *Nature*, **314**, 67–73.

Kidd, J. R. (1993). Population Genetics and Population History of Amerindians as Reflected by Nuclear DNA Variation. Unpublished PhD Dissertation, Yale University.

Kidd, J. R., Black, F. L., Weiss, K. M., Balazs, I. & Kidd, K. K. (1991). Studies of three Amerindian populations using nuclear DNA polymorphisms. *Human Biology*, **63**, 775–94.

Kidd, J. R., Pakstis, A. J. & Kidd, K. K. (1993). Global levels of DNA variation. *Proceedings of the Fourth International Symposium on Human Identification* 1993. pp.21–30. Madison: Promega.

Kidd, J. R., Pakstis, A. J., Black, F. L., Goldman, D., Weiss, K. M. & Kidd, K. K. (1996). Molecular haplotypes at seven loci in twelve populations. (In preparation).

Kidd, K. K. & Cavalli-Sforza, L. L. (1974). The role of genetic drift in the differentiation of Icelandic and Norwegian cattle. *Evolution*, **28**, 381–95.

Kidd, K. K. & Ruano, G. (1995). Optimizing PCR. In *PCR2: A Practical Approach*, ed. M. J. McPherson, B. D. Haines and G. R. Taylor, pp. 1–22. Oxford: Oxford University Press.

Kidd, K. K. & Sgaramella-Zonta, L. A. (1971). Phylogenetic analysis: concepts and methods. *American Journal of Human Genetics*, **23**, 235–52.

Lu, R. -B., Ko, H. -C., Chang, F. -M., Castiglione, C. M., Schoolfield, G., Pakstis, A. J., Kidd, J. R. & Kidd, K. K. (1996). No association between alcoholism and multiple polymorphisms at the dopamine D2 receptor gene (DRD2) in three distinct Taiwanese populations. *Biological Psychiatry*, in press.

Malaspina, P., Persichetti, F., Novelletto, A., Iodice, C., Terrenato, L., Wolfe, J., Ferraro, M. & Prantera, G. (1990). The human Y chromosome shows a low level of DNA polymorphism. *Annals of Human Genetics*, **54**, 297–305.

Murphy, P. D., Kidd, J. R., Castiglione, C. M., Lin, P. F., Ruddle, F. H. & Kidd, K. K. (1986). A frequent polymorphism for the cytosolic thymidine kinase gene, TK1, (17q21-q22) detected by the enzyme TaqI. *Nucleic Acids Research*, **14**, 438–1.

Murphy, P. D., Lin, P. F., Ruddle, F. H. & Kidd, K. K. (1987). A second useful polymorphism for the cytosolic thymidine kinase gene (TK1) with the enzyme BstEII which will allow haplotyping at this locus on chromosome 17 (q21–q22). *Nucleic Acids Research*, **15**, 7212–13.

Pagnier, J., Mears, J. G., Dunda-Belkhodia, O., Schaefer-Rego, K. E., Beldjord, C.. Nagel, R. L. & Labie, D. (1984). Evidence for the multicentric origin of the sickle cell haemoglobin gene in Africa. *Proceedings of the National Academy of Sciences, USA*, **81**, 1771–3.

Poloni, E. S., Excoffier, L., Mountain, J. L. & Langaney, A. (1995). Nuclear DNA polymorphism in a Mendenka population from Senegal: comparison with eight other human populations. *Annals of Human Genetics*, **59**, 43–61.

Ruano, G., Deinard, A. S., Tishkoff, S. & Kidd, K. K. (1994). Detection of DNA sequence variation via deliberate heteroduplex formation from genomic DNAs amplified *en masse* in 'Population Tubes'. *PCR Methods and Applications*, **3**, 225–31.

Sambrook, J., Fritsch, E. F. & Maniatis, T. (1989). In *Molecular Cloning: A Laboratory Manual, 2nd edn.*, ed. N. Ford, C. Nolan and M. Ferguson. Cold Spring Harbor: Cold Spring Harbor Laboratory Press.

Scozzari, R., Torroni, A., Semino, O., Sirugo, G., Brega, A. & Santachiara-Benerecetti, A. S. (1988). Genetic studies on the Senegal population. 1. Mitochondrial DNA polymorphisms. *American Journal of Human Genetics*, **43**, 534–44.

Slatkin, M. (1995). A measure of population subdivision based on microsatellite allele frequencies. *Genetics*, **139**, 457–62.

Tishkoff, S. A., Dietzsch, E., Speed, W., Pakstis, A. J., Cheung, K., Kidd, J. R., Bonne-Tamir, B., Santachiara-Benerecetti, A. S., Moral, P., Watson, E., Krings, M., Pääbo, S., Risch, N., Jenkins, T. & Kidd, K. K. (1996). Global patterns of linkage disequilibrium at the CD4 locus and modern human origins. *Science*, **271**, 1380–7.

Tishkoff, S. A., Ruano, G., Kidd, J. R. & Kidd, K. K. (1995*b*). Distribution and frequency of a polymorphic *Alu* insertion at the PLAT locus in humans. *Human Genetics* (In press).

Wainscoat, J. S., Hill, A. V. S., Boyce, A. J., Flint, J., Hemandez, M., Thein, S. L., Old, J. M., Lynch, J. R., Galusi, A. G., Weatherall, D. J. & Clegg, J. G. (1986). Evolutionary relationships of human populations from an analysis of nuclear DNA polymorphisms. *Nature*, **319**, 491–3.

Weber, J. L. & Wong, C. (1993). Mutation of human short tandem repeats. *Human Molecular Genetics*, **2**, 1123–8.

Weiss, K. M., Kidd, K. K. & Kidd, J. R. (1992). Human genome diversity project. *Evolutionary Anthropology*, **1(3)**, 80–2.

Weissenbach, J., Gyapay, G., Dib, C., Vignal, A., Morissette, J., Millasseau, P., Vaysseix, G. & Lathrop, M. (1992). A scond-generation linkage map of the human genome. *Nature*, **359**, 794–801.

18 *Contrasting gene trees and population trees of the evolution of modern humans*

N. SAITOU

A gene tree is an essential descriptor of any evolutionary process, for the semi-conservative replication of the DNA double helix automatically produces a bifurcating gene tree. It should be emphasized that the genealogical relationship of genes is independent of the mutation process, especially when neutral evolution (Kimura, 1983) is considered. The former is a direct product of DNA replication, while the latter may or may not happen within a certain time period and DNA region. Therefore, even if many nucleotide sequences happened to be identical, there must be a genealogical relationship for those sequences. However, it is impossible to reconstruct the genealogical relationship without mutational events. In this respect, extraction of mutations from genes and their products is critical for reconstructing phylogenetic trees.

We can, therefore, best estimate a gene tree according to the mutation events realized on its expected gene tree (see Fig. 18.1(*a*)). We call this ideal reconstruction of the gene tree the realized gene tree (see Fig. 18.1(*b*)), while the reconstructed one from observed data is called the 'estimated' gene tree (Saitou, 1995*b*). Branch lengths of realized and estimated genes tree are proportional to mutational events. These mutational events are not necessarily proportional to physical time. Because of limitations in information, estimated gene trees are often unrooted. By definition, expected gene trees are strictly bifurcating, while realized and estimated gene trees may be multifurcating. This is because of the possibility of no mutation at a certain interior branch, such as branch X of Fig. 18.1(*a*).

Ideally, branch lengths of a phylogenetic tree are proportional to physical time. We call this type of tree the 'expected tree'. It is a rooted tree. Both species/population trees and gene trees have their expected trees, but their properties are somewhat different from each other. An expected gene tree directly reflects the history of DNA replications. In contrast, a

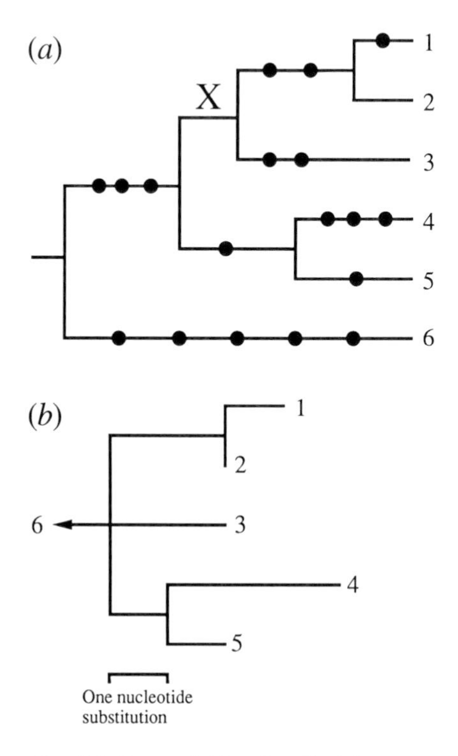

Fig. 18.1. Expected (*a*) and realized (*b*) gene tree (From Saitou (1995*b*)). Full circles on the expected gene tree denote nucleotide substitutions. Because no substitution occurred at branch X of the expected gene tree (*a*), the corresponding branch does not exist in the realized gene tree (*b*).

species/population tree is only a simplified view of a nexus of gene trees. Therefore, the speciation time (or time of population diversification) is not always clear, in contrast to the clear DNA replication event. A species/population tree reconstructed from observed data is called an 'estimated' species/population tree, while there is no realized species/population tree.

There are several other important differences between gene trees and species/population trees. Even when orthologous genes are used, a gene tree may be different from the corresponding species/population tree. This difference arises from the existence of a gene genealogy in ancestral species/population. A simple example is illustrated in Fig. 18.2. A gene sampled from species A has its direct ancestor at the speciation time T_1 generations ago, and so does a gene sampled from species B. Thus the divergence time between the two genes sampled from the different species always overestimates that of the species. The amount of overestimation corresponds to the coalescence time in the ancestral species, and its

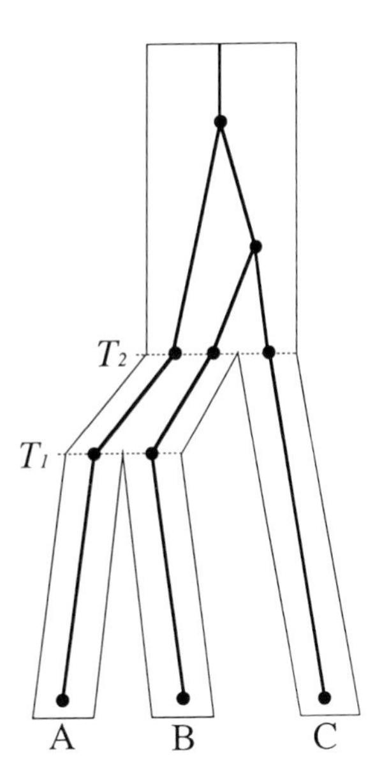

Fig. 18.2. Difference between a gene genealogy and species tree (From Saitou, 1996). Full circles and thick lines denote the genealogical relationship while thin lines (outlining the gene tree) denote the species tree. A, B, and C denote three genes each sampled from extant species, while X and Y denote ancestral genes. T_1 and T_2 denote the two speciation times.

expectation is $2N$ for neutrally evolving nuclear genes of a diploid organism, where N is the population size of the ancestral species. Therefore, if the two speciation events (T_1 and T_2) are close enough, the topological relationships of the gene tree may become different from those of the species tree, as shown in Fig. 18.2. Although species A and B are more closely related than to C, genes sampled from species B and C happen to be more closely related to each other than to that sampled from species A. The probability (P_{error}) of obtaining an erroneous tree topology is given by $P_{error} = (2/3)e^{-T/2N}$, where $T = T_2 - T_1$ generations (Nei, 1987). For example, P_{error} is 0.404 when $T = 50\,000$ and $N = 50\,000$. Therefore, a species tree estimated from a single locus may not be correct even if the gene tree has been correctly estimated; we should use more than one locus. Saitou and Nei (1986) computed the probabilities of obtaining the correct species tree from a number of gene trees for the case of a three species tree. They considered a trinomial distribution, and the topology supported by the largest number of loci was regarded as the correct one. Under this condition, we need only one locus when $T/2N$ is 4, but 7 loci when $T/2N$ is 1, if we want the probability to be larger than 0.95.

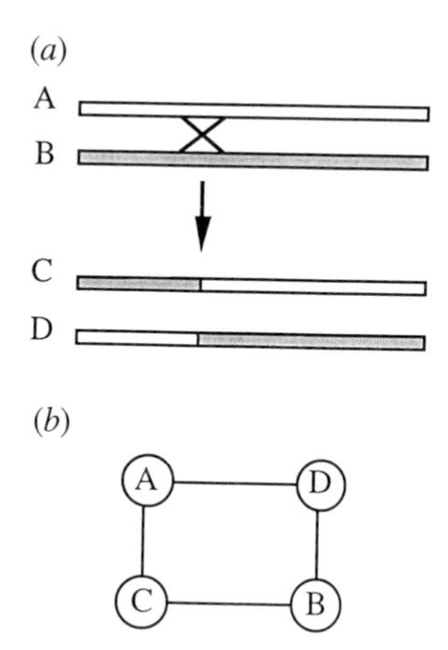

Fig. 18.3. (*a*) A recombination event creates new alleles C and D from existing alleles A and B. (*b*) A network of four alleles caused by the recombination described in (*a*).

When gene conversion and/or recombination has occurred within the gene region under consideration, a tree structure may no longer exist. Fig. 18.3 shows this situation schematically. When a recombination occurs between alleles A and B that diverged some time ago, new recombinant alleles C and D are produced (Fig. 18.3(*a*)). If we consider the relationship among those four alleles, the resultant graph is not a tree but a network (Fig. 18.3(*b*)). The distance (measured in terms of nucleotide difference) between alleles A and C is the same as that between B and D, and it is smaller than that between A and D (or between B and C) if we consider the location of the recombination event. It should be noted that the mutations are assumed to have accumulated uniformly over the sequence. This example clearly shows the limitation of a tree representation in some cases. Bandelt (1994) recently proposed a method for constructing such networks from sequence data.

A gene tree for HTLV-I DNA

Human T-lymphotropic virus type I (HTLV-I) has been found in Japan, Africa, and the Caribbean Islands, but it has also been found in Melanesia (see Yanagihara & Garruto, 1992 for a review and Yanagihara *et al.*, 1995). Nerurkar *et al.* (1993) sequenced parts of the HTLV-I genome found in Melanesians (Papua New Guineans and Solomon Islanders), and Song *et al.* (1994) determined several simian T-lymphotropic virus type I (STLV-I)

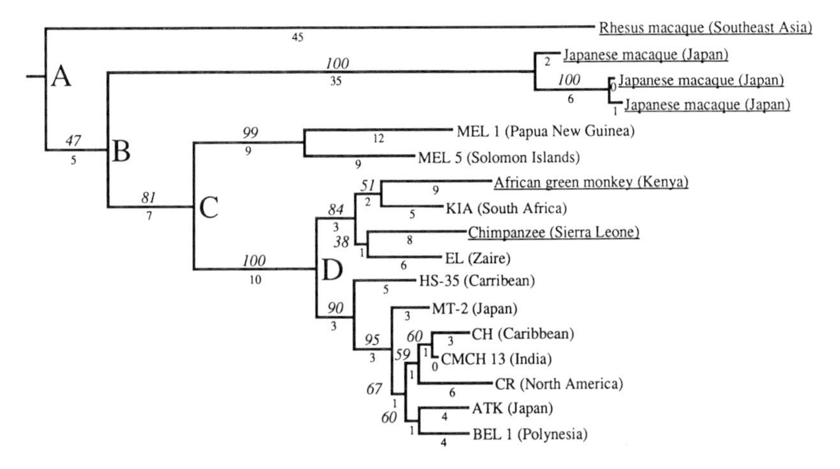

Fig. 18.4. A neighbour-joining tree of HTLV-I and STLV-I sequences (modified from Song *et al.*, 1994). The tree is rooted by including an HTLV-II sequence. STLV-I sequences are underlined. Numbers below branches are estimated numbers of nucleotide substitutions at corresponding branches, and those above internal branches are bootstrap probabilities (%).

sequences. A phylogenetic tree of HTLV-I and STLV-I sequences is shown in Fig. 18.4. HTLV-II which is remotely related to HTLV-I and STLV-I was used as an outgroup to locate the root of the tree. Because we are interested in reconstructing the realized gene tree, all the estimated branch lengths (integers below branches) are numbers of nucleotide substitutions that have occurred in the compared DNA region.

First of all, it is evident that the resultant gene tree does not correspond to the phylogenetic tree of species involved in the comparison (human, chimpanzee, African green monkey, Japanese macaque, and rhesus macaque). Because the two macaque species and African green monkey are Old World monkeys, they should be monophyletic in the real species tree. Accordingly, the branching point (node) A of Fig. 18.4 is unlikely to correspond to the separation time of rhesus and Japanese macaques, nor node B to the separation time of Japanese macaques and humans. Alternatively, it may be more reasonable to assume that node C corresponds to the time of the first human migration into Melanesia (*ca.* 50 000 BP) from Sunda land. If so, node D, the coalescent point for the so-called cosmopolitan HTLV-I strains, corresponds to about 25 000 BP under the assumption of a rough constancy of the evolutionary rate. Nodes A and B are also roughly dated as *ca.* 90 000 BP and 70 000 BP, respectively. Both human and macaques cohabited in the Asian Continent around that time, and it is conceivable that an interspecific transmission of the virus occurred. Similar interspecific transmissions evidently occurred much more recently in Africa, for African green monkey and chimpanzee STLV-I strains clustered with some human sequences from Africa (see Fig. 18.4).

In spite of some initial enthusiasm, there are doubts as to the utility of the phylogenetic tree of the HTLV-I virus for elucidating modern human evolution. As shown above, the interspecific transmission of this virus between human and other primates seems to occur rather frequently. If so, horizontal transfer of the virus among humans may easily occur. Thus the HTLV-I/STLV-I gene trees should not be used without due consideration.

Gene trees and population trees for mitochondrial DNA

Genetic polymorphism of human mitochondrial DNA (mtDNA) has been extensively studied by using both restriction enzymes and direct sequencing. We studied both gene trees and population trees based on mtDNA polymorphisms detected by restriction enzymes (Harihara & Saitou, 1989; Saitou & Harihara, 1996). A summary of the results is given in this section.

Published mtDNA data were collected for a total of 885 individuals from 15 human populations (see Harihara & Saitou, 1989 for details). The restriction enzymes used were *Ava* II, *Bam* HI, *Hpa* I, and *Msp* I. Each enzyme produces various patterns of restriction fragments, and the sets of restriction sites deduced from such patterns are called mtDNA morphs. Fig. 18.5 shows the relationship of 26 mtDNA morphs found by using *Ava* II. It is clear that morph 1 is at the centre of radiation, followed by morph 5. When there is more than one possibility of connecting different morphs with more than one restriction site difference (e.g. morphs 1 and 20), a loop is created. This is reminiscent of the phylogenetic network.

A combination of each mtDNA morph for different restriction enzymes is called a mtDNA type. A total of 57 mtDNA types (or haplotypes) were found by this procedure. Although we observed a network structure for mtDNA morphs, the real evolutionary history of mtDNA molecules must be a phylogenetic tree, for mtDNA is considered to undergo no recombination. We therefore produced the mtDNA gene tree (Fig. 18.6) using the maximum parsimony method (Fitch, 1977). It should be mentioned that the tree shown in Fig. 18.6 is only one of many equally parsimonious trees.

MtDNA type 1 was the most frequent (670 individuals out of 885 had this type), and was found in all 15 populations. This cosmopolitan mtDNA type is shown as the large ellipse in Fig. 18.6, and is at the centre of radiation. Most of the mtDNA types found in African populations (Bantu and Bushman) form a monophyletic cluster at the left side of branch α, although some mtDNA types found in Arabian (P) and Roman (R) populations are also included in this 'African' cluster. This clear distinction

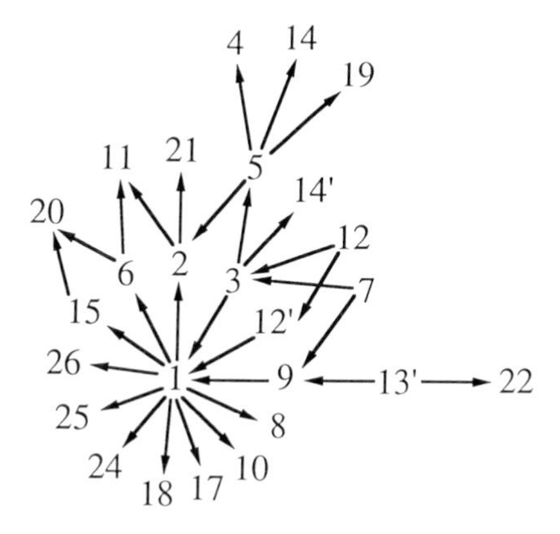

Fig. 18.5. A phylogenetic network for 26 mitochondrial morphs produced by using restriction enzyme *Ava* II (modified from Fig. 18.1a of Harihara & Saitou, 1989). Arrows indicate the direction of restriction site loss.

of mtDNA types found in African populations is a basis of the 'Out-of-Africa' hypothesis championed by Cann, Stoneking & Wilson (1987) using their restriction site data, and later extended by Vigilant *et al.* (1991) using nucleotide sequence data. However, re-analysis of these data (Hedges *et al.*, 1991; Madisson, Ruvolo & Swofford, 1992) showed that there is still uncertainty for the support of this hypothesis.

Because the gene tree of Fig. 18.6 does not have a root, we assigned a root to mtDNA type I (cosmopolitan type) for the following four reasons (Saitou & Harihara, 1996). (1) Under neutral evolution, the most frequent allele (mtDNA type 1) is likely to be the oldest, with a probability equal to its frequency (Watterson & Guess, 1977). In this case, the frequency of mtDNA type 1 is 670/885 = 0.76. (2) All 15 populations had mtDNA type 1. (3) mtDNA type 1 is the centre of radiation of the remaining mtDNA types; there were 23 branches connected to mtDNA type 1. (4) If we use the midpoint rooting method assuming a rough constancy of evolutionary rate (Farris, 1972), mtDNA type 1 becomes the root.

A rooted tree of 56 mtDNA types was thus obtained. We then estimated the evolutionary rate of mtDNA as follows: the average number of restriction site differences between the ancestral mtDNA type (type 1) and present-day individuals was computed to be 0.514. Because the mtDNA type 1 was a combination of *Ava* II morph 1 (8 restriction sites), *Bam* HI morph 1 (1 restriction site), *Hpa* I morph 2 (3 restriction sites), and *Msp* I morph 1 (23 restriction sites), the total number of nucleotides assayed by these four restriction enzymes is $8 \times 5 + (1 + 3) \times 6 + 23 \times 4 = 156$. Thus the average number of nucleotide differences per nucleotide site between the ancestral mtDNA type and present-day individuals was

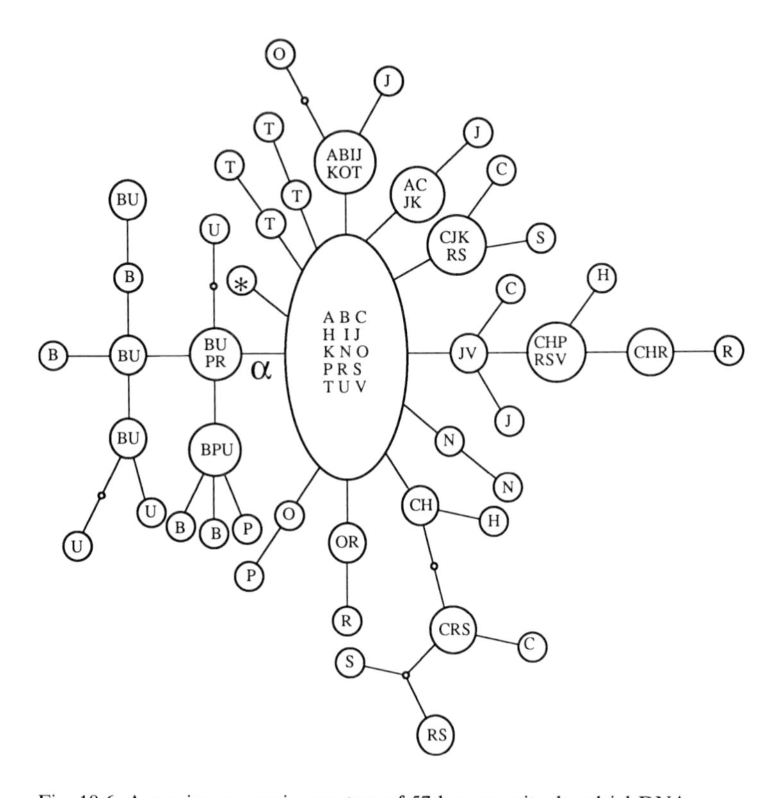

Fig. 18.6. A maximum parsimony tree of 57 human mitochondrial DNA haplotypes found in 15 human populations (modified from Saitou & Harihara, 1996). Letters denote the populations in which each mitochondrial DNA haplotypes were found. Abbreviations of the populations are; A: Ainu, B: Bantu, C: Caucasian, H: Jewish, I: Amerindian, J: Japanese, K: Korean, N: Negritos, O: Oriental, P: Arab, R: Roman, S: Sardinian, T: Tharu, U: Bushman, and V: Vedda. A circle with an asterisk represents 12 different mtDNA types that are one site apart from mtDNA type 1. Small open circles designate intermediate mtDNA types not found but necessary to explain the relationship of known mtDNA types.

estimated to be $0.154/156 = 0.00329$. This is equivalent to the 'sequence divergence' of 0.66% ($= 0.00329 \times 2 \times 100$); this is not very different from the corresponding value (0.57%) estimated by Cann *et al.* (1987).

The situation in which the root of a gene tree exists in the most common type or allele is not restricted to mtDNA types, but is a general pattern of genealogy for closely related genes. Fig. 18.7(*a*) shows a hypothetical gene genealogy for 14 genes with the common allele (C) and six variant alleles (V1–V6). Because only 7 mutational events were extracted (designated as full circles) and no mutation was observed along the lineages to all the common allele genes, the ancestral gene (the root) is identical with the

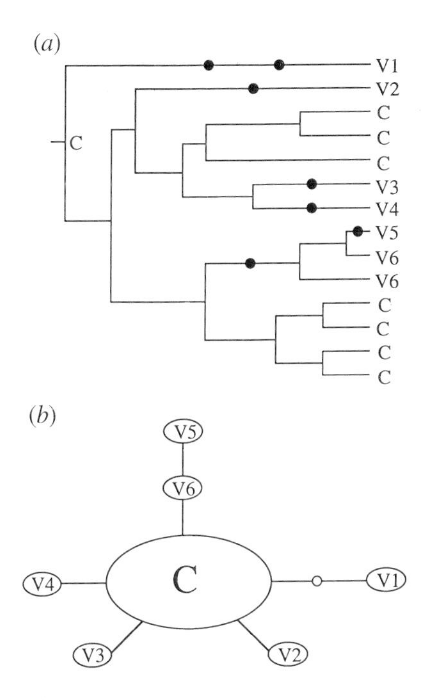

Fig. 18.7. (*a*) A schematic expected tree for 14 genes. Full circles denote mutational events. (*b*) An unrooted gene tree reconstructed by using the observed mutations as shown in the tree above.

common allele. When we reconstruct the gene tree, even the ideal reconstruction (see Fig. 18.7(*b*)) is a gross simplification of the real genealogy. Although the seven C genes are not monophyletic (see Fig. 18.7(*a*)), this cannot be extracted from the reconstructed gene tree of Fig. 18.7(*b*). We should, therefore, be careful in interpreting the branching pattern of a gene genealogy.

We estimated genetic distances among the 15 human populations based on information about the number of nucleotide differences between all the possible pairs of mtDNA types and the frequency of each mtDNA type (Harihara & Saitou, 1989). Fig. 18.8 is an unrooted population tree constructed from that distance matrix data using the neighbour-joining method (Saitou & Nei, 1987). There are some branches with negative lengths in that tree, designated as broken lines. Although this is annoying, the appearance of negative branches is inevitable for non-metric measures such as this genetic distance. This is because the triangle inequality is sometimes violated when there are a number of parallel changes in the allele frequency. For example, let us consider three hypothetical populations (α, β, and γ) and assume that the observed distances were 0.03, 0.05, and 0.01

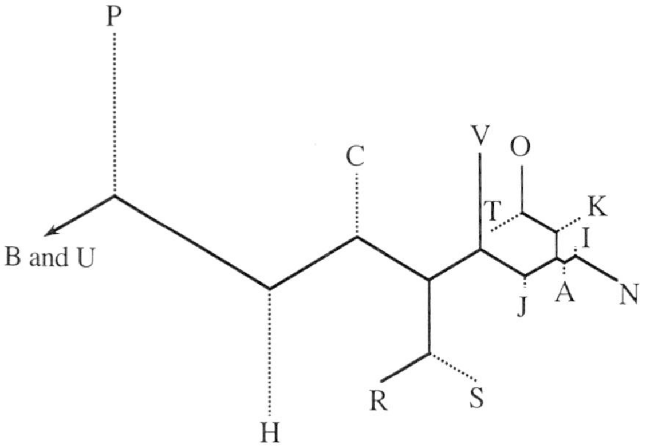

Fig. 18.8. An unrooted neighbour-joining tree for 15 human populations (modified from Saitou & Harihara, 1996). Abbreviations of the populations are the same as those of Fig. 18.6.

for population pairs α–β, β–γ, and γ–α, respectively. In this example, the triangle inequality is violated, for the distance (0.05) between β and γ is larger than the sum ($0.04 = 0.03 + 0.01$) of distances for β–α and α–γ. The estimates of three branch lengths between the internal node X and three populations then become -0.005, 0.035, and 0.015 for branch αX, βX, and γX, respectively. In fact, branch αX was estimated to be negative.

In any case, let us examine this population tree. Two African populations (Bantu and Bushman) are far apart from the remaining 13 non-African populations. This is clearly because of the distinct clustering of African mtDNA types observed in the gene tree (see Fig. 18.6). Seven *circum*-Pacific populations (Ainu, Amerindian, Japanese, Korean, Negritos, Oriental, and Tharu) are tightly clustered, and the Vedda of Sri Lanka are clustered with these *circum*-Pacific populations. Middle Eastern populations (Arabians and Jews) are located between African populations and the remaining populations. Because the genetic distance data used for producing this population tree were based on the single locus (mtDNA), there are large standard errors in the distance estimates, and the resulting tree may not be completely reliable. We need to study many independent genetic loci in nuclear DNA.

Ancient mitochondrial DNA

Recently so-called ancient DNA studies are have become popular and many ancient human mtDNA sequences have now been published. I will briefly discuss a new aspect of utilizing this ancient DNA data combined

Table 18.1. *Association between burial style and genetic relationship at the Takuta–Nishibun site of Kyushu, Japan*

Burial type	*Number of individuals in mtDNA type*		
	A	non-A	Total
Kamekan (buried in jar-coffin)	6	3	9
Dokoubo (direct burial)	3	14	17
Total	9	17	26

From Oota *et al.*, 1995.

with archaeological data. Oota *et al.* (1995) extracted and amplified a part of the mtDNA D loop region for 26 human bones and teeth found from an archaeological site in the southern part of Japan, dated at *ca.* 2000 BP. Two regions of nucleotide sequences were determined, and phylogenetic trees were constructed (not shown). Nine individuals belonged to the most frequent mtDNA sequence (type A), and the remaining 17 individuals belonged to the other 10 mtDNA sequences, in which 7 of them radiated directly from type A.

There were two types of burial at the site: Kamekan (burial in earthenware jar-coffins) and Dokoubo (direct burial in the earth). To investigate the possibility of a correlation between burial style and genetic relatedness, we computed the probability of obtaining the observed frequency distribution (see Table 18.1). The resultant probability was 0.028 (Fisher's exact test), thus the null hypothesis (of no association) was rejected at the 5% level. This raised two hypotheses about the relationship between burial style and mtDNA type. One assumes that the two burial styles were used at the same period, and the people at the site were buried according to their genetic background (probably kinship). The other hypothesis assumes that these two burial styles were used at different periods, and the genetic constitution of the populations might have been somewhat different between the different periods. This implies an inflow of people with a different genetic background, together with a different culture at least in relation to burial style.

Although ancient DNA data are often used for phylogenetic reconstruction, comparison with archaeological evidence is an important field, as explained above. In this respect, Kurosaki, Matsushita & Ueda (1993) amplified not only mtDNA but also nuclear DNA (short-VNTR loci) from bones of two female individuals (sexing was morphologically determined) buried about 2000 years ago in Japan. These females (mature and juvenile) were buried side by side on the same hill, and both had about 20 cone-shell bracelets on their arms. Because of these characteristics, some archaeologists considered that they were members of the same family, probably mother and daughter who had ruled over the area as shaman or leader. Kurosaki *et al.* (1993) clearly showed, however, that these females were not genetically related. This finding was contrary to the archaeological conjecture.

Gene trees for nuclear DNA

There are few studies on the reconstruction of gene trees for nuclear DNA, with the exception of the β-globin gene cluster. DNA variation of this region has been extensively studied using restriction enzymes (e.g. Chen *et al.* 1990), and recently Fullerton *et al.* (1994) determined 3-kilobase β-globin sequences for 72 chromosomes. This kind of sequence data will become the standard for future studies of nuclear DNA variation.

Thanks to its extremely high mutation rate and high genetic variation, examination of microsatellite (short-VNTR or STR) loci is becoming popular in human population studies. Bowcock *et al.* (1994) examined 30 microsatellite loci for 148 individuals from 14 human populations, and constructed a colourful tree of 'individuals'. It is not clear from the text whether Bowcock *et al.* considered that tree to be a gene tree. Although a tree of 'individuals' is equivalent to a tree of 'genes' in the case of mtDNA, this does not apply to nuclear DNA data when unlinked loci are used. Since dozens of unlinked loci were used in Bowcock *et al*'s tree of 'individuals', that tree should not be considered as a gene tree in the usual sense. In reality, it presents the relationship of different combinations of unlinked alleles, not the genealogy of individuals. Long branches to extant 'individuals' in the tree of Bowcock *et al.* (1994) do not, therefore, mean long evolutionary times but are merely a reflection of recombination events, and it is erroneous to put a time scale to such a tree.

Population trees for nuclear genes

In contrast to the relatively few studies on nuclear gene trees, those on population trees based on nuclear gene data are abundant. Edwards and Cavalli-Sforza (1964) pioneered the construction of phylogenetic trees of human populations using allele frequency data. Nei and Roychoudhury (1974) estimated the divergence of three major races, and Negroid (African) was estimated to diverge first. This may be the first indication of the 'Out-of-Africa' hypothesis from genetic data.

Saitou, Tokunaga & Omoto (1992) applied the neighbour-joining method for the first time to genetic distance data. Recently, Nei and Roychoudhury (1993) and Bowcock *et al.* (1994) both used the neighbour-

(*a*)

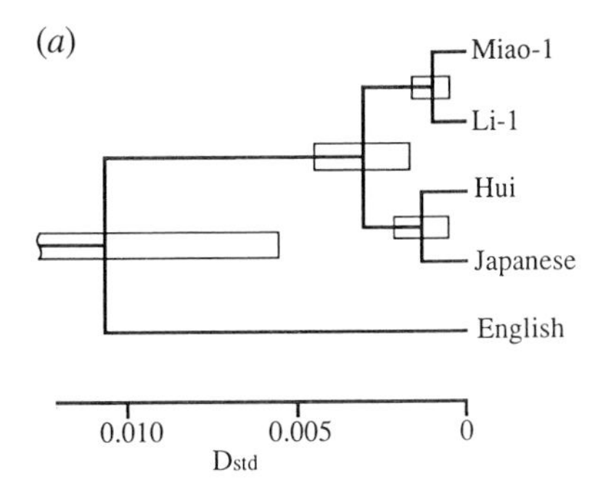

(*b*)

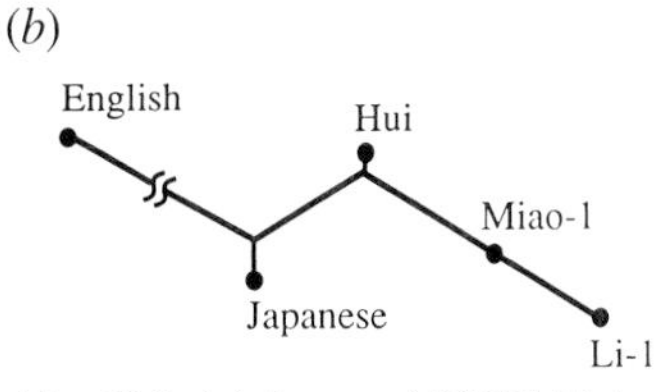

Fig. 18.9. (*a*) A rooted UPGMA tree and (*b*) an unrooted neighbour-joining tree for six human populations (from Saitou *et al.*, 1994).

joining method for reconstructing human population trees. Although they used different datasets (classical markers and microsatellite loci, respectively), similar relationships were obtained.

Theoretically, there is no qualitative difference between a species tree and a population tree. Because a population tree usually means the relationship between populations within a species, however, there is always a chance for intraspecific populations to have high gene flow with each other. Therefore, a rooted tree, in which populations are always assumed to differentiate, may be misleading. In this sense, an unrooted tree representation is more appropriate. Fig. 18.9 shows rooted and unrooted trees for the same genetic distance data of five populations (Saitou *et al.*, 1994). Although Hui, a Muslim population at Hainan Island, is clustered with Japanese in the UPGMA rooted tree (Fig. 18.9(*a*)), it is located between Japanese and Miao-1, another population on Hainan Island. This unrooted population tree suggests that there has been some gene flow between Hui and the surrounding populations of Hainan Island.

Saitou (1995*a*) examined allele frequency data from 12 polymorphic nuclear loci for 30 human populations and constructed an unrooted tree by using the neighbour-joining method (Fig. 18.10). Current human populations are more or less clustered according to their geographical locations; Africa, West and East Eurasia, North and South America, and Sahul land. Sahul land, or the Sahul shelf, existed until about 10 000 years ago, and later

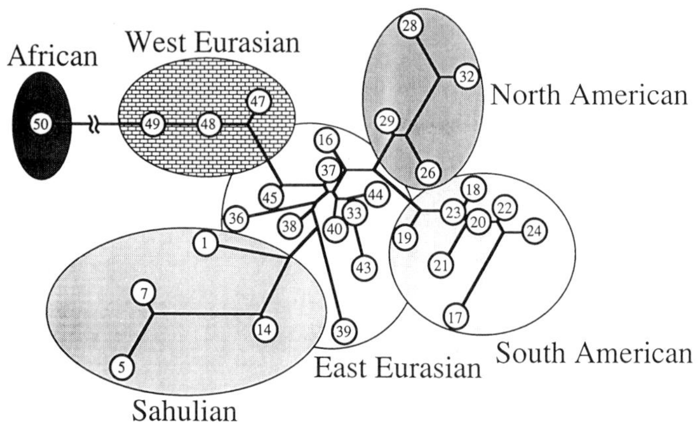

Fig. 18.10. An unrooted neighbour-joining tree for 30 human populations (modified from Saitou, 1995*a*). Population IDs are as follows. Sahulian: 1 = Australian Aborigines (Northern Territory), 5 = Papua New Guinean (North Central Highland), 7 = Papua New Guinean (East Highland), 14 = Micronesian (East Caroline Island). South American: 17 = Yanomama 18 = Makiritare, 19 = Aymara, 20 = Baniwa, 21 = Cayapo, 22 = Macushi, 23 = Wapishana, 24 = Ticuna. North American: 26 = Eskimos (North Alaska), 28 = Athabaskan Indian, 29 = Eskimos (Canada), 32 = Dogrib Indian. East Eurasian: 16 = Polynesian (Samoa Island, now living in New Zealand), 33 = Japanese, 36 = Balinese, 37 = 'Mais, 38 = Filipino, 39 = Negritos, 40 = Han, Northern China, 43 = Ainu, 44 = Korean, 45 = Nepali. West Eurasian: 47 = Indian (South India), 48 = Iranian, 49 = English. African: 50 = Yoruba (Nigeria).

separated into Australia and Papua New Guinea. This continent-wide clustering apparently reflects the history of human population dispersal in the last 100 000 years. Although this pattern is somewhat blurred because of the great human movements particularly within the last 10 000 years, we can still extract the ancient course of human dispersal using genetic data from current populations. Saitou (1995*a*) thus proposed a new classification of human populations based on this genetic affinity tree, as shown in Fig. 18.10. It should be noted that the classification was not meant for the current human populations. It was for those at around the end of Pleistocene, i.e. *ca.* 10 000 BP. The great movement of Polynesian people occurred much later, and the centre of the Pacific was not yet populated at that time. Thus the four clusters surrounding the Pacific (East Eurasian, Sahulian, North American, and South American) can be further grouped to form a '*circum*-Pacific' supercluster. This supercluster corresponds to the 'pan-Mongoloid' cluster of Saitou *et al.* (1992).

Importance of finiteness for evolutionary studies

As the real world is always finite, the course of evolutionary history should also be treated in this finite framework. Random genetic drift caused by the finiteness of the population size is a good example. I would like to emphasize three other aspects of the evolutionary process in which finiteness should be taken into account.

The first is the number of ancestors for one individual. There are 2^n ancestors for a diploid organism such as humans when we go back n generations. This number exceeds the current world population (*ca.* 5×10^9) when $n = 30$ or larger. Of course, the number of individuals at that time (about 6000 years ago if we consider one generation to be 20 years) must have been much smaller than the current level, and inevitably there are many redundancies among the ancestors, i.e. inbreeding. Let us look at this parent–offspring relationship from a different point of view. It is clear that the number of ancestors for a particular mitochondrial DNA is always one, for the circular molecule is inherited without recombination. It immediately follows that the number of ancestral individuals who actually contributed a part of their genetic material is the number of non-recombining units in the genome. Unfortunately, we do not know this number at present. However, the upper limit is the number of nucleotides for a genome, and this is about 3×10^9 for the human genome. Therefore, if the total number of ancestors exceeds this number in a certain generation, there will be ancestral individuals who did not contribute to the genetic composition of a particular present-day descendant. Let us call this ancestor a 'null' ancestor. For example, all male ancestors are null when we consider mtDNA.

The second aspect of finiteness is the number of 'genes' in a genome. Probably the best current estimate of the total number of genes in the human genome is *ca.* 60 000–70 000 (Fields *et al.*, 1994). All the genes from those with housekeeping activities to those involved in complicated ethological characters are in this finite set. If we consider the yet unknown enzymes and proteins expressed in various tissues, the total number of typical genes responsible for biochemical pathways may easily exceed 10 000. Therefore, it is possible that many morphological characters attributed to hereditary factors may be non-hereditary. The same applies to the complicated nature of human brain functions. Unless some unexpected structures that were previously considered to be merely junk are found to be functional, we may be able to map all the functions of genes in the human genome.

The third finiteness is in the number of nucleotides in a non-recombining unit. In theory, the number can be infinite, and it is better to have longer sequences for obtaining better reconstruction of gene trees from a statistical viewpoint (see e.g. Saitou & Nei, 1986). However, there is always a limit to growth. Horai *et al.* (1995) compared the entire 16.5 kilobase mtDNA genomes of five hominoid species (human, common chimpanzee, pygmy chimpanzee, gorilla, and orang-utan). There will be no need for more study of the mtDNA gene tree of these species, except for intraspecific variation.

Many people are often interested in evolution of a particular gene. In this case, the possible number of nucleotides to be compared is usually much smaller than the entire mtDNA genome. Specific nucleotide changes responsible for the creation or loss of gene function may be delineated, but the estimation of the time frame can be difficult. When the evolutionary time is expected to be quite large, it will be almost impossible to estimate the divergence time of two remotely related genes. In any case, we should be cautious in reconstructing gene trees because of this finiteness.

Acknowledgements

This study was partially supported by a Grant-in-Aid for Scientific Research on Priority Areas (Molecular Evolution) of Ministry of Education, Science and Culture (Japan).

References

Bandelt, H. -J. (1994). Phylogenetic networks. *Verhandlungen des Naturwissenschaftlichen Vereins in Hamburg (NF)*, **34**, 51–71.

Bowcock, A. E., Ruiz-Linares, A., Tomfohrde, J., Minch, E., Kidd, J. R. & Cavalli-Sforza, L. L. (1994). High resolution of human evolutionary trees with polymorphic microsatellites. *Nature*, **368**, 455–7.

Cann, R. L., Stoneking, M. & Wilson A. C. (1987). Mitochondrial DNA and human evolution. *Nature*, **325**, 31–6.

Chen, L. Z., Easteal, S., Board, P. G. & Kirk, R. L. (1990). Evolution of beta-globin haplotypes in human populations. *Molecular Biology and Evolution*, **7**, 423–37.

Edwards, A. W. F. & Cavalli-Sforza, L. L. (1964). Reconstruction of evolutionary trees. *Systematics Association Publication*, **6**, 67–76.

Farris, S. J. (1972). Estimating phylogenetic trees from distance matrices. *American Naturalist*, **106**, 645–68.

Fields, C., Adams, M. D., White, O. & Venter J. C. (1994). How many genes in the human genome? *Nature Genetics*, **7**, 345–6.

Fitch, W. M. (1977). On the problem of discovering the most parsimonious tree. *American Naturalist*, **111**, 223–57.

Fullerton, S. M., Harding, R. M., Boyce, A. J. & Clegg, J. B. (1994). Molecular and population genetic analysis of allelic sequence diversity at the human β-globin locus. *Proceedings of the National Academy of Sciences, USA*, **91**, 1805–9.

Harihara, S. & Saitou, N. (1989). A phylogenetic analysis of human mitochondrial DNA data. *Journal of the Anthropological Society of Nippon*, **97**, 483–92.

Hedges, S. B., Kumar, S., Tamura, K. & Stoneking, M. (1991). Human origins and analysis of mitochondrial DNA sequences. *Science*, **255**, 737–9.

Horai, S., Hayasaka, K., Kondo, R., Tsugane, K. & Takahata, N. (1995). Recent African origin of modern humans revealed by complete sequences of hominoid mitochondrial DNAs. *Proceedings of National Academy of Sciences, USA*, **92**, 532–6.

Kimura, M. (1983). *The Neutral Theory of Molecular Evolution*. Cambridge: Cambridge University Press.

Kurosaki, K., Matsushita, T. & Ueda, S. (1993). Individual DNA identification from ancient human remains. *American Journal of Human Genetics*, **53**, 638–43.

Maddison, D. R., Ruvolo, M. & Swofford, D. L. (1992). Geographic origins of human mitochondrial DNA: phylogenetic evidence from control region sequences. *Systematic Biology*, **41**, 111–24.

Nei, M. (1987). *Molecular Evolutionary Genetics*. New York: Columbia University Press.

Nei, M. & Roychoudhury, A. K. (1974). Genic variation within and between the three major races of man, Caucasoids, Negroids, and Mongoloids. *American Journal of Human Genetics*, **26**, 434–6.

Nei, M. & Roychoudhury, A. K. (1993). Evolutionary relationships of human populations on a global scale. *Molecular Biology and Evolution*, **10**, 927–43.

Nerurkar, V. R., Song, K. -J., Saitou, N., Mallan, R. R. & Yanagihara, R. (1993). Interfamilial and intrafamilial genomic diversity of human T lymphotropic virus type I strains from Papua New Guinea and the Solomon Islands. *Virology*, **196**, 506–13.

Oota, H., Saitou, N., Matsushita, T. & Ueda, S. (1995). A genetic study of 2,000-year-old human remains of Japan (Yayoi period) using mitochondrial DNA sequences. *American Journal of Physical Anthropology*, **98**, 133–45.

Saitou, N. (1995*a*). A genetic affinity analysis of human populations. *Human Evolution*, **10**, 17–33.

Saitou, N. (1995*b*). Methods for building phylogenetic trees of genes and species. In *Molecular Biology: Current Innovations and Future Trends*, ed. H. Griffin and A. Griffin., pp. 115–35. Wymondham: Horizon Scientific Press, in press.

Saitou, N. (1996). Reconstruction of gene trees from sequence data. In *Computer Methods for Macromolecular Sequence Analysis*, ed. R. Doolittle, pp. 427–49. Orlando: Academic Press.

Saitou, N. & Harihara, S. (1995). Gene phylogeny and population phylogeny reconstructed from human mitochondrial DNA data. In *Human Evolution in the Pacific Region*, ed. C. K. Ho, G. Kranz and M. Stoneking, Washington: Washington State University Press, in press.

Saitou, N. & Nei, M. (1986). The number of nucleotides required to determine the branching order of three species, with special reference to the human–chimpanzee–gorilla divergence. *Journal of Molecular Evolution*, **24**, 189–204.

Saitou, N. & Nei, M. (1987). The neighbour-joining method: a new method for reconstructing phylogenetic trees. *Molecular Biology and Evolution*, **4**, 406–25.

Saitou, N., Tokunaga, K. & Omoto, K. (1992). Genetic affinities of human populations. In *Isolation and Migration*, ed. D. F. Roberts, N. Fujiki, and K. Torizuka, pp. 118–129. Cambridge: Cambridge University Press.

Saitou, N., Omoto, K., Du, C. & Du, R. (1994). Population genetic study in Hainan

Island, China. II. Genetic affinity analyses. *Anthropological Science*, **102**, 129–47.

Song K. -J., Nerurkar, V. R., Saitou, N., Lazo, A., Blakeslee, J. R., Miyoshi, M. & Yanagihara, R. (1994). Genetic analysis and molecular phylogeny of simian T-cell lymphotropic virus type I: evidence for independent virus evolution in Asia and Africa. *Virology*, **199**, 56–66.

Vigilant, L., Stoneking, M., Harpending, H., Hawkes, L. & Wilson, A. C. (1991). African populations and the evolution of human mitochondrial DNA. *Science*, **253**, 1503–7.

Watterson, G. A. & Guess, H. A. (1977). Is the most frequent allele the oldest? *Theoretical Population Biology*, **11**, 141–160.

Yanagihara, R. & Garruto, R. (1992). Serological and virological evidence for human T-lymphotropic virus type I infection among the isolated Hagahai of Papua New Guinea. In *Isolation and Migration*, ed. D. F. Roberts, N. Fujiki and K. Torizuka, pp. 143–153. Cambridge: Cambridge University Press.

Yanagihara, R., Saitou, N., Nerurkar, V.R., Song, K. J., Bastian, I., Franchini, G. & Gajdusek, D. C. (1995). Molecular phylogeny and dissemination of human T-cell lymphotrophic virus. *Cellular and Molecular Biology*, **41**, S145–61.

19 *Methods and models for understanding human diversity*

H. HARPENDING, J. RELETHFORD AND S. T. SHERRY

Introduction

The study of modern human origins and the study of human variation are the same, since they both refer to processes by which contemporary human diversity developed. There is no shortage of discussion in the literature about competing models of human origins, so we will summarize current viewpoints in a cursory way, then discuss several testable hypotheses suggested by these viewpoints. After that, we describe several relatively unexplored ways of looking at human molecular data and evaluate the merits and possibilities of new and older methods.

We try to distinguish between what we call population perspectives and phylogenetic perspectives. These perspectives descend from the population genetics of the last several decades, in which distinct anthropological and biological traditions developed. The anthropological tradition, summarized, for example, in Crawford and Mielke (1980), was concerned with local differentiation of populations driven by gene flow and drift. The biological tradition, summarized for example in Nei (1987), was concerned with species differences and mutations, especially neutral mutations. Our current interest in global human origins and dispersions calls for a blend of these traditions both in theory and in techniques of data analysis and presentation. Anthropologists should learn how to put mutations in their models, and biologists should stop treating human groups as if they were species.

Models of human origins

Modern humans are very different from their immediate precursors, at least in the African and European fossil and archaeological records. The difference seems less marked in Asia. Contemporary theories of the difference between modern and archaic humans can be placed along a continuum:

1. Modern humans were bearers of a new gene complex.
2. Modern humans were a new race.
3. Modern humans were a new species.

Hypothesis (1) corresponds to the multiregional theory of human origins in which new advantageous genes spread through established archaic populations. In this scenario, neutral or nearly neutral loci in our species today existed in hundreds of thousands of copies in our ancestors in the Middle Pleistocene.

Hypothesis (2) suggests that modern humans evolved somewhere as a variant of archaic humans and then spread with partial incorporation of genetic material from other races of the species. There are many examples known of this kind of process in historical times such as the European spread into the Americas and the Bantu expansion in sub-Saharan Africa. Under this model there were fewer copies in the Middle Pleistocene of our neutral loci, but since there was partial incorporation of existing populations the number was at least several tens of thousands to several hundreds of thousands.

Hypothesis (3) is the Noah's Ark or Garden of Eden model of human origins, in which modern humans were the product of a speciation-like event. They then expanded over and beyond the range of antecedent archaic populations who went extinct from competition or worse. Since this new species arose from an isolated archaic population distinct from others, the number of copies of today's loci could have been very small, on the order of hundreds or thousands.

We describe the familiar positions in the debate in these terms in order to emphasize that the differences among these models are essentially demographic. When we look at the ancestry of today's genetic material, do we see a water glass or a funnel? If we see a funnel, how narrow is the neck? Does the funnel have one neck or several passages through the neck? Understanding human origins is precisely understanding the demographic history of our ancestors, including total population size and the extent to which the ancestral populations were united by gene flow or separated into isolated subpopulations or races.

Genetic data suggest that the effective size of our species is about 10 000 breeding individuals. What this means is not so clear in view of the compliations described below, but this estimate of the effective size of humanity is not new (Nei, 1987) and it has been confirmed by recent tabulations (Gibbons, 1995). Obviously, there are many more than 10 000 humans today, and there have been for a long time, so some time within the 'memory' of genetic systems (i.e. the reciprocal of the mutation rate or so) the total number of our ancestors for some long period was certainly of this order. Since 10 000 individuals could hardly have been spread over the temperate Old World, the genetic evidence with little reservation supports hypothesis (3) above.

Although the debate drags on, it is obvious that the genetic evidence is clearly in favour of hypothesis (3), the Noah's Ark model. This evidence is simply that the effective size of our species is of the order of several thousand to several tens of thousands of individuals. Estimates from mitochondrial DNA sequences suggest that the recovery from our near extinction started during the last interglacial or during the early part of the last glaciation (Harpending *et al.*, 1993).

We will proceed by accepting that our history is best described by model (3) or, perhaps, something between (2) and (3). Our demographic history is a funnel, and we should now try to refine our knowledge of the shape and structure of the funnel. In particular, we will discuss ancient subdivision of our species into partially isolated demes during the bottleneck, the timing of the recoveries that occurred, and the mechanisms leading to today's differences between major regional populations of humans.

All our loci have been through the same demographic history, but they do not all record that history in the same way. Our inferences below depend on differences between loci in effective number, in mutation rate, and in the extent to which they conform to the infinite sites assumption in which every mutation is to a new and different allele. Classical markers and nuclear RFLP (restriction fragment length polymorphism) loci have low mutation rates compared to mitochondrial DNA sites, craniometric traits (Lynch, 1988), and short tandem repeat markers (Bowcock *et al.*, 1994). Among the high mutation systems, we expect that the infinite sites or infinite alleles assumptions will be violated a lot by mtDNA sequences and perhaps less by craniometric traits.

Important concepts and models

Signatures of ancient demographic events remain in the diversities within and the differences between contemporary populations. We use *diversity* to mean genetic differences between individuals within populations, usually measured by heterozygosity. We use *difference* to mean genetic differences between populations, usually described by a statistic called F_{ST}. Estimating and interpreting these quantities from data is a treacherous business, so we discuss these quantities and how they are measured before going on to the substantive matters.

Diversity

When we speak of diversity we really mean heterozygosity, but only the theoretical heterozygosity that would be observed under random mating. In other words, we are usually not concerned with real genotype distributions but rather with properties of the gene frequencies. In the case of mitochondrial DNA, which is a haploid system, we use the same statistic but the heterozygosity is, of course, virtual.

In the case of sequences it is convenient to extend the notion of heterozygosity to the sum of heterozygosities over the sites in the sequence. This sum is also the mean number of differences between all pairs of sequences, or mean sequence divergence (MSD).

For a neutral genetic locus the diversity is determined by the mutation rate μ and the effective number of the population N. In all that follows N will refer to the effective number of loci transmitted each generation: for diploid loci $N = 2N_e$, twice the conventional effective size. For mitochondrial DNA $N = N_f$, the effective number of females, ordinarily the probability that two women have the same mother.

For neutral genetic loci the amount of diversity in a population is determined at equilibrium by the balance between mutation, which increases diversity, and finite size, which leads to loss of diversity every generation. There are three common formulae that show these relationships.

First, if a population has been the same size for a very long time, long enough that the equilbrium between loss of diversity by drift balances the gain of diversity by new mutation, the equilbrium diversity is approximately

$$H = \frac{2N\mu}{1 + 2N\mu}$$

under the infinite alleles model, which postulates that every mutation is to a new and unique allele. If we are dealing with a sequence and measuring MSD rather than sitewise heterozygosity the analogous formula is

$$MSD = 2N\mu$$

where the mutation rate in this case is the sum of the rates at each nucleotide position in the sequence.

Second, if a population has gone through a severe bottleneck, a reduction in size, then the initial recovery of diversity follows:

$$H = MSD = 2\mu T$$

where T is the time in generations since the recovery. In this sense, diversity can measure the 'age' of a population if age means time since expansion to a very large size from a very small size.

What do 'very large' and 'very small' mean? Diversity approaches its equilibrium value exponentially with parameter $-(2\mu + 1/N)$ so that 'very large' means $N \gg 1/2\mu$ in which case the rate of recovery is determined by the mutation rate. This expression for the rate shows another important property of diversity changes. If a population falls to a very small size, $1/N$ becomes relatively very large, and diversity is lost rapidly. After recovery, N is large and the $1/N$ term is small, so the rate of recovery of diversity is determined by the mutation rate and may be very slow.

Since diversity is lost much faster than it is gained when population size fluctuates over time, the diversity that we observe in nature reflects minimum population sizes over time rather than average sizes. This is usually expressed by saying that the effective number of a population is the harmonic mean of the numbers over time, but this harmonic mean has to be discounted backwards in time by the effect of mutations.

Another important influence on effective number is population subdivision into partially isolated demes: subdivision increases the overall effective number of a population. The effective number of a species divided into two demes that exchange, on average, one-fourth of an individual per generation is twice as large as the species effective number would be if it were panmictic. This means that an ancient effective number of 10 000 could reflect a species subdivided into two isolated demes each with only 2500 or so members. This would be like the population structure of common chimpanzees today – the western chimpanzees are different enough from those in central and east Africa that some would call them a different species.

Difference

The difference between populations in the simplest cases is relatively independent of the diversities within them. The customary way to describe genetic differences among populations is the statistic F_{ST}, the ratio of variation between groups to the total variation of a trait. It is computed by comparing heterozygosity within populations to what heterozygosity would be if there were no population structure. Write H_0 for the heterozygosity of a trait within populations, averaged over all the populations. In the case of nuclear loci this is usually taken to be one less the sum of squares of allele frequencies at a locus, i.e. it is the heterozygosity that would exist if mating within the population were random. In the case of mitochondrial DNA it is 'virtual heterozygosity', since we are mitochondrial haploids. Write H_e for heterozygosity if all the populations were to mate at random. Then:

$$F_{ST} = \frac{H_e - H_0}{H_0}$$

There is no good simple formula for the equilibrium value of F_{ST} in a subdivided population, but there is a complex formula with several simple special cases. We need the complex formula in this chapter because we propose below that human differences reflect ancient population structure with isolated demes, so isolated that the rate of migration (gene exchange) among them was of similar order to the mutation rate. Standard textbook treatments of F_{ST} assume that the number of new mutants per generation, $N\mu$, is much less than the number of migrants Nm, but we cannot assume this for present purposes, particularly in view of loci like short tandem repeats (STRs) that have high mutation rates.

By following the style of Crow and Aoki (1984) but writing differential equations instead of difference equations, we can find the value of F_{ST} in an island model of population structure at equilibrium. While this is a special kind of structure in which all populations exchange migrants with all the others at the same rate, it is a good guide to intuition and simulations show the results to be quite robust. The model is that there are L populations, each of effective number N. In the time interval δt an individual disappears to be replaced by a copy of a random member of the same population with probability δt and an individual disappears to be replaced by a copy of a random individual from one of the other $L - 1$ populations with probability $m\delta t$. The genetic locus we are studying has K possible alleles, and the locus mutates to one of $K - 1$ other alleles with probability $\mu\delta t$ in the time interval δt. When the equilibrium of drift, mutation, and migration is reached,

$$F_{ST} = \frac{1}{1 + 2mN\left(\frac{L}{L-1}\right)^2 + 2N\mu\frac{KL}{(K-1)(L-1)}}$$

(Sherry, Stoneking & Harpending, manuscript). The textbook expression for F_{ST} at equilibrium, $1/(2Nm + 1)$, can be recovered from this by setting the mutation rate to zero and letting the number of populations L be very large. In our discussion of human races below, the number of populations will usually be 3 or 4, and the terms in $L/(L - 1)$ cannot be ignored. The lesson is that we cannot compare a published estimate of F_{ST} from three populations with another published estimate from 15 populations – they are simply incommensurable.

The effective number N for a diploid locus is four times N for a haploid locus if the sex ratio is unity. Thus F_{ST} estimates from nuclear loci are not commensurable with estimates from mitochondrial DNA without correction for the fourfold difference in N.

Mutation decreases F_{ST}. In the case of infinite alleles K is very large and the coefficient of the number of new mutants $2N\mu$ in the denominator is effectively unity, while in the case of two alleles the coefficient is two. Most mtDNA differences in human populations are transitions, and this formula shows that the forward and backward mutations that occur decrease F_{ST}. In other words high mutation rates should *decrease* differences among populations, and they should decrease them the most at loci where the infinite alleles or infinite sites model fails.

Just as there is a sequence measure of diversity, there is a version of F_{ST} computed from sequence data. We simply replace heterozygosity with mean sequence diversity in the formula for computing F_{ST}. Thus, H_e is the mean pairwise number of differences between all sequences pooled together, while H_0 is the mean pairwise difference within populations. We have not been able to derive a general expression for the equilibrium value of this statistic under the case of reversible mutation, but under the infinite sites model where every mutation is at a new and different site, the statistic has a simple equilibrium form:

$$F_{ST,MSD} = \frac{1}{1 + 2Nm\left(\frac{L}{L-1}\right)^2}$$

This form is derived from the model described in Rogers and Harpending (1992) in which the number of pairs of sequences that differ at i sites is increased by a mutation in a pair that differ at $i - 1$ sites and is decreased by reproduction in a finite population where the pair may be replaced by a pair that differs at no sites. In this special case, the mutation rate does not appear in the denominator. This means that *MSD* estimates of F_{ST} should be higher than corresponding sitewise estimates, and Table 19.1 shows that this is indeed true for human mtDNA. This is not a good model for human mtDNA sequences, since on other grounds it is known that there are backward and forward transitions in the histories of mtDNA sequences, but the qualitative insight that F_{ST} estimated from *MSD* is expected to be greater than the sitewise estimate is important.

F_{ST} estimated from nuclear loci, either classical markers or nuclear RFLP sites, is about 10–12% among major regional groups of humans. It is the same estimated from craniometric differences among regions (Relethford & Harpending, 1994; 1995). There are several ways to compute

Table 19.1. *F_{ST} estimates from three mitochondrial DNA data sets*

		HVS I sequences	HVS I and II sequences	RFLP sites
MSD	F_{ST}	0.13	0.06	0.06
Sitewise	F_{ST}	0.03	−0.02	0.01

The first column shows F_{ST} estimates from 103 segregating sites in HVS I of the control region from 242 Africans, 146 Asians, 150 Europeans, and 33 Sahulians. The second column shows estimates from 97 segregating sites in HVS I and HVS II combined from 81 Africans, 34 Asians, 11 Europeans, and 13 Sahulians. The number of segregating sites is less than in the first column since the number of individuals sampled is smaller. The negative sitewise estimate of F_{ST} arises from the bias correction, which is large in this case of a small number of individuals per population. The third column shows corresponding estimates from RFLP sites over the whole molecule from 20 Africans, 34 Asians, 140 Sahulians, and 45 Europeans. The data in column three are the same data analysed by Stoneking *et al.* (1990).

an estimate of F_{ST} from sequences, and these have to be adjusted for the different effective number in the two cases. The equivalent F_{ST} from mitochondrial DNA after correction is about three percent. In other words mitochondrial DNA diversity is only a third or a fourth as great as it ought to be if it reflected the same processes as classical markers. Table 19.1 shows estimates of F_{ST} from three different mtDNA sets of data, with both sitewise and *MSD* versions of the statistic.

An interesting aside is that Stoneking *et al.* (1990) reported a value of about 0.30 for this statistic computed from RFLP marker frequencies in several major races. Whittam *et al.* (1986) estimated F_{ST} to be only 0.063 using essentially the same data set. The fivefold difference in these two estimates reflects differences in the statistical method used to estimate F_{ST}, and the lower figure is the correct one since it agrees with F_{ST} estimated from sequences. We (Relethford & Harpending, 1994) and others (e.g. Takahata, 1992) have uncritically used the high figure of 0.30 and drawn completely inappropriate conclusions. Stoneking *et al.* used an unusual method of estimating the statistic (Takahata & Palumbi, 1985) that, at least in this case, gives an apparently incorrect figure.

Details of the funnel

We will describe two characteristics of the funnel about which competing models have been proposed, and we will show how current data seem to support one model over the other in each case rather clearly. The first characteristic is whether the funnel opened up once or whether there were several funnels in different places at different times. The underlying issue is the greater genetic diversity observed today in African populations in some marker systems but not others. The second characteristic of the funnel is how much gene exchange there has been at different levels of the funnel between populations ancestral to today's major regional populations, which we will call 'races' using the word in its colloquial sense. (There is no technical meaning of race in human biology.) Different models of the history of race differences lead to testable hypotheses about the genetic data.

African diversity

There is greater diversity in sub-Saharan African populations in mitochondrial DNA sequences and in craniometric traits (Relethford & Harpending, 1994; 1995). On the other hand, there is no greater diversity of African populations in classical markers like blood groups and enzymes, nor in nuclear RFLP sites. It has been proposed that the failure of classical markers and RFLPs to show higher African diversity is due to ascertainment bias: since they were ascertained mostly in European populations, systems most diverse in Europeans were most likely to be discovered. A recent analysis of this hypothesis suggests that it cannot explain all the observed differences (Rogers & Jorde, manuscript).

Two scenarios have been proposed to explain greater African diversity. These are:

1. The population of our ancestors was subdivided in the neck of the funnel. The subdivision ancestral to Africans was larger than the other subdivisions, so diversity was greater then. African diversity today reflects ancient differences in effective size.
2. The funnel opened at different times for different ancestral populations. African diversity is greater because ancestors of Africans expanded from the neck of the funnel earlier than the ancestors of other major racial groups. Diversity, then, has been accumulating longer in African populations. This is presumably what people have in mind when they speak of an 'African origin' of modern humans.

Diversity today in our species should be thought of as the sum of two components: diversity that was present in the neck of the funnel, when effective size was small, and diversity that has accumulated since the funnel opened and the effective size of our species has been large. The relative contribution of these two components at a locus should depend on the mutation rate for the locus: systems with very low mutation rates have not had enough time since the funnel opened (perhaps 60 000 years) for diversity to accumulate, and race differences in diversity may reflect differences in the ancient structured population in the neck of the funnel. Systems with high mutation rates, on the other hand, have accumulated diversity since the funnel opened. Diversity today in these systems would reflect in large part the accumulation of mutations in a large population in the last 60 000 years or so. Mitochondrial DNA control region sequences are clearly of the latter sort, and almost all diversity today has accumulated since the funnel opened (Harpending *et al.*, 1993). A third possibility is that systems like short tandem repeats with extrememly high mutation rates have 'forgotten' the funnel and reflect recent group effective sizes.

With these differences between loci in mind, the two models of African diversity make very different predictions about contemporary diversity.

Hypothesis (1) predicts that African populations should be more diverse in all systems, but that this ancient difference in diversity should be obscured to some extent by mutations accumulated since the neck opened in high mutation systems. Thus, excess African diversity should be somewhat *less* in mitochondrial DNA control region sequences, craniometric traits, and other long DNA sequences (since the mutation rate of a non-recombining sequence is the sum of the rates at each nucleotide position). Excess African diversity should be clearest in low mutation sytems like nuclear RFLP loci, classical markers, and other single site systems.

Hypothesis (2), earlier African expansion, makes different predictions. Under this model markers with low mutation rates should reflect ancient relative subdivision size, so race differences should be minimal if effective sizes in the neck of the funnel were equivalent. There has not been enough time for these sites since the bottleneck opened to accumulate much diversity, and they should not reflect differences between races in when the funnel opened. High mutation sites, on the other hand, should show the greatest diversity in those groups that expanded earliest. If Africans expanded earlier, African diversity should be highest in systems with high mutation rates but not in systems where the time since the funnel opened, in generations, is much greater than the reciprocal of the mutation rate.

The available data support hypothesis (2) over hypothesis (1). African diversity is higher in mtDNA sequences (Harpending *et al.*, 1993), in craniometric traits (Relethford & Harpending, 1994), and in short tandem repeat polymorphisms (Bowcock *et al.*, 1994). African diversity is not higher in classical markers nor in nuclear RFLP polymorphisms (Nei & Roychoudhury, 1974; Bowcock *et al.*, 1994). It appears that the ancient expansion of African ancestors predated expansions of the ancestors of other major regional groups by at least several tens of thousands of years. African diversity is higher in high mutation rate systems but has diversity comparable to other regions in systems with low mutation rates. The inference seems solid enough that it should be a central hypothesis for archaeological research into human origins.

Race differences

Two models of race differences

The 10% difference among major human races is too great to have arisen in only a few hundred thousand years by drift in a species with a million or so members. Two different models have been proposed to account for these differences, the model of successive fissions and the model of frozen ancient structure.

1. The model of successive fissions posits that the history of races is a history of successive bifurcations of large populations (Nei & Roychoudhury, 1974; Cavalli-Sforza *et al.*, 1988). After the splits there was no gene exchange between the daughter populations, and differences between them accumulated by mutation. In their explicit formulation of the model, Nei and Roychoudhury assume that the populations were very large, effectively infinite, so that extant differences are due to mutation only. With this assumption, differences should be proportional to the product of the mutation rate and time since separation. Given estimates of mutation rates, time since separation can be estimated.
2. The model of frozen ancient structure ascribes today's differences to mutation, migration, and genetic drift in a subdivided ancient population of our ancestors with small subdivisions, each with only several thousands of members (Harpending *et al.*, 1993; Relethford & Harpending, 1994). When our species emerged from the funnel and the great population expansions occurred in the last interglacial and during the last glaciation these differences were 'frozen' since genetic drift among large subdivisions would be so slow that it could hardly change the ancient pattern.

These two models make contrasting predictions about race differences at different kinds of loci. The model of successive fissions posits that differences reflect the accumulation of mutations, so differences among groups, i.e. F_{ST}, should be greatest for systems with high mutation rates. Since mutation lowers F_{ST} in a structured population at equilibrium when the infinite sites assumption fails (see above), the ancient structure model predicts that F_{ST} should be lower rather than higher for high mutation rate systems like mtDNA sequences.

Under the successive bifurcation model race differences should be greatest for mtDNA sequences, for short repeats, and for craniometric traits, since difference is predicted to be proportional to the product of mutation rate and time since separation. Race differences should be lowest for classical markers and for nuclear RFLP markers.

Evidence so far is clearly in favour of the ancient structure model since differences among races are much lower for mitochondrial DNA control region sequences than they are for classical markers, and they are about the same for craniometric traits. (Craniometric traits are a high mutation rate system that may conform to the infinite sites assumption.) Equivalent F_{ST} between major races is three to four times greater for classical markers than it is for mtDNA sequences, the reverse of the pattern predicted by the bifurcating model.

The model of successive bifurcations does not fit the data. This means that today's differences between populations cannot be used to estimate separation times of races, since there are no separation times. Population trees do not represent the history of human races, and maps of the world with snake-like arrows repeatedly undergoing mitosis do not portray the history of human differentiation in the right way.

There are many more data in the literature to be analysed from the perspective of these specific models, and details of our largely verbal hypothesis tests ought to be made formal and explicit. We do not claim that the issues that we have posed are decided by the evidence so far, but we offer these formulations and preliminary answers as examples of the way that genetic data can and ought to be applied to issues about our past that are more interesting and more complex than the tired question of multi-regionalism or not.

New kinds of method

In this section we describe several promising approaches to understanding our history from genetic marker data. These approaches have not yet been well explored in anthropology but they may become important.

Sites where the ancestral state is known

Batzer *et al.* (1994) describe polymorphic frequencies in 16 populations of 4 *Alu* insertion polymorphisms that are not present in any of the great apes. The remarkable property of these data is that the ancestral state is known, that is in ancestral hominids the frequency must have been zero. We have no good way of knowing when the mutations occurred, of course, but nevertheless incorporation of the root reveals interesting aspects of human

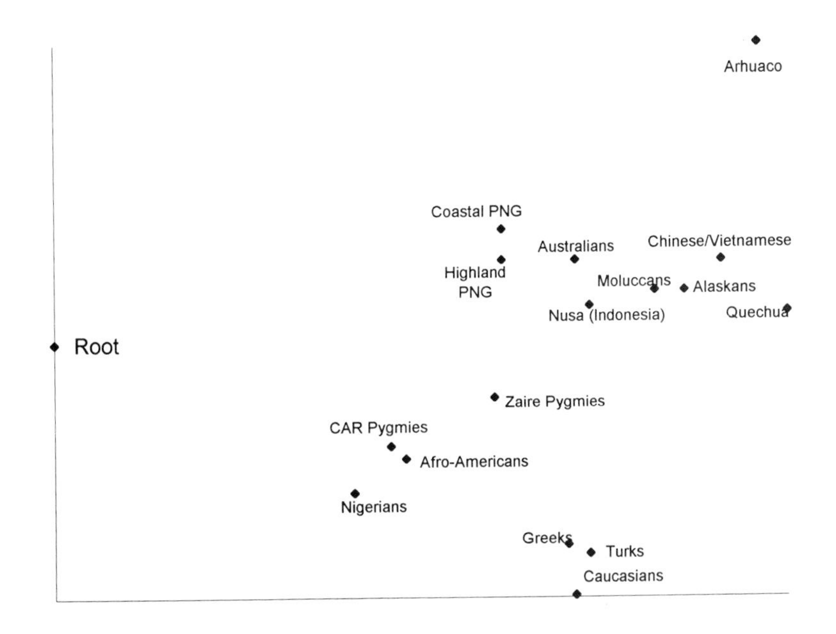

Fig. 19.1. Principal coordinates plot of populations typed for 4 *Alu* insertion polymorphisms. The scale of the axes is arbitrary, and the Root should be even further away from the extant populations than the Figure shows. This is simply a visual portrayal of the differences among the populations.

differences that are not apparent when comparing extant groups. The authors of this paper discuss this point, but then they proceed, unfortunately, to analyse differences under the bifurcating tree model of human differences. They portray their results as a tree of populations, completely obscuring the simple pattern in their data.

A sensible way to look at these data is simply to look at the distances among extant populations and between them and the root. Fig. 19.1 shows a principal coordinates plot of the populations with the root included. The pattern is clearly one in which large tropical populations of our species are closest to the root, while more peripheral modern humans like Europeans, north Asians, and Amerindians are further away. This is a sensible pattern reflecting, we suggest, the earlier expansion of the ancestors of these tropical populations.

More loci ought to be examined in this way. In particular we can infer the ancestral state of most nuclear RFLP populations and DNA markers with a good degree of confidence. If our hypothesis that the ancestors of Africans expanded earlier than the populations ancestral to other major regions is correct, then over many systems Africans ought to be closest to the root.

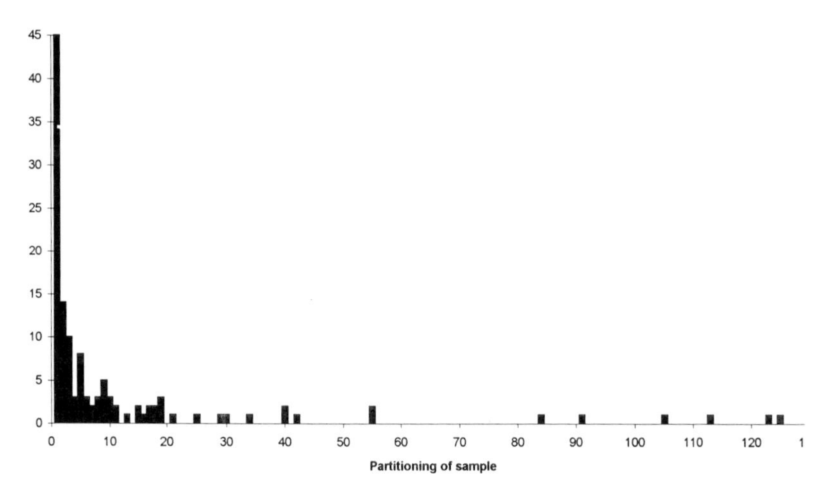

Fig. 19.2. Spectrum of site partitions in a sample of 303 mtDNA sequences at 410 sites. An entry in the histogram at, for example, 30 on the X axis means that one site was in one state in 30 of the 303 sequences and in the other state in 273 of them. All the differences are transitions.

Spectrum of sequence differences

Rogers and Harpending (1992) followed Slatkin and Hudson (1991) and Di Rienzo and Wilson (1991) in using the distribution of all pairwise sequence differences, the mismatch distribution, as a summary of important properties of a sample of sequences. The rationale is that, while there is a tree of descent of the sequences, it would be best if the tree could be reconstructed, but methods for reliable tree reconstruction do not exist. Therefore, we ought to seek summary statistics that contain useful information about the tree. The mismatch distribution is one such summary, and the site frequency spectrum is another.

Fig. 19.2 shows the frequency spectrum of 410 sites in a sample of 303 human mtDNA control region sequences. This is a histogram with one entry for each of the 126 segregating sites of the 410. Since there are only transitions in these data, any particular site can be in one of two states. If a variable site is in one state in one person and in the other state in the 302 other people, then it adds a mark to the histogram at 1. If it is in one state in two people and the other in 301, it adds a 2, and so on. Most of the sites in Fig. 19.2 are at the left of the histogram, meaning that they partition the sample into one or two people versus all the others. A few sites are further right on the X-axis, meaning that they partition the sample into two large subsets.

What is the difference between these two kinds of sites? A site that partitions the sample as (1:302) reflects a mutation that happened uniquely in the line of descent to a single individual. A site that partitions the sample into two large blocks of people reflects, under the infinite sites assumption, a mutation that occurred early in the tree of descent of the sample so that each branch of the tree, the branch with the mutation and the branch without, has a large number of descendants. In a perfect star phylogeny under the infinite sites model, all the mass of the histogram would be concentrated at unity since every mutation would have exactly one descendant.

The data shown in Fig. 19.2 are partly consistent with a star phylogeny except for a few sites. These sites in reality must be of two types, those that reflect mutations early in the tree of descent as under the infinite sites assumption, and mutational hot spots. It is occasionally possible to tell these two categories apart, but it is an art. It is important to note that sites to the right on the X-axis of Fig. 19.2 are precisely those sites that are informative when attempting to reconstruct the tree, and these informative sites are some unknown mixture of hot spots and genuine informative sites.

Most of the sites, on the other hand, suggest a star phylogeny, in agreement with the smoothness and peakedness of the mismatch distributions discussed by Rogers and Harpending (1992), Harpending *et al.* (1993), and the recent review by Rogers and Jorde (1995). We are currently exploring whether properties of match distributions or properties of spectra give redundant or independent information about the demographic history of a sample (Gurven, Sherry, and Harpending, manuscript). Spectra from simulated populations with waterglass rather than funnel histories never look like the spectrum in Fig. 19.2 with its concentration of mass at the left. We suggest that site spectra, along with mismatch distributions, should be studied routinely in samples of non-recombinant sequences like mtDNA sequences from human populations.

Summary and conclusions

We have tried to suggest in this chapter that there are interesting demographic issues in the prehistory of our species and that genetic data and properties of genetic systems can be of value in testing predictions from models of these phenomena. We described two such issues: two models of the genesis of greater genetic diversity in African populations and two models of the genesis of race differences in our species. We found support for the model of early African expansion, in the first case, and for frozen ancient population structure in the second.

We also suggest that there are new techniques and deductions from population genetic models that are still not thoroughly understood but that may provide us with new insights and new approaches to old problems. Differences between two estimates of group differentiation measured by F_{ST} on non-recombinant sequences suggest a new way of estimating mutation rates. Our models used the infinite sites assumption, but forward and back mutation can easily be incorporated, at least numerically. Polymorphisms where the ancestral or root state is known provide historical information that other loci do not, and our portrayal of group differences in *Alu* insertion frequencies showed support for our model of early African expansion. It also suggested a relatively early expansion of Sahulian ancestors that has not been apparent in other data. Finally, the spectrum of site frequencies for a sample of mtDNA control regions sequences is another way to summarize the properties of the tree of descent of a set of sequences without reconstructing the tree. The human spectrum of mtDNA control region sequences confirms the match distribution in portraying a star-like phylogeny of sequences reflecting a major ancient expansion of our ancestors during the last interglacial and during the first part of the last major glaciation.

Acknowledgements

We are grateful for help and suggestions from Andy Clark, Mike Gurven, Lynn Jorde, Alan Redd, Alan Rogers, Mark Stoneking, and Tom Whittam. Henry Harpending wishes to acknowledge support from the Human Diversity Project of the King's College Research Centre, Cambridge, UK.

References

Batzer, M. A., Stoneking, M., Alegria-Hartman, M., Bazan, H., Kass, D. H., Shaikh, T. H., Novick, G. E., Ioannou, P. A., Scheer, W. D., Herrera, R. J. & Deininger, P. L. (1994). African origin of human-specific polymorphic Alu insertions. *Proceedings of the National Academy of Sciences, USA*, **91**, 12288–92.

Bowcock, A. M., Ruiz-Linares, A., Tomfohrde, J., Minch, E., Kidd, J. R. & Cavalli-Sforza, L. L. (1994). High resolution of human evolutionary trees with polymorphic microsatellites. *Nature*, **368**, 455–7.

Cavalli-Sforza, L. L., Piazza, A., Menozzi, P. & Mountain, J. (1988). Reconstruction of human evolution: bringing together genetic, archaeological, and linguistic data. *Proceedings of the National Academy of Sciences, USA*, **85**, 6002–6.

Crawford, M. & Mielke, J. (ed.) (1980). *Current Developments in Anthropological Genetics II*. New York: Plenum Press.

Crow, J. F. & Aoki, K. (1984). Group selection for a polygenic behavioral trait: Estimating the degree of population subdivision. *Proceedings of the National Academy of Sciences, USA*, **81**, 6073–7.

Di Rienzo, A. & Wilson, A. (1991). Branching pattern in the evolutionary tree for human mitochondrial DNA. *Proceedings of the National Academy of Sciences, USA*, **88**, 1597–1601.

Gibbons, A. (1995). The mystery of humanity's missing mutations. *Science*, **267**, 35–6.

Harpending, H. C. (1994). Gene frequencies, DNA sequences, and human origins. *Perspectives in Biology and Medicine*, **37**, 385–94.

Harpending, H. C., Sherry, S. T., Rogers, A. R. & Stoneking, M. (1993). The genetic structure of ancient human populations. *Current Anthropology*, **34**, 483–96.

Lynch, M. (1988). The rate of polygenic mutation. *Genetical Research, Cambridge*, **51**, 137–48.

Nei, M. (1987). *Molecular Evolutionary Genetics*. New York: Columbia University Press.

Nei, M. & Roychoudhury, A. K. (1974). Genetic variation within and between the three major races of man, Caucasoids, Negroids, and Mongoloids. *American Journal of Human Genetics*, **26**, 421–43.

Relethford, J. & Harpending, H. C. (1994). Craniometric variation, genetic theory, and modern human origins. *American Journal of Physical Anthropology*, **95**, 249–70.

Relethford, J. & Harpending, H. C. (1995). Ancient differences in population size can mimic a recent African origin of modern humans. *Current Anthropology*, **36**, 667–74.

Rogers, A. R. & Harpending, H. C. (1992). Population growth makes waves in the distribution of pairwise genetic differences. *Molecular Biology and Evolution*, **9**, 552–69.

Rogers, A. R. & Jorde, L. B. (1995). Genetic evidence on modern human origins. *Human Biology*, **67**, 1–36.

Slatkin, M. & Hudson, R. R. (1991). Pairwise comparisons of mitochondrial DNA sequences in stable and exponentially growing populations. *Genetics*, **129**, 555–62.

Stoneking, M., Jorde, L. B., Bhatia, K. & Wilson, A. C. (1990). Geographic variation of human mitochondrial DNA from Papua New Guinea. *Genetics*, **124**, 717–33.

Takahata, N. & Palumbi, S. R. (1985). Extranuclear differentiation and gene flow in the finite island model. *Genetics*, **109**, 441–57.

Whittam, T. S., Clark, A. G., Stoneking, M., Cann, R. L. & Wilson, A. C. (1986). Allelic variation in human mitochondrial genes based on patterns of restriction site polymorphism. *Proceedings of the National Academy of Sciences, USA*, **83**, 9611–15.

Index